VOLUME SIX HUNDRED AND THIRTY FOUR

METHODS IN ENZYMOLOGY

Neutron Crystallography in Structural Biology

METHODS IN ENZYMOLOGY

VOLUME SIX HUNDRED AND THIRTY FOUR

METHODS IN ENZYMOLOGY

Neutron Crystallography in Structural Biology

Edited by

PETER C.E. MOODY

Department of Molecular and Cell Biology and Leicester Institute of Structural and Chemical Biology, University of Leicester, Leicester, United Kingdom

ELSEVIER

ACADEMIC PRESS

An imprint of Elsevier

Academic Press is an imprint of Elsevier
50 Hampshire Street, 5th Floor, Cambridge, MA 02139, United States
525 B Street, Suite 1650, San Diego, CA 92101, United States
The Boulevard, Langford Lane, Kidlington, Oxford OX5 1GB, United Kingdom
125 London Wall, London, EC2Y 5AS, United Kingdom

First edition 2020

Notices

Knowledge and best practice in this field are constantly changing. As new research and experience broaden our understanding, changes in research methods, professional practices, or medical treatment may become necessary.

Practitioners and researchers must always rely on their own experience and knowledge in evaluating and using any information, methods, compounds, or experiments described herein. In using such information or methods they should be mindful of their own safety and the safety of others, including parties for whom they have a professional responsibility.

To the fullest extent of the law, neither the Publisher nor the authors, contributors, or editors, assume any liability for any injury and/or damage to persons or property as a matter of products liability, negligence or otherwise, or from any use or operation of any methods, products, instructions, or ideas contained in the material herein.

ISBN: 978-0-12-819214-6
ISSN: 0076-6879

For information on all Academic Press publications
visit our website at https://www.elsevier.com/books-and-journals

Publisher: Zoe Kruze
Acquisitions Editor: Ashlie M. Jackman
Editorial Project Manager: Shellie Bryant
Production Project Manager: James Selvam
Cover Designer: Alan Studholme

Typeset by SPi Global, India

Contents

Contributors

Paul D. Adams
Molecular Biophysics and Integrated Bioimaging Division, Lawrence Berkeley National Laboratory (LBNL); Department of Bioengineering, University of California Berkeley, Berkeley, CA, United States

Pavel V. Afonine
Molecular Biophysics and Integrated Bioimaging Division, Lawrence Berkeley National Laboratory (LBNL), Berkeley, CA, United States

Jacob T. Andring
Department of Biochemistry and Molecular Biology, College of Medicine, University of Florida, Gainesville, FL, United States

Giuseppe Aprigliano
European Spallation Source ERIC, Lund, Sweden

Kaira Beltran
Department of Natural Sciences, Tennessee Wesleyan University, Athens, TN, United States

Matthew P. Blakeley
Large Scale Structures Group, Institut Laue–Langevin, Grenoble, France

Monika Budayova-Spano
Université Grenoble Alpes, CEA, CNRS, IBS, Grenoble, France

Toshiyuki Chatake
Institute for Integrated Radiation and Nuclear Science, Kyoto University, Kumatori, Osaka, Japan

Leighton Coates
Neutron Scattering Division, Oak Ridge National Laboratory, Oak Ridge, TN, United States

Jacob E. Combs
Department of Biochemistry and Molecular Biology, College of Medicine, University of Florida, Gainesville, FL, United States

Matthew J. Cuneo
Neutron Scattering Division, Oak Ridge National Laboratory, Oak Ridge, TN, United States

Steven Dajnowicz
Department of Chemistry and Biochemistry, University of Toledo, Toledo, OH; Neutron Scattering Division, Oak Ridge National Laboratory, Oak Ridge, TN, United States

Victoria Drago
Department of Chemistry and Biochemistry, University of Toledo, Toledo, OH, United States

Zoë Fisher
Department of Biology, Lund University; Scientific Activities Division, European Spallation Source ERIC, Lund, Sweden

Satoru Fujiwara
Institute for Quantum Life Science, National Institutes for Quantum and Radiological Science and Technology, Tokai, Ibaraki, Japan

Oksana Gerlits
Department of Natural Sciences, Tennessee Wesleyan University, Athens, TN, United States

John R. Helliwell
Manchester University, Manchester, United Kingdom

Samuel John Hjorth-Jensen
Department of Molecular Biology and Genetics—DANDRITE, Aarhus University, Aarhus C, Denmark

Takaaki Hosoya
Domain of Materials Science and Engineering, Graduate School of Science and Engineering, Ibaraki University, Hitachi; Divisions of Neutron Beamline and Research, Frontier Research Center for Applied Atomic Sciences, Tokai, Ibaraki, Japan

Anna Jennings
Neutron Scattering Division, Oak Ridge National Laboratory; Oak Ridge Associated Universities, Oak Ridge, TN, United States

David A. Keen
ISIS Facility, Rutherford Appleton Laboratory, Harwell Campus, Didcot, United Kingdom

Tatiana Kerber
School of Molecular Sciences, University of Western Australia, Perth, WA, Australia

Katarina Koruza
Department of Biology, Lund University, Lund, Sweden

Andrey Kovalevsky
Neutron Scattering Division, Oak Ridge National Laboratory, Oak Ridge, TN, United States

Katsuhiro Kusaka
Divisions of Neutron Beamline and Research, Frontier Research Center for Applied Atomic Sciences, Tokai, Ibaraki, Japan

Hanna Kwon
School of Chemistry, University of Bristol, Bristol, United Kingdom

Le Li
Neutron Scattering Division, Oak Ridge National Laboratory, Oak Ridge, TN, United States

Dorothee Liebschner
Molecular Biophysics and Integrated Bioimaging Division, Lawrence Berkeley National Laboratory (LBNL), Berkeley, CA, United States

Derek T. Logan
Section for Biochemistry and Structural Biology, Department of Chemistry, Lund University, Lund, Sweden

John M. Louis
Laboratory of Chemical Physics, National Institute of Diabetes and Digestive and Kidney Diseases, National Institutes of Health, DHHS, Bethesda, MD, United States

Márton Markó
Institute for Solid State Physics and Optics, Wigner Research Centre for Physics, Budapest, Hungary

Robert McKenna
Department of Biochemistry and Molecular Biology, College of Medicine, University of Florida, Gainesville, FL, United States

Flora Meilleur
Neutron Scattering Division, Oak Ridge National Laboratory, Oak Ridge, TN; Department of Molecular and Structural Biochemistry, North Carolina State University, Raleigh, NC, United States

Peter C.E. Moody
Department of Molecular and Cell Biology and Leicester Institute of Structural and Chemical Biology, University of Leicester, Leicester, United Kingdom

Nigel W. Moriarty
Molecular Biophysics and Integrated Bioimaging Division, Lawrence Berkeley National Laboratory (LBNL), Berkeley, CA, United States

Timothy C. Mueser
Department of Chemistry and Biochemistry, University of Toledo, Toledo, OH, United States

Dean A.A. Myles
Neutron Scattering Division, Oak Ridge National Laboratory, Oak Ridge, TN, United States

Gergely Nagy
Institute for Solid State Physics and Optics, Wigner Research Centre for Physics, Budapest, Hungary; European Spallation Source ERIC, Lund, Sweden

Nobuo Niimura
Divisions of Neutron Beamline and Research, Frontier Research Center for Applied Atomic Sciences, Tokai, Ibaraki, Japan

Poul Nissen
Department of Molecular Biology and Genetics—DANDRITE, Aarhus University, Aarhus C, Denmark

Esko Oksanen
European Spallation Source ERIC; Department of Biochemistry and Structural Biology, Lund University, Lund, Sweden

Andreas Ostermann
Heinz Maier-Leibnitz Zentrum (MLZ), Technische Universität München, Garching, Germany

Joshua Pierce
Neutron Scattering Division, Oak Ridge National Laboratory, Oak Ridge, TN, United States

Emma L. Raven
School of Chemistry, University of Bristol, Bristol, United Kingdom

Tobias E. Schrader
Forschungszentrum Jülich GmbH, Jülich Centre for Neutron Science (JCNS) at Heinz Maier-Leibnitz Zentrum (MLZ), Garching, Germany

Thomas Lykke-Møller Sørensen
Department of Molecular Biology and Genetics—DANDRITE, Aarhus University, Aarhus C, Denmark

Brendan Sullivan
Neutron Scattering Division, Oak Ridge National Laboratory, Oak Ridge, TN, United States

Ichiro Tanaka
Domain of Materials Science and Engineering, Graduate School of Science and Engineering, Ibaraki University, Hitachi; Divisions of Neutron Beamline and Research, Frontier Research Center for Applied Atomic Sciences, Tokai, Ibaraki, Japan

Susan S. Taylor
Department of Chemistry and Biochemistry, University of California at San Diego, Lo Jolla, CA, United States

Alexandre G. Urzhumtsev
Centre for Integrative Biology, Institut de Génétique et de Biologie Moléculaire et Cellulaire, Illkirch; Faculté des Sciences et Technologies, Université de Lorraine, Nancy, France

Gianluigi Veglia
Department of Chemistry; Department of Biochemistry, Molecular Biology and Biophysics, University of Minnesota, Minneapolis, MN, United States

Alice Vrielink
School of Molecular Sciences, University of Western Australia, Perth, WA, Australia

Irene T. Weber
Department of Biology; Department of Chemistry, Georgia State University, Atlanta, GA, United States

Kevin L. Weiss
Neutron Scattering Division, Oak Ridge National Laboratory, Oak Ridge, TN, United States

Taro Yamada
Divisions of Neutron Beamline and Research, Frontier Research Center for Applied Atomic Sciences, Tokai, Ibaraki, Japan

Naomine Yano
Divisions of Neutron Beamline and Research, Frontier Research Center for Applied Atomic Sciences, Tokai, Ibaraki, Japan

Jinkui Zhao
Neutron Scattering Division, Oak Ridge National Laboratory, Oak Ridge, TN, United States

Preface

Neutron crystallography applied to structural biology has expanded enormously over the last few years. What was a relatively obscure technique is now known about and properly discussed as a very important tool, with a unique contribution to the understanding how biomolecules interact and how enzymes work. The realization that without a knowledge of hydrogen positions, extrapolation of the role of these atoms is frequently guesswork (inspired and informed though it may be) has led to a recognition of the importance of the technique. Recent improvements and developments in all aspects have contributed to increasing its scope, making neutron crystallography much more accessible, and there is more to come. When assembling this volume, it was assumed that the reader will come from a background that is comfortable with X-ray crystallography. I have been very fortunate in being able to get expert contributions to this volume. After a personal introduction describing the fundamentals of neutron crystallography, this volume of *Methods in Enzymology* covers the practical imperative of getting the large crystals required and then describes key examples of the facilities for data collection, along with exciting new developments. The interactive interpretation of maps and refinement is described, as well as the techniques quality and limitations. Several detailed chapters describe specific examples of what can be achieved and the insights possible when we can see hydrogen atoms.

PETER C.E. MOODY
Leicester, United Kingdom

CHAPTER ONE

Fundamentals of neutron crystallography in structural biology

John R. Helliwell*
Manchester University, Manchester, United Kingdom
*Corresponding author: e-mail address: john.helliwell@manchester.ac.uk

Contents

Abstract

This chapter introduces this topic for the whole volume. It is not a review, rather it presents the basics, the key considerations and forward references to the other chapters. This starts by setting the scene of principles and overall strategy, moves onto planning an experiment including its feasibility and then outlines practicalities with options for the experiment. The crystal structure that results will lead to publication and associated with it, Protein Data Bank deposition.

Methods in Enzymology, Volume 634
ISSN 0076-6879
https://doi.org/10.1016/bs.mie.2020.01.006

1. Introduction

As a diffraction probe for biological crystallography neutrons have delightful properties. Without radiation damage to the crystal sample room temperature diffraction data collection is routine thus placing the structural details obtained especially relevant to physiological temperature and processes. A switch to deuterium from hydrogen in the biological macromolecule and in the bound waters is possible by various means and yields as strong scattering for the deuteriums as for carbons, revealing what is hidden from X-ray studies. Biological mechanisms that are ambiguous from other structural methods harness these properties of neutron diffraction.

Despite the advantages of neutrons as a diffraction probe, there are the practical challenges. The neutron fluxes available are relatively weak and so source and instrumentation optimization is very important. Specialist detection media have required development. Software for diffraction data processing continues to evolve and benefits from compute engine improvements. Model refinement has developed with a number of styles being used, and each style resting on the necessary platform of an already determined X-ray crystal structure, ideally also at room temperature. For metalloproteins the precise identification of metal type and occupancy may have been made by tuned wavelength X-ray anomalous dispersion. The neutron scattering of metals can vary in curious ways with neutrons (e.g., manganese is negative!). Finally, validation tools tailored to neutron studies have appeared in recent times which are proving highly valuable to researcher and the Protein Data Bank alike.

There are also special experiments. Cryo-freezing (vitrification) even with relatively large crystals has been proven to be possible (Blakeley, Kalb, Helliwell, & Myles, 2004) and thereby trapping of labile enzyme intermediates is viable (see the chapter "Heme peroxidase—Trapping intermediates by cryo neutron crystallography" by Kwon et al., in this volume). The spin polarization property of neutrons has the possibility to improve the measurements that can be made in biological neutron scattering including protein crystallography (Stuhrmann, 2004). Instruments based on this for neutron macromolecular crystallography are under consideration in the Europe, Japan and the USA and have been demonstrated in the USA (Pierce et al., 2019). These developments are discussed further in the chapter by Ichiro Tanaka describing JPARC and in depth in the chapter by Pierce and others on the instrument at Oak Ridge (ONRL). Whereas the most common diffraction resolution limits reached by the available image plate instruments are between 1.3 and 2.5 Å, higher resolutions can be possible at the Institut Laue Langevin, beam line D19 with its microstrip detector technology. Reactor sources are also now being used to study larger and larger macromolecules though through use of longer wavelengths either in the monochromatic or quasi Laue options, the latter also using very careful bandpass filtering and improved neutron guides. The time-of-flight instruments at spallation sources can potentially offer both the best signal to noise measurements and use of the wide emitted spectrum of neutrons, thereby allowing effective data collection from large molecular complexes in big unit cells and/or use of hydrogenated rather than deuterated crystal samples. In a final example, of a special experiment, is the first ever direct determination of a protein crystal structure from neutron data alone based on cadmium anomalous dispersion (Cuypers et al., 2016).

In the development of neutron biological crystallography these last decades, there have been some important hurdles to overcome and strategical questions to evaluate. While a neutron structure can resolve a mechanistic question that settles a whole domain of biochemistry, the use of neutron crystallography has been limited to a relatively small number of systems due to the limited resources and technical challenges. A comprehensive book on the topic up to 2011 is available (Niimura & Podjarny, 2011). As well as joining in with the technical developments, the provision of software and methods of analysis and examples it is necessary to consider some strategic issues. Firstly; can the protonation states of ionizable amino acids be predicted by computational chemistry methods? Our evaluations (Fisher, Wilkinson, Henchman, & Helliwell, 2009) based on neutron, ultra-high resolution X-ray and NMR titration structural studies as test cases show that experimental determination, such as from crystallography, remains paramount—not least to validate better prediction methods. Also, prediction methods do not predict which atom of an ionizable group has the hydrogen atom in the case of a singly protonated histidine or a protonated carboxyl. Furthermore, the state of individual bound waters as either hydroxyl, neutral or hydronium cannot be predicted. Secondly; even for crystallography, there are possible pitfalls in the final linking of structure to mechanistic interpretation. The chemical conditions of crystallization may, for example, be at a pH remote from the functioning pH. Also, kinetic and resonance effects of a key hydrogen atom may change upon deuteration, including a pKa shift. To address these points many datasets will need to be recorded and in this increased demand in turn would require faster data collection.

A concerted effort in the source, instrumentation and software development will be essential across the global neutron facilities to address this. To that end sharing expertise for example via neutron Facility Advisory Committees would be very important. Thirdly: does deuteration affect the structure, separate from the kinetics which it obviously will? Fisher and Helliwell (2008) investigated the effect of deuteration on structure by data mining of the Cambridge Structural Database and the Inorganic Crystal Structure Database, for deuterated and hydrogenated pairs of small-molecule structures analyzed by neutron and X-ray crystallography. Fisher and Helliwell (2008) concluded that at the diffraction resolutions typically employed in neutron biological crystallography, up to 1.5 Å resolution, this is not a significant worry.

Another, wider, strategic consideration is if the information desired could be more easily obtained with other technologies than neutron crystallography? X-ray free electron lasers allow "diffraction before destruction" and at room temperature, and the possibility of looking at the structures of relatively short-lived intermediates, giving new capabilities for macromolecular crystallography; a recent example of this is the structure of the 30S ribosomal subunit at ambient temperature by Dao et al. (2018). However, these still involve the use of X-rays and do not give the exquisite sensitivity that neutrons as a probe offer for the placement of hydrogens by neutron crystallography. Electrons are capable of revealing electrostatic surfaces and thus hydrogens either by micro electron diffraction (microED) or cryo electron microscopy (cryoEM) single particle approaches. The latter now offer higher resolution capabilities than ever before and are being for biological complexes that cannot be crystallized. However, it seems though that there is no chance to ever undertake such electron microscopy or diffraction studies at room temperature.

2. Basics of neutrons as a diffraction probe

2.1 Neutrons as waves

Wave particle duality is a cornerstone of the theory of quantum mechanics. In one's everyday laboratory usage of an X-ray diffractometer we perhaps do not realize that we are comfortable with this unusual "duality" feature of the atomic world but we happily use phrases such as: *does the crystal diffract?* And: *how many photons are in that reflection from the crystal?* Likewise, for neutrons we can both consider their diffraction as waves and the strength of a crystal reflection intensity in terms of "counts." De Broglie's equation relates these two worlds of particle and wave:-

$$p = h/\lambda \tag{1}$$

where p is the momentum of the neutron as a particle, λ is its wavelength as a wave and h is Planck's constant.

2.2 Scattering signatures of the elements and their isotopes

The wavelength of the neutrons to satisfy Bragg's Law of reflection from a crystal plane of atoms must be identical to that of an X-ray for a given inter planar spacing, d, and scattering angle, θ. The same point applies to the diffraction of an electron from a crystal plane of atoms. Recall that Bragg's Law, where n is an integer, is:-

$$n\lambda = 2d\sin\theta \tag{2}$$

For a neutron beam to have wavelengths suitable for diffraction, i.e., in the range of 1–6 Å the neutrons must pass through a moderator (see below). At these wavelengths their energy is only meV, this is a lot less than the energy of X-rays used in diffraction. Specifically for X-rays with wavelengths in the range of 1–2 Å the energies are between 12 and 6 keV (three orders of magnitude greater than the first ionization energy of atoms which is in the order of eV).

Neutrons are also neutral that is they have no charge. The scattering interaction of a neutron with atoms is via the nuclei, whereas the predominant interaction of electrons or photons is with the electrons of the atoms. This is shown in the way neutrons are scattered differently by different isotopes of the same element. The interaction of the neutral neutron with the nuclei means that they are scattered weakly compared with electrons or X-rays which interact with an atom's electron charge cloud and so in general neutrons show relatively small variations from element to element (with some notable exceptions). This latter property has "unusual" benefits such as is exploited in chemical crystallography structural studies of uranyl hydrides where the scattering of uranium and hydrogen are comparable in magnitude (Grant et al., 2012); for uranium the scattering length for neutrons is 8.4 fm relative to hydrogen of −3.74 fm. If X-rays are used the massive electron density of the uranium atom overwhelms the adjacent space.

The big advantages of neutron scattering for biological crystal samples are then as follows:-

- the non-damaging nature of their interaction with the samples, including no photo-reduction of metal atoms in the active centers of enzymes
- the particular scattering details from element to element
- the different scattering of neutrons by isotopes of the same element
- the scattering of a deuterium atom is as large as that of a carbon atom

These properties are exploited in biological crystallography. They allow routine use of ambient temperatures in data collection, a point which I mention again for due emphasis. The crystallized molecule is unaffected by radiation damage. Thirdly a substitution of a different isotope will not change the chemical mechanism but will lead to contrast between isotopes, a method known as contrast variation and exploited with hydrogen to deuterium exchange in biology experiments. However, the different masses of hydrogen to deuterium naturally alter the speed of an atom's movements and thereby the kinetics of chemical reactions. This latter is known as the kinetic isotope effect (KIE).

In biological crystallography the scattering lengths for neutrons of the most relevant isotopes (extracted from Sears, 1992) are listed in Table 1. For other elements, as might be encountered for metalloenzymes, a comprehensive tabulation is available in Dianoux and Lander (2003) and also at https://www.ncnr.nist.gov/resources/n-lengths/ which includes a neat periodic table graphic display as a means to access the various scattering data for each isotope separately and overall for each element taking account of their isotopic compositions.

Table 1 Neutron-scattering lengths.

Atom type (isotope) with atomic number in brackets	**Neutron scattering length b (10 fm)[a]**
H (D) (1)	−0.374 (0.667)
C (6)	0.665
N (7)	0.936
O (8)	0.580
P (15)	0.513
S (16)	0.285

[a]The use of 10^{-12} cm for a neutron scattering length is also common in the field and this is the same as fm the SI unit of length.

The overwhelming purpose of the neutron biological crystallography experiment is to determine protonation states of ionizable amino acids and the nature of bound waters, as neutral or charged species, and their orientations. In my own neutron crystallography studies of the manganese containing protein concanavalin A I was also introduced to the oddity, to me, of a negative neutron scattering signature of manganese. Among metals this is an unusual effect. Manganese has a neutron scattering length of −0.373 (Table 1) (similar to hydrogen) due to a phase shift during the scattering process. Another case is lithium, which can be used in crystallizations as a salt (McPherson and Gavira (2014)), and where lithium has a neutron scattering length of −0.19. This will lead to negative nuclear density being observed for manganese or lithium binding sites consistent with the negative neutron scattering factors for these elements. Li^+ with only two electrons is barely visible by X-rays. It is also worth noting that sulfur, as can be seen in Table 1, has a lower neutron scattering length than it would have for X-rays relative to a carbon (16 electrons vs 6 electrons). This can also surprise some users of neutron facilities. Since a neutron crystallography experiment follows on from X-ray crystallography, and likely in the future from cryoEM, these neutron scattering factors must of course be rigorously taken account of in the model refinement but identification of these metals and sulfurs will generally not be in doubt from the X-ray anomalous dispersion.

That neutrons interact with the nuclei has an important further consequence. The nucleus of an atom is very much smaller in size than the profile of its electron distribution. This means that the scattering fall off with angle from an atom that is seen with X-rays or electrons (which interact with the electron charge clouds of an atom) would hardly be seen at all by the neutrons for a stationary atom. Atoms are not stationary of course and their movement does produce a fall off of the scattering of the neutron but it is less than that seen for X-rays or electrons. This means that neutron maps appear better defined than equivalent resolution electron density maps.

Finally, I mention that you as the user may well be encouraged to have a sample with deuteriums fully replacing hydrogen atoms in your sample. Why is that? First, any unexchanged hydrogens on (for example) methylene carbon atoms will lead to cancellation in that nuclear density. Second, hydrogen has a relatively large incoherent scattering property which deuterium has not, by about 40 times. The incoherent scattering just adds noise to the single crystal diffraction pattern and, since there are as many hydrogen atoms in a biological molecule as carbon, nitrogen and oxygen, this is a significantly detrimental effect on the diffraction resolution of your neutron dataset.

You can at least undertake heavy water liquid or vapor exchange of your enzyme crystal. This will replace the bound light waters by heavy waters and lead to replacement of a large proportion of the exchangeable hydrogens. This is discussed more in the chapters that describe specific examples.

2.3 Neutron energies and phonon interactions

There are some other properties of neutrons and although they might not be of direct interest for structural biologists I think they need describing for completeness. A crystal can have so called phonon traveling waves going through it which lead to concerted atom to atom movements. The speed of these waves is the speed of sound. The neutrons interacting with the crystal sample have meV energies, as mentioned above, and this is well matched to the energy in these traveling phonon waves. This leads to inelastic interactions of the neutrons with the atoms in the sample. When you are at your experiment using the biological neutron diffractometer of your choice, you will become aware of a quite large number of inelastic neutron scattering instruments of varying types that use this property.

2.4 Neutrons: Their spin and their magnetic moment

A neutron has a spin of a ½ and a magnetic moment of -1.9 nuclear magnetons. The magnetic moment of the neutron means that neutrons can then be used to probe a crystal sample containing isotopes that have an unpaired number of spins in order to understand their magnetic properties. As with the inelastic scattering described above you will become aware of quite a large number of other neutron crystallography instruments which are used to study the magnetic properties of crystals at the neutron facility. They include being able to add large cryostats. Thus the neutron facility has great experience with cryostats and this bodes well for the spin polarization type of neutron biological crystallography experiments that are upcoming for us (see immediately below). As with synchrotron facilities, this exposure to a wide range of science I have also found to be highly educational and horizon broadening in my work overall as a scientist.

Exploiting the spin property of neutrons is also envisaged as a means to improve biological neutron crystallography using dynamic nuclear polarization methods. In this approach, described in the chapter by Pierce et al. (2019), a spin labeled protein crystal is cooled below 1 K and subjected to a magnetic field of >2.5 T, to polarize electrons in the sample, and then irradiated with microwaves (at the Larmor frequency) to transfer this spin polarization to protons. With the neutron beam spin polarized and the protons in the sample polarized the scattering varies significantly when the spins are aligned parallel or antiparallel. This can increase the neutron scattering strength of a crystal by a near order of magnitude, thus reducing the crystal sample volume needed especially for larger molecular weight macromolecules or complexes.

3. Types of neutron sources

Neutrons used in our research are generated in two types of source; nuclear reactors and spallation neutron sources. While reactors produce neutrons continuously, spallation sources produce pulses of neutrons due to the pulsed nature of the particle accelerators used. The generated neutrons have energies of MeV and thus have wavelengths that would be extremely short for our diffraction purpose. The emitted neutrons are therefore passed through materials known as "*moderators*" to reduce the neutron energies to the meV range. According to the type of moderator the neutron beams have wavelengths that are called "cold" having wavelengths of ~5 Å or "thermal" or "hot" neutrons having wavelengths of ~1 Å.

4. Neutron macromolecular crystallography instrument types

At the nuclear reactor neutrons are generated continuously and the instruments are fed either a monochromatic beam or a polychromatic beam. There is a conflict between Laue methods, that use a wide spectrum of the reactor's emitted neutrons (see Chapter "IMAGINE: The neutron protein crystallography beamline at the high flux isotope reactor" by Meilleur et al., in this volume and Blakeley et al., 2010 for a description of LADI-III at the Institut Laue Langevin) for efficient data collection, and monochromatic methods as used at BIODIFF at FRM-II (Ostermann & Schrader, 2015) which optimize the signal to noise of the measurements. The monochromatic data collection will use the Arndt and Wonacott (1977) rotation method (as is normal for X-ray work), with "frames" collected over a total range of rotation appropriate for the crystal symmetry and orientation so as to ensure a high diffraction data completeness.

The use of the polychromatic beam involves a crystal held stationary during the exposure and is the classic "Laue method" approach; the different wavelengths in the beam then stimulate the different Bragg reflections.

This is relatively rarely used for X-rays and the details may be unfamiliar to those new to neutron crystallography. The width of the neutrons' wavelength bandpass is selected to be, e.g., 2–3 Å from a full range of 1–6 Å. This restriction is to reduce the background scatter into the diffraction pattern, and this leads to the name *quasi Laue*. The wavelength/intensity profile of the incident beam will not be flat and so different Bragg reflections will be stimulated by different intensities, and this has to be calibrated for in software for data processing, which is readily done. A large enough number of these stationary crystal exposures are required with the sample oriented at different angles so as to obtain as complete as possible diffraction dataset. The nature of the Laue geometry and the need for a restricted bandpass leads to a lower completeness in the low resolution shell. The most obvious difference between neutron and X-ray crystallography beam lines is the nature of the detectors. The longer wavelength of neutrons means that higher angles are needed for equivalent resolution, and the image-plate detectors are constructed into large cylinders, completely surrounding the crystal.

At the spallation neutron source, the neutrons are generated at the moment a pulsed proton beam hits the target, such as mercury. The pulsed nature of the source allows determination of the wavelength of the neutrons arriving at the detector by recording their time-of-flight. Unlike X-ray photons that travel at the speed of light, neutrons are nuclear particles that have a mass. Therefore by applying Eq. (1) we can calculate their momentum—and hence velocity as we know the mass—from the wavelength and vice versa. The velocities of "cold" neutrons are sufficiently slow such that their times-of-flight over some tens of meters are in the range of tens of milliseconds. This makes it practical to determine the wavelength of each neutron arriving at the detector with the required precision. The Laue method is employed but with the detector of the diffraction pattern being sensitive to not only the position of the neutrons striking it in each Bragg reflection but also the time of their arrival after the pulse. This immediately avoids the build-up of background on the Bragg reflection intensity as this noise can be electronically "gated out." This also means that the full wavelength spectrum of the neutrons can be harnessed, say from 1 to 6 Å. There will be Bragg reflections that arrive at the same position on the detector, called multiples, and these can also be separated by their time of arrival. The time of flight method at neutron sources offers significant advantages over reactor sources provided as high an average flux is achieved as with the nuclear reactor derived beams. This latter requirement has been a major hurdle in the technological development of spallation neutron sources. The planned European Spallation

Source at 5 MW is expected (see the Chapter "Neutron macromolecular crystallography at the European spallation source" by Oksanen and colleagues, in this volume) to deliver about two to three times the average flux of neutrons as a neutron beam from the 58 MW fission nuclear reactor at the Institut Laue Langevin. Reliable operation of the USA's Oak Ridge Spallation Neutron Source, a 1.4 MW source, has however, taken about a decade to achieve.

Examples of each of the above types of measuring instrument and neutron diffraction beam lines are described in subsequent chapters.

5. Sample preparation

Preparation of recombinant protein in microbial hosts fed on deuterated carbon sources allows proteins to be produced which are fully deuterated ("perdueterated"). The facilities for these microbiological sample preparation procedures have not always been routinely available, but now exist alongside most of the neutron sources. Without this full deuteration the hydrogen atoms affixed to carbon atoms that have nuclear negative scattering signatures (listed in Table 1) will almost cancel out the scattering from carbon in $-CH_2-$, at least at moderate resolution (as mentioned above). The first time the user sees this cancellation effect in a nuclear density map is somewhat unnerving, but the corresponding X-ray derived electron density map superimposed on the nuclear map provides a full picture. With a fully deuterated (perdeuterated) crystal sample the cancellation effect in the nuclear density map of a carbon with its hydrogens does not occur. The first demonstration of this and the benefits that full deuteration gave is the study of myoglobin by Shu, Ramakrishnan, and Schoenborn (2000). The impact on life of heavy water is that it is toxic to animals, presumably due to the kinetic isotope effect, but algae can survive and indeed be used as a laboratory tool of purifying perdeuterated proteins for further physical chemistry investigation (Crespi, Archer, & Katz, 1959) such as neutron protein crystallography (https://www.ill.eu/users/support-labs-infrastructure/deuteration-laboratory/).

Growth of a crystal from pure protein can be guided by the conditions found for the prior X-ray crystallography research but may be slightly different in heavy water conditions. For example, a difference between pH and pD may need to be accounted for. Also while crystal volumes that one needs can be radically smaller with fully deuterated vs partially deuterated protein the larger the crystal is, the better. The crystals of a few tens of micrometers that are now routinely used for X-ray data collection at synchrotrons will not suffice. Methods to achieve suitable crystals are described in the chapter "Large crystal growth for neutron protein crystallography" by Fisher and colleagues (in this compendium).

6. Planning an experiment with neutrons

The most likely question to arise will be whether your crystal sample unit cell volume, solvent content and crystal sample size will combine to yield good diffraction. Although sample scattering efficiency can be calculated directly as was done when evaluating feasibility of a diffraction experiment at a synchrotron X-ray source for microcrystals (Andrews et al., 1988), experience of successful neutron data collections at various instruments built up over the years suggests an empirical approach is better and this is conveniently displayed in the "Wilkinson plot" (see Fig. 1 of Habash et al., 1997; updated by Blakeley, 2009). In the early days of neutron protein crystallography there were data collections that took around 1 year. This is now transformed to much shorter times and we are able to use much weaker scattering samples. Despite the remarkable developments in the field,

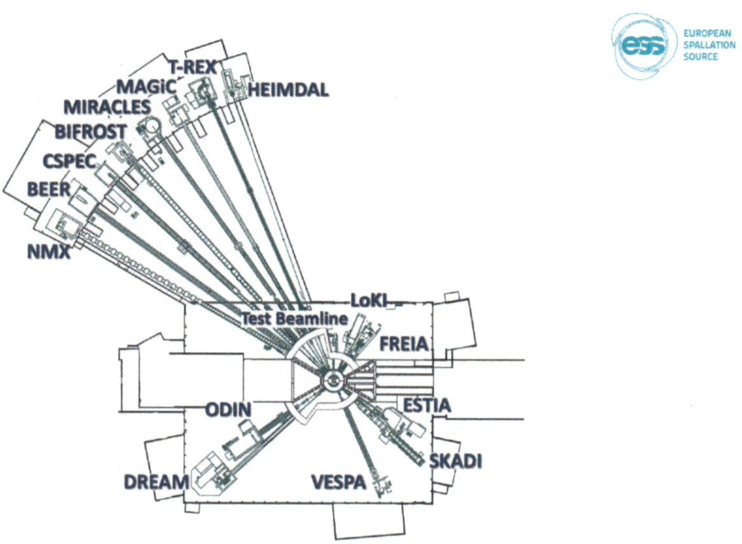

Fig. 1 Layout of the 16 instruments currently under construction at ESS. SANS: LoKI and SKADI. Reflectometry: ESTIA and FREIA. Imaging: ODIN. Engineering Diffraction: BEER. Single-Crystal Diffraction: MAGiC and NMX (Neutron Macromolecular Crystallography). Powder Diffraction: DREAM and HEIMDAL. Direct-Geometry Spectroscopy: CSPEC and T-REX. Indirect-Geometry Spectroscopy: BIFROST, MIRACLES and VESPA. Moderator Characterization and Support: Test Beamline. *This figure is kindly provided by the ESS (Dr Ken Andersen).*

it must be appreciated that the time-scales involved are still considerably longer than for modern synchrotron X-ray experiments, a typical data collection run might be 10 days long.

Before submitting a beamtime proposal to a neutron facility, I believe it essential that a discussion is had by the user with the relevant instrument scientist. The complexity and specialist knowledge required to conduct neutron crystallography experiments means that the research will necessarily be a collaboration with the beam-line scientists. Before being awarded enough beam time to collect a full dataset it will usually be necessary to apply for a short time to test that the crystals diffract well enough to collect the data that will answer the biological question being addressed.

7. Measuring your diffraction data

A neutron crystallography project will follow on from an X-ray crystal structure most likely measured at cryo temperature ca. 100 K. A triumph though this structure may be, nevertheless there will be changes from a room temperature structure (see, e.g., Deacon et al., 1997). As well as changes in the cell dimensions there are a higher proportion of split occupancy amino acid side chains than seen at room temperature and the thermal parameters are the "wrong side" of the glass transition for a protein which occurs in a narrow temperature range of 180–210 K (Parak, Knapp, & Kucheida, 1982). As the neutrons used do not damage the crystal, it is clearly better to collect neutron data at the temperature that is as close to physiological as possible, the default is to collect at room temperature. Clearly, X-ray data that is to be used for joint refinement should be collected at the same temperature as the neutron data, ideally from the sample that will have been used for your neutron data collection. To contain X-ray radiation damage at room temperature shorter exposure times than used for the cryo crystal structure will be needed, but as the resolution of neutron data will be less than what can be obtained with X-rays, these shorter exposures can provide a good match to the neutron diffraction resolution. Of course, if the neutron experiment is to take place at cryogenic temperatures, it will be appropriate to collect the complementary X-ray data similarly.

The monochromatic neutron instrument concept means that longer measuring times are needed per project than with reactor Laue or time-of-flight spallation sources. Comparisons can be made from near identical samples studied at different facilities, due allowance being made for the usually different $<I/\sigma(I)>$ values at the diffraction pattern "edge."

A detailed set of comparisons have been made, for example for galectin C at various neutron macromolecular crystallography instruments (i.e., monochromatic, reactor Laue and spallation time-of-flight; see table 1 of Manzoni et al., 2016 for summaries of those diffraction datasets).

Diffraction data completeness must be attained where very fine details are being sought, such as a split occupancy protonation state, and where ultra-careful evaluations are to be undertaken. A direct repeat of an experiment is a simple means to consider the replicability of a result, although it may be difficult to convince a beam allocation panel to grant the time in preference to new work. Monochromatic datasets can attain 100% completeness through all the resolution ranges although may require re-orientation of the crystal to avoid blind region gaps that may be significant at the longer wavelengths used (Arndt & Wonacott, 1977). Laue geometry cannot attain 100% diffraction data completeness at low resolution due to its nature (see fig. 1A of Cruickshank, Helliwell, & Moffat, 1987). However in practice ~80% Laue diffraction data completeness is routinely achievable with neutrons and is as effective for establishing protonation states and bound water orientations as a monochromatic dataset. It is not possible to set a hard rule about the observability of a structural feature however, as visibility depends on various factors notably the B factor of the relevant atom. Overall, the neutron derived crystal structure, more of which below, is tethered to an underpinning X-ray crystal structure and gives a robustness to the neutron based analyses of particular hydrogenation features that the X-ray determination does not give. If the result is a structural feature "never seen before" caution is paramount and again a direct repeat of an experiment, if possible, is a simple means to test the replicability of a result.

The chapter by Dorothee Liebschner and colleagues explores in detail the considerations of quality and completeness of neutron diffraction data.

8. Refining the molecular model against neutron data

There is not yet a consensus of the *best* way to do this, and in any case each way used and published, provided thorough peer review of article with data has been done, will have worked satisfactorily. In principle a joint refinement (X-ray data with the neutron data) should be used to achieve the best molecular model, as it has the best possible diffraction data to parameter ratio. This assumes that the diffraction dataset measurements are from isomorphous (ideally identical) crystals, and matching the diffraction resolution is important to prevent one set dominating the refinement.

A pragmatic and simple method for model refinement that I have employed, is to use the X-ray crystal structure model with its "heavy" atoms held fixed and then use the neutron diffraction data to refine those atoms that the neutrons are sensitive to, namely the deuterium and hydrogen atoms. Even this is not totally simple because the deuteriums may not have fully replaced the hydrogens on the exchangeable sites when heavy water exchange is used. For these exchangeable H/D sites it is necessary to refine the relative occupancies of the hydrogens and deuteriums. This of course yields information on the solvent accessibility of parts of a macromolecule. If the H:D ratio is about 2:1 their scattering will cancel, meaning it is not possible to confidently establish those particular protonation (deuteration) states.

Refinement in PHENIX is described in the chapter by Pavel Afonine and colleagues. The overall process of refinement and model building is described in the chapter by Derek Logan.

It is important to note that the Protein Data Bank deposition and/or the subsequent curation of neutron macromolecular crystal structures has not been consistently done over the decades (Liebschner, Afonine, Moriarty, Langan, & Adams, 2018), and there have been problems with neutron diffraction data being "lost" when both X-ray and neutron data have been deposited, but a path forward to improving the annotation, validation and deposition of neutron models and hybrid X-ray and neutron models is suggested by Liebschner et al. (2018). A particularly simple solution to this challenge is for a publication to also have the finalized coordinates along with the X-ray and neutron diffraction data files on which they are based available with the article.

9. Validating your model

The considerable structural knowledge held in the databases, the PDB and the Cambridge Structural Database, provide the information for the core checks needed in validation. The *phenix.refine* software package has appropriate ongoing validation tools to judge the progress of the X-ray crystal structure refinement and incorporates neutron derived details in its specific to neutron routines (Liebschner et al., 2018). Also, the SHELXE package has various validation tools and these are described by Gruene, Hahn, Luebben, Meilleur, and Sheldrick (2014); these involve hydrogen atom restraints for each amino acid type and a cross validation method. Following these procedures the PDB's own validation should be straightforward, any queries from the PDB's evaluator being dealt with prior to article submission of course. A PDB

Validation Report is normally required by reputable journals on article submission, and this makes the referee's task considerably easier. I would personally also encourage the submission to the journal of the PDB's to-be-released files of the final model and the processed diffraction datasets, X-ray and neutron, with the article and PDB validation report so as to allow direct calculations by editor and referees. However the pool of reviewers experienced at looking at neutron structures is limited, and you might have to explain why expectations normal for synchrotron-derived X-ray data and structures are not always appropriate. The data science checks to be performed for biological X-ray crystallography have been described by Helliwell (2018), and provide a guide for evaluation of neutron structures with the appropriate caveats. If the structural details revealed have "never been seen before," and rely on a repeated diffraction dataset for verification (as mentioned above), it is vital that those repeat experiment files are also included with the article submission. It is likely that the depositor will choose just one of these to be available at the PDB but the extra dataset should be linked to the article as a supplementary file.

10. Data reuse

Software for diffraction data processing and for model refinement are both still improving. The use of profile fitting has been highlighted recently for the raw diffraction data measured at MANDI on the SNS (Sullivan et al., 2018). Such profile fitting is routinely applied in the Daresbury Laue software suite (Nieh et al., 1999). The procedures for model refinement coupled with validation are also clearly evolving as outlined above (Liebschner et al., 2018). This potential for improving the measurement of diffraction data and the resulting structures as methods improve, means raw diffraction data should be retained and made available. The raw diffraction data needs to be Findable, Accessible, Interoperable and Reusable, known as FAIR, and is the best means possible of getting the very best out of the neutron facility investment in individual projects. A new policy for access to raw diffraction data underpinning macromolecular crystallography has been published recently simultaneously in IUCrJ, Acta Cryst D, Acta Cryst F and J Appl Cryst (see Helliwell et al., 2019).

11. Concluding remarks

Neutron macromolecular crystallography offers a wonderful world of capabilities for structural studies, capable of resolving controversies left by other probes of structure. There are neutron facilities across all the continents that offer neutron macromolecular crystallography instruments and expertize to assist in sample preparation, experiments, processing the measured data, refining the molecular model and finally deposition and publication. A range of enzyme structural studies using neutrons are described in the other chapters of this volume of Method of Enzymology and will serve as examples for future studies, showing what can be achieved. Like the synchrotron facilities, the neutron facilities involve both a very wide range of the sciences and engineering of the sources and the instruments; these synergies are reflected for example, in the programme for the European Neutron Scattering Conference with joint synchrotron and neutron instrumentation sessions. The layout of the planned instruments at the European Spallation Source illustrates the diversity of instruments and scientific scope of a neutron facility (see Fig. 1). So, please do enjoy this diversity of scientific life during your user visits to the neutron facility.

Acknowledgments

I am grateful to the University of Manchester for support. I am grateful to the EPSRC for research grant funding for ongoing developments of the Daresbury Laue processing suite (PI Dr. M.M. Harding, CoPI Prof. J.R. Helliwell) and the Daresbury Laboratory for beamtime at the Synchrotron Radiation Source over many decades. I am very grateful to Keith Moffat and Durward Cruickshank (1924–2007) for many discussions and collaborations on synchrotron Laue diffraction. I am grateful to the Institut Laue Langevin (ILL) for neutron beamtime and ILL PhD studentships for M.P. Blakeley and S. Fisher, now both Drs. I am grateful to the numerous colleagues who I have served with on neutron advisory committees in Europe, Japan and the USA with whom it has been a pleasure to work alongside the facilities namely at the ILL Genoble, JAERI Tokai, SNS and HIFR Oak Ridge National Laboratory. This book chapter is dedicated to Dr. Clive Wilkinson and Dr. Mogens Lehman, who both passed away in recent years, and who encouraged my interest in neutron protein crystallography in the mid-1990s. I thank the following for their comments on my manuscript drafts: Peter Moody, Andreas Osterman, Esko Oksanen, Dean Myles, and M.P. Blakeley. I thank Ken Anderson for providing Fig. 1. I heartily thank Matthew Blakeley and Trevor Forsyth at the Institut Laue Langevin, Esko Oksanen and Ken Andersen at ESS, Derek Logan at Lund University, Andreas Osterman and Tobias Schrader at FRM2, Leighton Coates, Flora Meilleur and Dean Myles at Oak Ridge National Laboratory and Nobuo Niimura, Japan for many ongoing and useful discussions.

References

Andrews, S. J., Papiz, M. Z., McMeeking, R., Blake, A. J., Lowe, B. M., Franklin, K. R., et al. (1988). Piperazine silicate (EU-19)—The structure of a very small crystal determined with synchrotron radiation. *Acta Crystallographica. Section B, 44*, 73–77.

Arndt, U. W., & Wonacott, A. J. (1977). *The rotation method in crystallography: Data collection from macromolecular crystals*. published by North Holland: Elsevier.
Blakeley, M. (2009). Neutron macromolecular crystallography. *Crystallography Reviews*, *15*, 157–218.
Blakeley, M. P., Kalb, A. J., Helliwell, J. R., & Myles, D. A. (2004). The 15-K neutron structure of saccharide free concanavalin A. *Proceedings of the National Academy of Sciences of the United States of America*, *101*, 16405–16410.
Blakeley, M. P., et al. (2010). Neutron macromolecular crystallography with LADI-III. *Acta Crystallographica Section D*, *66*, 1198–1205.
Crespi, H. L., Archer, S. M., & Katz, J. J. (1959). Culture of algae and other micro-organisms in deuterium oxide. *Nature*, *184*, 729–730.
Cruickshank, D. W. J., Helliwell, J. R., & Moffat, K. (1987). Multiplicity distribution of reflections in Laue diffraction. *Acta Crystallographica. Section A*, *43*, 656–674.
Cuypers, M. G., Mason, S. A., Mossou, E., Haertlein, M., Forsyth, V. T., & Mitchell, E. P. (2016). Macromolecular structure phasing by neutron anomalous diffraction. *Scientific Reports*, *6*, 31487.
Dao, E. H., Poitevin, F., Sierra, R. G., Gati, C., Rao, Y., Ciftci, H. I., et al. (2018). Structure of the 30S ribosomal decoding complex at ambient temperature. *RNA*, *24*, 1667. https://doi.org/10.1261/rna.067660.118.
Deacon, A., Gleichmann, T., Kalb, A. J., Price, H., Raftery, J., Bradbrook, G., et al. (1997). The structure of concanavalin A and its bound solvent determined with small–molecule accuracy at 0.94Å resolution. *Faraday Transactions*, *93*(24), 4305–4312.
Dianoux, A.-J., & Lander, G. (2003). *The neutron data booklet* (2nd ed.). Grenoble: Institut Laue Langevin published by the.
Fisher, S. J., & Helliwell, J. R. (2008). An investigation into structural changes due to deuteration. *Acta Crystallographica. Section A*, *64*, 359–367.
Fisher, S. J., Wilkinson, J., Henchman, R. H., & Helliwell, J. R. (2009). An evaluation review of the prediction of protonation states in proteins versus crystallographic experiment. *Crystallography Reviews*, *15*, 231–259.
Grant, D. J., Stewart, T. J., Bau, R., Miller, K. A., Mason, S. A., Gutmann, M., et al. (2012). Uranium and thorium hydride complexes as multielectron reductants: A combined neutron diffraction and quantum chemical study. *Inorganic Chemistry*, *51*, 3613–3624.
Gruene, T., Hahn, H. W., Luebben, A. V., Meilleur, F., & Sheldrick, G. M. (2014). Refinement of macromolecular structures against neutron data with *SHELXL2013*. *Journal of Applied Crystallography*, *47*, 462–466.
Habash, J., Raftery, J., Weisgerber, S., Cassetta, A., Lehmann, M., Hoghoj, P., et al. (1997). Neutron Laue diffraction study of concanavalin A: The proton of Asp28. *Faraday Transactions*, *93*(24), 4313–4317.
Helliwell, J. R. (2018). Data science skills for referees: Biological X-ray crystallography. *Crystallography Reviews*, *24*, 263–272.
Helliwell, J. R., Minor, W., Weiss, M. S., Garman, E. F., Read, R. J., Newman, J., et al. (2019). Findable accessible interoperable re-usable (FAIR) diffraction data are coming to protein crystallography. *Acta Crystallographica. Section D*, *75*, 455–457.
Liebschner, D., Afonine, P. V., Moriarty, N. W., Langan, P., & Adams, P. D. (2018). Evaluation of models determined by neutron diffraction and proposed improvements to their validation and deposition. *Acta Crystallographica Section D*, *74*, 800–813.
Manzoni, F., Saraboji, K., Sprenger, J., Kumar, R., Noresson, A.-L., Nilsson, U. J., et al. (2016). Perdeuteration, crystallization, data collection and comparison of five neutron diffraction data sets of complexes of human galectin-3C. *Acta Crystallographica Section D*, *72*, 1194–1202.
McPherson, A., & Gavira, J. A. (2014). Introduction to protein crystallization. *Acta Crystallographica. Section F*, *70*, 2–20.

Nieh, Y. P., Raftery, J., Weisgerber, S., Habash, J., Schotte, F., Ursby, T., et al. (1999). Accurate and highly complete synchrotron protein crystal Laue diffraction data using the ESRF CCD and the Daresbury Laue software. *Journal of Synchrotron Radiation, 6,* 995–1006.

Niimura, N., & Podjarny, A. (2011). *Neutron protein crystallography: Hydrogen, protons, and hydration in bio-macromolecules.* Published by Oxford: Oxford University Press.

Ostermann, A., & Schrader, T. (2015). BIODIFF: Diffractometer for large unit cells. *Journal of Large-Scale Research Facilities, 1,* A2.

Parak, F., Knapp, E. W., & Kucheida, K. (1982). Protein dynamics Mossbauer spectroscopy on deoxymyoglobin crystals. *Journal of Molecular Biology, 161,* 177–194.

Pierce, J., Crow, L., Cuneo, M., Edwards, M., Herwig, K. W., Jennings, A., et al. (2019). A prototype system for dynamically polarized neutron protein crystallography. *Nuclear Instruments and Methods in Physics Research A, 940*(2019), 430–434.

Sears, V. F. (1992). Neutron scattering lengths and cross sections. *Neutron News, 3*(3), 26–37. https://doi.org/10.1080/10448639208218770.

Shu, F., Ramakrishnan, V., & Schoenborn, B. P. (2000). Enhanced visibility of hydrogen atoms by neutron crystallography on fully deuterated myoglobin. *Proceedings of the National Academy of Sciences of the United States of America, 97*(8), 3872–3877.

Stuhrmann, H. B. (2004). Unique aspects of neutron scattering for the study of biological systems. *Reports on Progress in Physics, 67*(7), 1073–1115.

Sullivan, B., Archibald, R., Langan, P. S., Dobbek, H., Bommer, M., McFeeters, R. L., et al. (2018). Improving the accuracy and resolution of neutron crystallographic data by three-dimensional profile fitting of Bragg peaks in reciprocal space. *Acta Crystallographica Section D, 74,* 1085–1095.

Further reading

Afonine, P. V., Grosse-Kunstleve, R. W., Echols, N., Headd, J. J., Moriarty, N. W., Mustyakimov, M., et al. (2012). Towards automated crystallographic structure refinement with phenix.refine. *Acta Crystallographica Section D, 68,* 352–367.

CHAPTER TWO

Large crystal growth for neutron protein crystallography

Monika Budayova-Spano[a], Katarina Koruza[b,†], Zoë Fisher[b,c,*]

[a]Université Grenoble Alpes, CEA, CNRS, IBS, Grenoble, France
[b]Department of Biology, Lund University, Lund, Sweden
[c]Scientific Activities Division, European Spallation Source ERIC, Lund, Sweden
*Corresponding author e-mail address: zoe.fisher@esss.se

Contents

Abstract

The use of neutron protein crystallography (NPX) is expanding rapidly, with most structures determined in the last decade. This growth is stimulated by a number of developments, spanning from the building of new NPX beamlines to the availability of improved software for structure refinement. The main bottleneck preventing structural biologists from adding NPX to the suite of methods commonly used is the large volume of the individual crystals required for a successful experiment. A survey of deposited NPX structures in the Protein Data Bank shows that about two-thirds came from crystals prepared using vapor diffusion, while batch and dialysis-based methods all-together contribute to most of the remaining one-third. This chapter explains the underlying principles of these protein crystallization methods and provides practical examples that may help others to successfully prepare large crystals for NPX.

† Current address: Structural Biology Brussels, Vrije Universiteit Brussels, Brussels, Belgium.

Methods in Enzymology, Volume 634
ISSN 0076-6879
https://doi.org/10.1016/bs.mie.2019.11.015

1. Introduction

Neutron protein crystallography (NPX) is a rapidly growing field, with most structures in the Protein Data Bank (PDB) deposited in the last decade. There are multiple reasons for this recent growth, including: the addition of new macromolecular crystallography (MX) beamlines to the global suite available for users; improved software for model refinement; intense pulsed neutron sources with higher flux; optimization of existing instruments allowing smaller crystals to be used (Blakeley, Hasnain, & Antonyuk, 2015). Despite these advances, the number of successful NPX experiments resulting in a refined and deposited structures (142) in the PDB are still far fewer than those coming from X-rays (~137,000), solution nuclear magnetic resonance (NMR) (~12,500), and electron microscopy (EM) (~3400), and are just slightly better represented than solid-state NMR (104) and electron diffraction (120) (accessed May 2019). While isotope labeling of proteins and higher flux sources are helping some biological researchers become neutron users, the largest bottleneck to determining neutron crystal structures for novel or challenging proteins is still limited by crystal volume. In a review article by Blakeley and colleagues it was shown that the size range for most successful NPX experiments, where crystal volumes were actually reported, is ~1 mm^3 in volume (Blakeley et al., 2015). Some of the smallest crystals that have been successfully used in NPX experiments are in the 0.1–0.2 mm^3 range. If one assumes a crystal that is isometric, this means growing a crystal that is $0.53 \times 0.53 \times 0.53$ mm in all three dimensions. This is of a size that most X-ray crystallographers no longer try to produce (Blakeley, 2009).

Of the 142 bona fide neutron crystal structures deposited in the PDB 56 are from unique proteins (PDB accessed May 2019; Table 1). The rest come from mutants of the same protein, different ligand/inhibitor/substrate/product complexes, or different metal-substituted versions of metalloproteins. Out of the 142, 108 were from proteins that were H/D exchanged while 34 were perdeuterated. For the 56 unique structures, 44 were from H/D exchanged proteins and 12 perdeuterated. Unfortunately, crystal dimensions and total volumes are not consistently reported and often not even mentioned at all in the primary literature, so one cannot make definitive statements about crystal volume vs deuteration status.

Of the reported neutron structures, some came from extremely large crystals (20–50 mm^3) prepared from H/D exchanged protein, including

Table 1 Summary of deposited neutron crystal structures in the Protein Data Bank (accessed May 2019).

Total number of deposited neutron crystal structures	142
Unique	56
Unique—deuterated	12
Unique—H/D exchanged	44
Number of crystals prepared by method:	
Vapor diffusion	95
Batch	34
Dialysis	8
Counter-diffusion in capillary	5

sperm whale carbonmonoxy-myoglobin, human deoxyhemoglobin, ribonuclease A, hen egg white lysozyme, and xylose isomerase (Cheng & Schoenborn, 1990; Kovalevsky et al., 2008, 2010; Mason, Bentley, & McIntyre, 1984; Wlodawer & Sjolin, 1982). In contrast, the smallest crystals reported were of A-DNA (0.06 mm^3) and perdeuterated aldose reductase (0.1 mm^3) (Hazemann et al., 2005; Leal, Teixeira, Blakeley, Mitchell, & Forsyth, 2009). Crystal size is clearly still a major limitation and NPX is mostly confined to protein systems where it is possible to get large amounts of soluble protein for crystallization trials. A recent paper by Tanaka explains the relationship between unit cell volume, overall crystal volume, and diffraction intensity/resolution (Tanaka, 2019). It is noteworthy that while large crystals with small unit cells will diffract neutrons the best, in the case of large unit cells it is not only the ability to discern close reflections that determine success, but mostly how large the crystal is overall. It is suggested that the effective volume, defined as $V_{crystal}/(V_{cell})^2$ should instead be used to assess achievable diffraction and crystallographic resolution. In essence, data from a large unit cell can be successfully collected by most MX neutron beamlines provided the crystal volume is very large. With the high-flux European Spallation Source (ESS), currently under construction and expected to come online in 2023, the projected flux will enable data collection in a day from much smaller (~0.01 mm^3) crystals than what is possible today. Although the difficulty of obtaining diffracting crystals is also often the bottleneck in X-ray crystallography, with the high X-ray flux available at synchrotrons and advanced detector instrumentation there is little incentive to increase the crystal size much beyond 0.008 mm^3 (~200 μm on each edge). Instead, for X-ray work there has been development in parallel, automated, high-throughput methods that strive to use a minimal amount of protein per experiment. For NPX the challenges are different, as increasing the crystal volume while maintaining diffraction quality is the primary driver.

For deposited NPX structures, about two-thirds of structures came from crystals grown using vapor diffusion, while batch/dialysis/counter-diffusion/other all-together contribute the remaining one-third (Table 1; PDB information accessed May 2019). For vapor diffusion, several also report combining some form of seeding or crystal feeding with crystallization optimization (Gavira, 2016; Manzoni et al., 2016; Yamada et al., 2013). This chapter will expand on the underlying principles of protein crystallization and the strategies that have been successfully used to prepare large crystals for NPX, including examples for vapor diffusion, batch, and dialysis.

2. Crystallization conditions, nucleation, and growth

The conditions under which a given particular protein crystallizes cannot be inferred or predicted. Crystallization is a sensitive process that depends on a large number of physicochemical parameters (e.g., precipitant and protein concentration, equilibration rate, temperature) that are difficult to control in conventional setups. These variables are explored through different approaches: one is a systematic screening through varying different parameters in an incremental way, while the other is using commercial screens designed to cover a very large crystallization space in a relatively small number of conditions (so-called sparse-matrix screening) (Jancarrik & Kim, 1991; Luft et al., 2001). The parameters to be screened include precipitant concentration, protein concentration, pH, temperature, but also the rate of the actual process of vapor diffusion (a form of evaporation) or mass transfer. Evaporation of crystallization drops in vapor diffusion set-ups is difficult to control as there can be quite some variability in sealing the drops, depending on the materials and methods used. The crystallization process is influenced by both with thermodynamic and kinetic features and occurs in multi-dimensional phase spaces. Thermodynamic parameters are influenced by temperature, pH, choice of solvent, presence of impurities and so on. Kinetic trajectories define or control the nucleation and growth rate and the final properties of the crystal (Maeki et al., 2011).

There are two steps in the crystallization process: nucleation and growth. Nucleation is a stochastic event with an activation energy, whereby a microscopic nucleus is formed from the association of a few individual protein molecules (Candoni, Grossier, Hammadi, Morin, & Veesler, 2012). Once the nucleus is formed and exceeds the critical size, the crystal growth takes place and can be described by equilibrium thermodynamics. While the theoretical background of the crystallization process is well established and studied with model systems, the principles are often difficult to implement in practice as the level of supersaturation cannot be controlled effectively in crystallization setups using small enough amounts of precious protein (Vekilov & Vorontsova, 2014).

3. Protein solubility and the phase diagram

A very useful tool to explore the nucleation and crystal growth space is the phase diagram (Saridakis & Chayen, 2000). Construction of a crystallization phase diagram can be designed in almost any kind of plate and works very well with batch and vapor diffusion, as these can be set up using a relatively small amount of protein. Here the idea is to design the phase diagram to systematically explore the relationship of a particular set of parameters, such as precipitant concentration vs protein concentration or temperature. Visual inspection and scoring of the drops in a plate then allows the researchers to map out where the optimal set of conditions are to promote nucleation and subsequently to manipulate the conditions to promote crystal growth (Fig. 1). Using the knowledge gained through phase diagram mapping to fine-tune conditions helps the researcher gain control over the crystal nucleation vs growth trajectory and this has been a successful strategy in our hands (Koruza, Lafumat, Nyblom, Knecht, & Fisher, 2018).

An important factor for direct crystallization from solution is knowledge of the solid-liquid equilibrium of the protein in solution. Both nucleation and the crystal growth rate are directly correlated to supersaturation (i.e., the higher the supersaturation, the faster the nucleation and growth rate). Therefore, controlling the supersaturation level over time can impact the maximum crystal size achievable. The quantity often used to define the state of supersaturation is the metastable zone width. Since it is a direct measure of stability of the solution in its supersaturated state, the metastable zone width is an essential parameter affecting crystal size. Essentially, the larger the width of the metastable zone, the higher the stability of the solution in the supersaturated state (Saridakis & Chayen, 2000). The diagram in Fig. 1 can be described in terms of four distinct zones: (1) the stable zone of an undersaturated solution where no nucleation or crystal growth is possible,

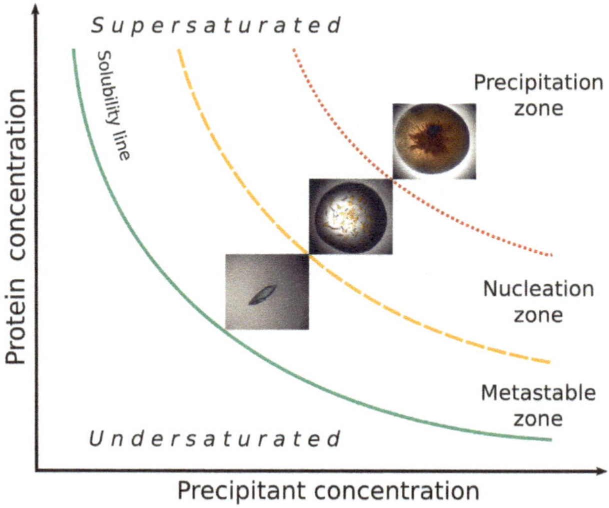

Fig. 1 A schematic representation of a phase diagram showing the solubility of a protein in solution as a function of the concentration of the precipitant.

(2) the supersaturated metastable zone where growth may occur but no spontaneous nucleation is observed, (3) the labile supersaturated zone of spontaneous and rapid nucleation, and (4) the precipitation zone where disordered structures (aggregates or precipitates) form faster than crystals.

4. Effect of temperature on protein crystallization

Temperature or precipitant gradients can be used to precisely and reversibly control the relative supersaturation levels of protein solutions. Temperature controls the balance between enthalpy and entropy effects on free energy and are typically comparable in magnitude. Depending on whether crystallization is enthalpy-driven or entropy-driven, proteins become either more soluble at higher temperatures (direct solubility) or less soluble at higher temperatures (reverse solubility) (Budayova-Spano, Dauvergne, Audiffren, Bactivelane, & Cusack, 2007; Oksanen et al., 2009).

The temperature influence is due to variation of the acid/base constants of the protein side chains (Chernov & Komatsu, 1995). In addition, the effective pK_a values of the ionizable groups are related to the ionic strength of the medium. As a result, the solubility increases with temperature when the ionic strength is low (i.e., solution contains low dielectric constant components) and the reverse. The temperature-solubility function is not

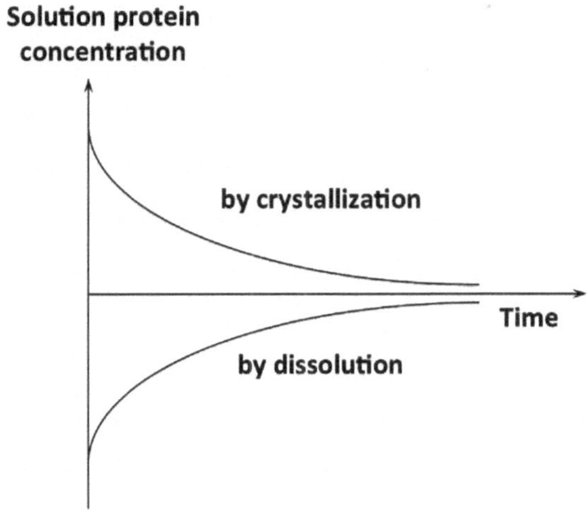

Fig. 2 Schematic diagram illustrating the principle of solubility measurements.

a property of the protein itself, but linked to the protein solution system. Temperature therefore influences nucleation, growth rate, protein solubility, and protein–protein interactions in solutions.

The determination of the solubility curve is carried out by measuring the protein concentration at the crystal-solution equilibrium at a given temperature (Fig. 2) (Cacioppo, Munson, & Lee Pusey, 1991; Ducruix & Ries-Kautt, 1990). This can be performed by introducing a small crystal to a supersaturated or undersaturated protein solution with crushed microcrystals in a batch set-up (see Section 5.2). As the crystals grow or dissolve, the protein concentration is monitored and determined regularly by removing aliquots and measuring the UV absorbance from tyrosine and tryptophan residues at $\lambda = 280\,\text{nm}$. When the concentration of the solution reaches a constant value, the system has reached equilibrium and this final protein concentration corresponds to the solubility (Fig. 2). Depending on the volume of the set-up it takes a few days to a week to measure a point on the solubility curve. An example of a solubility curve measured as a function of temperature in the case of hydrogenated recombinant urate oxidase complexed with 8-azaxanthin in the presence of H_2O and D_2O is described in Budayova-Spano et al. (2007).

The gradient of precipitant concentration or temperature can be used to affect the solubility and to modify the supersaturation rates of the sample in order to carefully manipulate crystal nucleation and crystal growth (Candoni et al., 2012). This can be also used to induce formation of different phases and growth mechanisms that lead to solution-mediated phase transitions. Temperature also affects the quantity and size of the crystals. It can be used to dissolve smaller crystals for the benefit of larger ones. By analogy to what is known from small molecule crystallization, the kinetic ripening mechanism has been proposed to account for this phenomenon (Madras & McCoy, 2003). In contrast to phase transitions, it concerns only the equilibrium between crystals of the same composition and structure, i.e., crystals within the same phase.

5. Seeding

A common approach to large crystal growth for NPX is seeding, either repeated macroseeding, crystal feeding, or microseeding into large vapor diffusion drops (Stura & Wilson, 1991, 1992; Thaller et al., 1985). Microseeding can generate better-diffracting crystals since crystals grow in the metastable zone (Asherie, 2004). Macroseeding or feeding of crystals tends to induce damage from repeated transfers or the addition of solutions and it is often difficult to suppress further nucleation or crystal imperfections from developing. As spontaneous nucleation cannot occur in undersaturated solutions (Fig. 1), crushed crystals (microseeds) introduced into this phase will dissolve. The same applies along the saturation boundary, but crystal seeds added will not dissolve or increase in size. Spontaneous homogeneous nucleation occurs above the saturation curve and the metastable region, in the supersaturation zone. The metastable zone is ideal for seeding and crystal growth. In addition, seeding into the supersaturation zone above the metastable region can speed up the nucleation remarkably and provide more consistent results (Luft & DeTitta, 1999; Stura & Wilson, 1992). Finally, beyond the supersaturation zone seeding can result in excessive nucleation but also induce formation of amorphous precipitate (Asherie, 2004). The size and number of crystals must be manipulated by using a dilution series of the microseed stock solution (Luft & DeTitta, 1999).

5.1 Large volume, sitting drop vapor diffusion

In this section we will first focus on large volume, sitting drop vapor diffusion methods and approaches that have worked for three different proteins. Table 2 is a summary of crystal properties for the examples of carbonic anhydrase II (CA II), the carbohydrate-binding domain X-2 Leu100Phe mutant (X-2 L100F), and carbonic anhydrase IX surface variant (CA IX_{SV}). Both CA II and X-2 L100F were successfully crystallized to large volumes in our

Table 2 Protein and crystal properties.

	CA II	X-2 L100F mutant	CA IX_{SV}
[protein] mg/mL	10–20	13	17
Space group	$P2_1$	$P2_12_12_1$	$P2_1$
Unit cell parameters (Å, °)	$a=42$, $b=41$, $c=72$, $\beta=104°$	$a=73$, $b=50$, $c=45$	$a=44$, $b=65$, $c=47$, $\beta=115°$
Unit cell volume ($Å^3$)	~120,000	~164,000	~122,000
Crystal volume (mm^3)	2.0	1.6	1.4
X/N resolution (Å)	1.6/1.8	1.6/1.6	1.3/2.7[a]
Solvent content (%)	~40	~46	~43
PDB ID	4q49	5dpn	6rqn
Reference	Michalczyk et al. (2015)	Ohlin et al. (2015)	Koruza et al. (2019)

[a]Tested on iBIX at J-PARC (user program proposal no. 2019PX0020).

laboratory using the same approach. As such, only details for CA II are presented here. A different approach with seeding was used for the challenging protein CA IX_{sv}.

Besides offering an appropriate support for large volume sitting drop experiments, there are additional benefits to using the "sandwich box" (Hampton Research) system: these include the possibility to H/D exchange drops/crystals, pH control of drops/crystals through vapor acidification, and adding ligands (e.g., dry co-crystallization) (Gelin et al., 2015; Michalczyk et al., 2015).

5.1.1 Carbonic anhydrase II (CA II)

For this project the crystallization conditions were well known and described, however, systematic scale-up and optimization for large crystal volume had not been attempted. Early literature on CA II showed that large crystals in the 1960s were grown using dialysis and ammonium sulfate as precipitant (Tilander, Strandeberg, & Fridborg, 1965). The ammonium sulfate protocol was subsequently adapted for hanging drop vapor diffusion and this was successfully used for many decades. Screening in the 2000s revealed sodium citrate also to be a good precipitant for CA II (Budayova-Spano, Fisher, et al., 2006). However, crystals took longer to appear (several days instead of overnight) but actually yielded better crystals with less tendency to aggregate.

A B

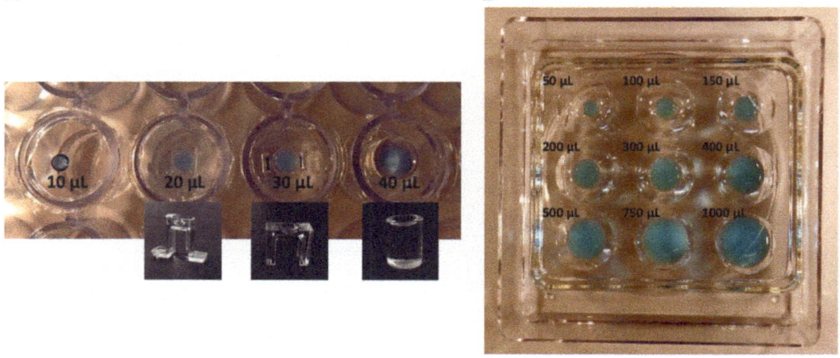

Fig. 3 The hardware used for scaling-up progression in vapor diffusion. (A) Hanging drop 10 μL of 0.01% methylene blue solution (left) and sitting drops in small to medium volumes (20, 30 and 40 μL) using different types of plastic microbridges and glass rod supports from Molecular Dimensions or Hampton Research (maximum volumes 40, 45, 100 μL, respectively) in a standard Linbro 24-well plate (Hampton Research). (B) Sitting drops in large volumes in sandwich box plate set-up available from Hampton Research. 9-Well siliconized Pyrex dish and the volumes from 50 μL to ~1000 μL of 0.01% methylene blue solution. The plate sits on the lid of standard petri dish over a reservoir (25 mL). The box is sealed with vacuum grease.

Fig. 3 illustrates the progression of scaling-up and the hardware used: hanging drops in small volumes, 2–10 μL (Fig. 3A), sitting drops in small to medium volumes (Fig. 3B), and finally sitting drops in very large volumes (Fig. 3B). However, our experiments showed that the crystal growth protein and precipitant concentration that was optimized for small hanging drops cannot be maintained through the scaling-up process, and the conditions require further optimization.

Fine screening in our lab showed that 1.1–1.3 M sodium citrate with 50–100 mM Tris pH 7.8 or higher was the best in small (10 μL) drops using conventional Linbro 24-well plates and hanging drops on 22 mm cover slides sealed with vacuum grease (Fig. 3A). It was also seen that the protein will crystallize at concentrations ranging from 10 to 50 mg/mL or even higher. This condition with a protein concentration of ~25 mg/mL was selected as the optimum for scaling-up in preparation for neutron studies. The first thing was to ensure that the same condition works for the same precipitant and protein combination and then switch to the same volume but in a sitting drop format. Once verified, systematic scale-up of drop volumes could begin, starting in microbridges and going from 10 to 40 μL. From here we scaled up to large drops (Fig. 3B) starting at 100 μL.

Our preferred hardware is the sandwich box plate set-up available from Hampton Research. In this set-up the drops are prepared in the 9-well siliconized Pyrex plates and the drop volumes can range from a few μL to ~1000 μL (Fig. 3B). The plate sits on the lid of petri dish over a reservoir that is recommended by Hampton Research to be 25 mL. The box is sealed with vacuum grease and left to undergo vapor diffusion. Our first attempts with 100 μL drop volumes (50:50 protein to precipitant ratio), performed even at higher protein concentration (~35 mg/mL), did not yield any crystals in more than 6 months and the drops remained clear. But when sodium citrate salt was directly added to the reservoir in attempt to speed up evaporation and hopefully nucleation, within a week or so crystals started appearing and grew to maximum size within a month. It was calculated that about 30% more precipitant was present in the reservoir than in the drop with the protein. The final set of conditions that were used to give consistently large, well-diffracting crystals were: 1.10–1.15 M sodium citrate, 50 mM Tris pH 8.5 in the drop, and 1.3–1.4 M sodium citrate, 50 mM Tris pH 8.5 in the reservoir (Fig. 4). We found we could use these conditions to grow crystals from more dilute protein sample (10–20 mg/mL), enabling scale-up to 500–600 μL drops. Using this approach consistently gave crystals between 1.0 and 2.0 mm^3 (Table 2). Numerous neutron crystal structures were solved from CA II crystals grown in this way (Aggarwal et al., 2016; Fisher, Aggarwal, Kovalevsky, Silverman, & McKenna, 2012; Fisher et al., 2010, 2011; Koruza et al., 2019; Kovalevsky et al., 2018; Michalczyk et al., 2015). The same approach of increasing the precipitant concentration in the reservoir

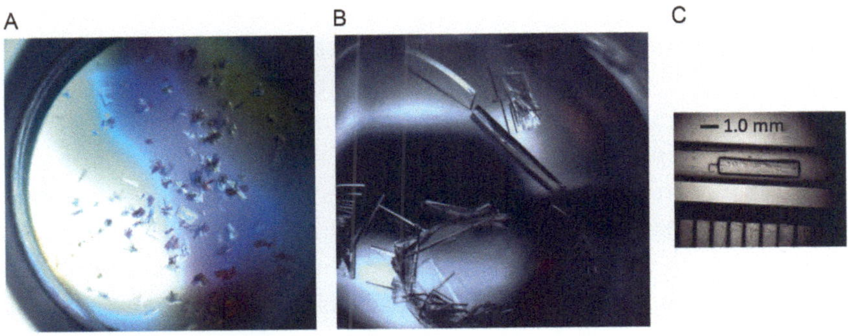

Fig. 4 Sitting drop vapor diffusion set-ups for crystallization of CA II. (A) The drop with initial hits and (B) the optimized condition that was used to give large crystals (1.10–1.15 M sodium citrate, 50 mM Tris pH 8.5 in the drop, and 1.3–1.4 M sodium citrate, 50 mM Tris pH 8.5 in the reservoir). (C) Crystal (~2 mm^3) mounted in quartz capillary.

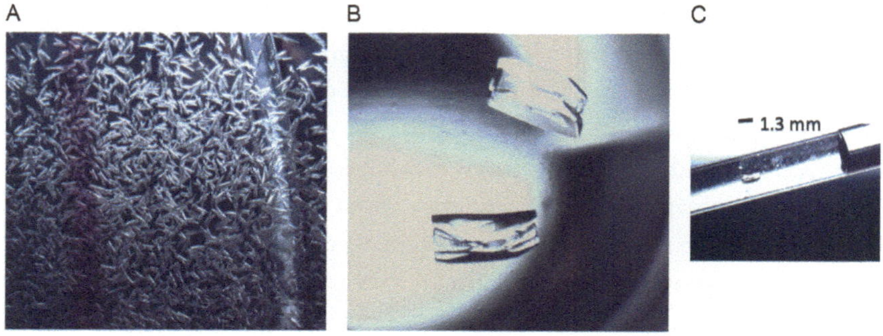

Fig. 5 Sitting drop vapor diffusion set-ups for crystallization of carbohydrate-binding domain Leu100Phe mutant of X-2. (A) Photograph of the drop with initial hits and (B) the optimized condition that was used to give large crystals (25% PEG 1500, 0.1 M MES pH 5.5 in the drop, and 28% PEG 1500, 0.1 M MES pH 5.5 in the reservoir). (C) Crystal (~1.6 mm^3) mounted in quartz capillary.

by around 30% worked precisely the same way for phosphate binding protein (PBP), X-2 L100F (Fig. 5; Table 2), and KDN9P phosphatase (Bryan et al., 2013; Ohlin et al., 2015; Sippel, Bacik, Quiocho, & Fisher, 2014).

The main challenge was that conditions that were very successful in small hanging or sitting drops did not necessarily scale to large volumes, especially when it came to drops larger than 100 μL. Some care and thought has to go into what the main constraint is, and then adjust the conditions to adapt to the larger volumes. In the case of the proteins mentioned above it was sufficient to simply add more precipitant to the reservoir and in effect to drive vapor diffusion a little faster.

5.1.2 Human carbonic anhydrase IX surface variant (CA IX_{SV})

CA IX_{SV} crystallization started with sparse-matrix screening using commercial kits and high-throughput robotics. Identification of small crystals by optical microscopy allowed for the set-up of multiple identical drops, and once crystals were formed they were subsequently used for seed stock preparation. Seed stock is prepared by crushing small crystals in mother liquor and diluted with precipitant solution. The diluted seed stock is then added to crystallization set-ups and provides nucleation sites (D'Arcy, Mac Sweeney, & Haber, 2003).

We used seeding for growing large crystals of CA IX_{SV} as described in Koruza et al. (2018). Seed stocks were prepared by following the instructions from Seed Bead™ (www.hamptonresearch.com; accessed July 2019) and briefly described here. The tube containing a plastic bead was cooled down on ice while opening the 20 μL sitting drop well with crystals, that we previously identified for creating the seed stock. Crystals were crushed with a glass probe and 5 μL of reservoir solution was added from the reservoir to the drop well containing the crushed crystals. The mixture from the drop well was aspirated and dispensed several times in order to dislodge crystals stuck to the plate. 5 μL of the mixture was pipetted from the drop to the tube on ice and this step was repeated for total of 5–10 times, always adding 5 μL from the reservoir solution. The whole procedure was observed under the microscope in order to transfer all of the crushed crystals, especially the ones that might be sticking to the plate, into the tube with the bead. Finally, there was approximately 50 μL of solution containing crushed crystals in the tube. In last step of seed stock preparation, the tube was vigorously vortexed for approximately 3 min, stopping every 30 s to cool the tube on ice. In this case the suspension was used for microseed matrix screening (MMS), where seed crystals are systematically added to a crystallization screen. In an effort to reduce the number of crystals, increase crystal size, and slow the crystal growth in large crystallization drops, serial dilutions of seed stock up to 1 in 100,000 were prepared, starting by mixing 5 μL of undiluted seed stock with 45 μL of well solution. Five dilutions were prepared beside the undiluted seed stock (marked as 1×10^{-1}, 1×10^{-2}, 1×10^{-3}, 1×10^{-4}, 1×10^{-5}). Before pipetting the 5 μL from the previous dilution, the tube was cooled on ice and solution was mixed and vortexed for at least 30 s. All seed stocks were frozen immediately after at −80 °C, fresh seeds are better than old seeds when setting up drops for large crystal growth. The serial dilutions were tested by setting up the vapor diffusion and batch drops with one of the known crystallization conditions (Figs. 6 and 7) (0.1 M Tris pH 8.5, 30% PEG 4000, 0.2 M sodium acetate) in ratio 3:2:1 (protein:precipitant:seed stock) (D'Arcy, Villard, & Marsh, 2007) as described in detail in Koruza et al. (2018). Before pipetting the seed stock, it has to be vortexed in case the suspended crystals have settled in the tube. It was observed that seed dilution 1×10^{-4} consistently produced the largest crystals of CA IX_{SV}. Five known crystallization conditions were tested for large crystal growth by setting up the sandwich box plate (Hampton Research) (Fig. 3B) with 25 mL reservoir solution. Concentration of the protein was 17.5 mg/mL. Initially the ratio 3:2:1 was used for setting up 100 μL drops, but over-nucleation was observed after a week as in Fig. 6, therefore drops with 3-times more protein and 5-times more precipitant were set up in

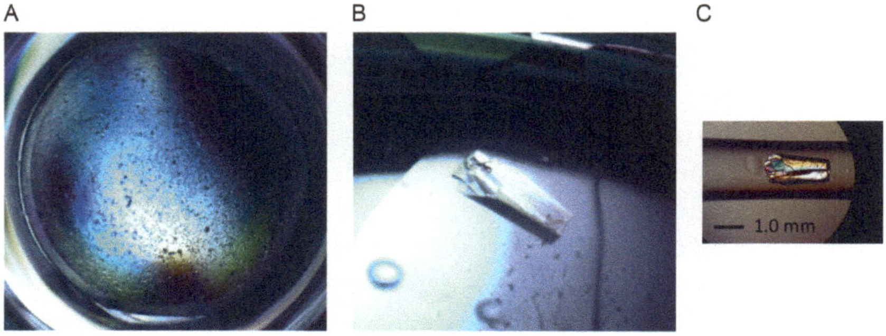

Fig. 6 Sitting drop vapor diffusion set-ups for crystallization of CA IX_{SV}. (A) The drop with initial hits and (B) the optimized condition that was used to give large crystals in ratio 3:2:1 (protein:precipitant:seed stock), where precipitant is 24% PEG 3350, 10% isopropanol, 0.1 M Tris pH 8.5 and seeds were used in dilution 1×10^{-4}. (C) Crystal (~1.4 mm^3) mounted in quartz capillary.

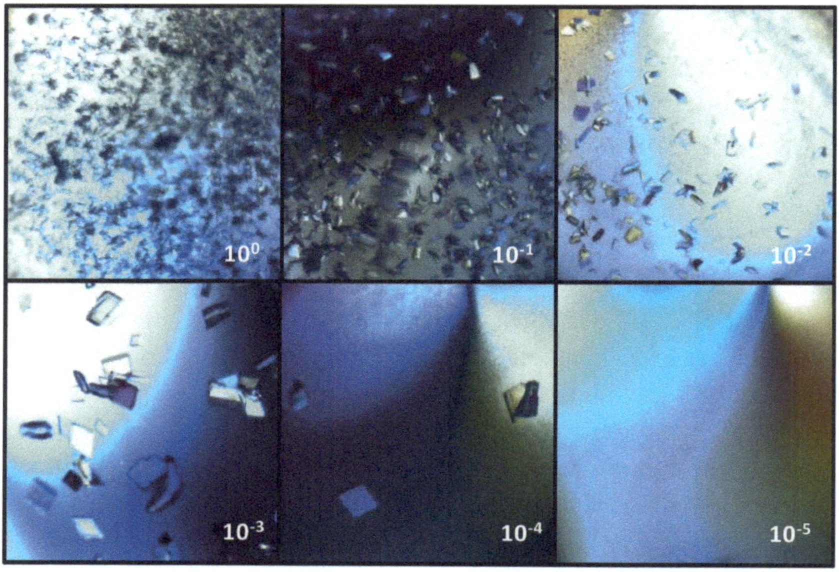

Fig. 7 The effect of diluting and testing the seed stock solution for optimal nucleation in vapor diffusion sitting drops. Drops were set up in a ratio of 3:2:1 (protein:precipitant: seed stock). Conditions used: 0.1 M Tris pH 8.5, 30% PEG 4000, 0.2 M sodium acetate, total drop volume was 24 μL.

150 μL (ratio 9:10:1) with seed stock dilution 1×10^{-4} (Fig. 7). The boxes were sealed with vacuum grease and left to undergo vapor diffusion for 2 months. The only condition that yielded crystals with volume more than 1 mm^3 (Fig. 6) was 24% PEG 3350, 10% isopropanol, 0.1 M Tris pH 8.5.

5.2 Batch

In conventional batch crystallization a closed vessel is used (e.g., a capillary, small container, or plate with a small reservoir such as the Macro-Store PlateTM from Hampton Research), and the sample is mixed with the crystallization agent at a concentration allowing the supersaturation to be instantaneously reached, creating a homogeneous crystallization medium requiring no equilibration with a reservoir through evaporation or diffusion (Rayment, 2002). Initial batch experiments can be used to determine the conditions where the protein crystallizes spontaneously. The components of the batch experiment are mostly identical to those in the original crystallization screen that have been used in vapor diffusion hanging/sitting drop. The main difference is the kinetic path that nucleation and crystal growth will follow on a schematic phase diagram (Fig. 8). The optimal initial supersaturation (protein and/or precipitant concentration) for batch crystallization technique and vapor diffusion will be different, always lower, and have to be optimized for a given protein (Chayen, 1998). The search for batch conditions is done in two stages: (1) discover what concentration of precipitant is required to obtain spontaneous crystal growth by direct mixing and (2) search for conditions that support crystal growth. Usually crystallization by direct mixing occurs at about 60–80% of the concentration of the precipitant required in a hanging drop vapor diffusion experiment (Chayen, 1998). Once the initial

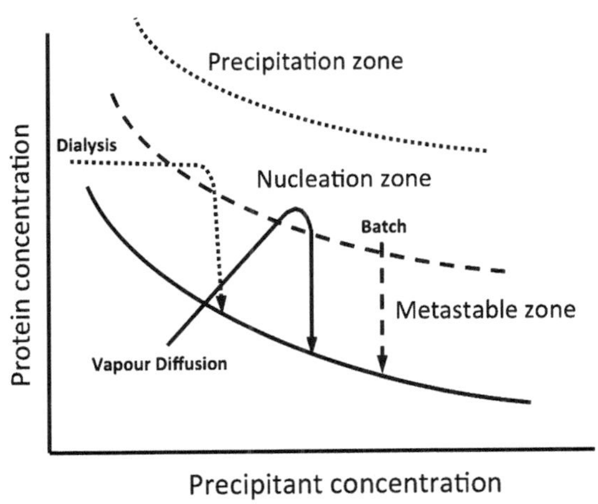

Fig. 8 Schematic view of a phase diagram as a function of the concentration of precipitant, with the kinetic trajectories illustrated by arrows in the case of different crystallization modes: comparison between vapor diffusion (solid line), batch (dotted line) and dialysis (dashed line) salting-out experiments.

conditions that yield crystals in batch have been defined, refinement of the crystallization protocol usually involves adjusting the protein and precipitant concentration, pH, salt, and additives. This is similar to vapor diffusion except it is possible to use larger crystallization volumes at lower protein or precipitant concentrations to control nucleation and crystal growth.

Batch crystallizations can be set-up in a large range of volumes with no upper limit in principle. In small volumes (0.5–1 μL) it is called microbatch, and drops are immersed in an inert oil (e.g., paraffin oil) to avoid evaporation. It is not very useful for growing large crystals and instead is used in initial stages to optimize the crystallization conditions for larger volumes. An alternative batch method without the use of oil is by thorough and direct mixing of protein and precipitant. Into a well-sealed container or in the presence of a reservoir with the same precipitant concentration to limit evaporative concentration. Under-oil set-ups can suffer from the slow evaporation of water from the drops, this can result in formation of salt deposits that interfere with protein crystal growth.

Traditional crystallization batch methods have been successfully used to grow large crystals for neutron studies of several proteins, including aspartic proteinase endothiapepsin, HIV-protease, porcine insulin, human deoxyhemoglobin, equine cyanomethemoglobin, D-xylose isomerase, beta-lactamase, and pyridoxal 5′-phosphate enzyme (Coates et al., 2008; Dajnowicz et al., 2016, 2017; Kovalevsky et al., 2010; Matsumura et al., 2008; Tomanicek et al., 2011; Wlodawer, Savage, & Dodson, 1989).

5.2.1 Urate oxidase (UO)

Urate oxidase (UO) was crystallized in complex with different substrate analogs with a temperature-controlled batch device and methodology developed specifically with NPX in mind (Fig. 9A) (Budayova-Spano et al., 2007; Oksanen et al., 2009). It combines temperature control with seeding in a batch set-up to drive the process of crystallization and allow the manipulation of the kinetics of the crystallization process, taking advantage of the information from the phase diagram.

The crystallization solution is kept in the metastable zone by regulating the temperature of the crystallization solution. The crystallization solution (100–500 μL) is poured into a specially designed quartz cell with an optically clear bottom covered with a quartz air-tight cap, which is attached to a brass support incorporating a single well or several wells of the crystal growth apparatus maintained at the same temperature. The accessible temperature range is 233–353 K, with the temperature controlled by Peltier elements

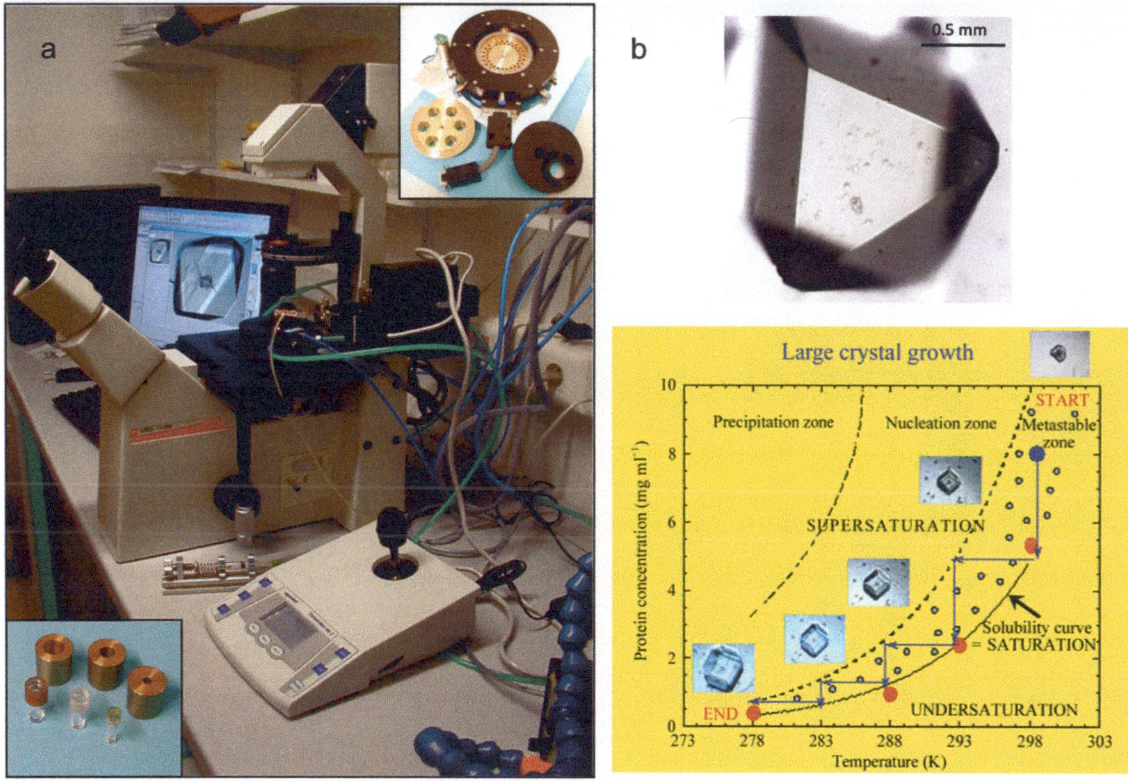

Fig. 9 (A) A semiautomated protein crystal growth setup combining the optical microscopy, video, and PC equiped with software for crystal visualization and temperature control. At the top right is shown crystal growth apparatus (a versatile carussel) incorporating several wells or a single well in thermal contact with Peltier elements. Bottom left, are represented quartz tubes of various volumes with optical bottom, housing crystallization solution, covered with quartz air tight cap and (B) large crystal of UO in complex with cyanide grown with device and methodology to approximate dimensions 1.2 × 1.2 × 1 mm. Scale bar = 0.5 mm. *Images reproduced with permission from Royal Society (Budayova-Spano, M., Dauvergne, F., Audiffren, M., Bactivelane, T., & Cusack, S. (2007). A methodology and an instrument for the temperature-controlled optimization of crystal growth.* Acta Crystallographica Section D, 63, *339–347; Oksanen, E., Blakeley, M. P., Bonneté, F., Dauvergne, M. T., Dauvergne, F., & Budayova-Spano, M. (2009). Large crystal growth by thermal control allows combined X-ray and neutron crystallographic studies to elucidate the protonation states in Aspergillus flavus urate oxidase.* Journal of the Royal Society Interface, 6, *599–610).*

to an accuracy of 0.1 K. The use of Peltier elements helps to improve temperature regulation. The microscope and computer allows visualization and temperature control in real time. Knowledge of the phase diagram for UO and the ability to control the temperature to drive crystallization allows for custom-made crystallization experiments. The final crystal volumes obtained in this way for Uox-8-azaxanthine complex, 9-methyl urate complex, urate complex with cyanide and urate complex were 4, 2, 1.5 and 1 mm^3, respectively, and yielded high-quality neutron and X-ray diffraction data (Oksanen et al., 2009). Fig. 9B shows the picture of a large crystal ($\sim$1.2 mm^3) crystal of UO in complex with cyanide that was prepared using the device and methods described here.

The same set-up was also used to grow large crystals for several perdeuterated proteins, such as yeast inorganic pyrophosphatase (PPase) with a volume of about 0.7 mm^3 (Budayova-Spano et al., 2007), and also wild-type haloalkane dehalogenase (DhaA) from *Rhodococcus rhodochrous* with a volume of about 0.25 mm^3 (Stsiapanava et al., 2011). PPase crystals are unstable and only yielded medium neutron diffraction resolution to 2.8 Å. In the case of DhaA, following a successful preliminary neutron diffraction experiment, a full data set to 1.85 Å resolution was collected at room temperature using the LADI-III beamline at the Institut Laue-Langevin (ILL).

5.3 Dialysis

In dialysis, a low molecular weight precipitant easily penetrates a semipermeable membrane and equilibrates into the protein solution, slowly approaching the concentration at which the macromolecule crystallizes by the salting-out effect (Figs. 1 and 8). The high molecular weight of the macromolecule prevents its diffusion through the membrane on the basis of size exclusion determined by the pore size of the membrane. The transfer of solutes and solvent through semipermeable membranes provides the gradual variation in conditions required for early crystal nucleation and for relatively undisturbed lattice formation of crystal growth (Zeppezauer, Eklund, & Zeppezauer, 1968).

In typical dialysis set-ups the protein solution is contained within a dialysis chamber (e.g., dialysis buttons from Hampton Research, Aliso Viejo, CA, USA), which is sealed with a semipermeable dialysis membrane. The dialysis button is then placed into a suitable container that holds a reservoir of precipitant solution against which the content of dialysis button will be equilibrated through the membrane. Equilibration against the precipitant in the surrounding solvent slowly drives the protein toward supersaturation within the dialysis chamber, eventually resulting in nucleation and crystal growth. Dialysis buttons can be bought in different sizes allowing one to set up drops up to 350 μL in volume.

The precipitating solution can also be varied, simply by removing the initial precipitant solution and exchanging it with another. The protein material can thus be reused until the correct conditions for nucleation and crystal growth are found, providing the protein is still soluble and has not precipitated and denatured.

Dialysis can also be used to exploit the salting-in region of the phase diagram where solubility increases with increasing precipitant concentration. Specifically, for a charge-neutral species (i.e., proteins at their pI) the salting-in effect dominates at low salt concentrations in that protein solubility increases as the addition of ions in solution disrupts attractive protein–protein interactions. Further increasing the salt concentration strengthens attractive protein–protein interactions as the salting-out effect begins to dominate and protein solubility decreases. The surface charge density of a protein has a strong effect on the salting out/in behavior. At a pH close to the pI or for a large-size protein with a small number of either positive or negative net charges (i.e., low surface charge density), only the monotonic salting-in behavior could be observed because the charge neutralization process is less dramatic. On the other hand, when a protein has high surface charge density due to being small or having a large number of positive charges, the anions might not completely neutralize the positive charges, therefore causing a decrease in protein solubility (Zhang, 2012). Accordingly, in the salting-in region the protein can be crystallized by lowering the precipitant concentration and forcing the protein out of solution. Dialysis also allows the supersaturation to be varied in a reversible manner, so that crystals can be grown and dissolved providing no precipitation or denaturation occurs.

Traditional methods of dialysis, including using commercially available micro-dialysis chambers, have been successfully applied for growing neutron diffraction quality crystals for a number of proteins, these include of D-xylose isomerase (Hanson et al., 2004), porcine insulin (Maeda, Chatake, Tanaka, Ostermann, & Niimura, 2004) and cytochrome *c* peroxidase (Casadei et al., 2014).

In recent years, the instrument and methods discussed in Section 5.2 and Fig. 9 has been adapted to incorporate a dialysis button in addition to batch, and is implemented in the new version of the instrument with automated flow and temperature control, the OptiCrys instrument is shown in Fig. 10A (Budayova-Spano et al., 2007; Junius et al., 2016). The protein solution is held in a specially designed stainless steel (or titanium) dialysis button with a polycarbonate (or quartz/glass window) at the bottom. It is separated from the precipitant solution by a dialysis membrane with the molecular weight cut-off chosen as appropriate for the size of the protein (Fig. 10B). The dialysis membrane is placed over the top of the dialysis chamber containing the sample (a variety of sizes of 25–200 μL are available) and is held in place with an elastic ring in a groove. A stainless steel reservoir holding 1 mL is placed over the dialysis button in order to ensure water tightness. The reservoir is sealed with vacuum grease and a glass cover slide or sealing tape. The dialysis setup is attached to a brass support incorporating one or several wells of the temperature-controlled crystal growth apparatus as shown in Fig. 9 for batch crystallization. With this setup, control of the temperature of the crystallization experiment is performed in an automated way, while the chemical composition of the reservoir solution must be changed manually by the user. In an updated version of the instrument (Fig. 10A) the dialysis button is replaced by a fluidic assembly functioning as a continuous flow cell (Fig. 10B). It is composed of the dialysis chamber with the semipermeable membrane, located on the bottom, and the reservoir chamber on the top connected to a pressure-driven flow control system (Fluigent) ensuring also automated regulation of the chemical composition (Junius et al., 2016). In both configurations, to recover the crystal, the reservoir is removed and a surgical blade is used to cut the membrane, creating an opening for crystal removal.

Using this system it is then possible to manipulate multiple parameters (e.g., temperature, precipitant concentration, pH) so that the state of the protein/crystal can be observed as it moves along a well-defined kinetic trajectory in the phase diagram. Any combination of precipitant concentration and temperature can be explored in a systematic manner by sampling a continuum of potential crystal-producing conditions without physically perturbing the mother liquor while the total volume of protein solution remains constant during the entire experiment. Due to the flexibility of the system, as little or as much sample can be used as desired.

To validate the method beyond model systems, like chicken white egg lysozyme, we have successfully tested the setup with a number of proteins for which large, single and well-diffracting crystals were previously not available or were difficult to obtain. These include: recombinant urate oxidase from *Aspergillus flavus* (Budayova-Spano, Bonnete, et al., 2006), fluorescent protein EosFP from *Lobophyllia hemprichii* (Wiedenmann et al., 2004), lactate

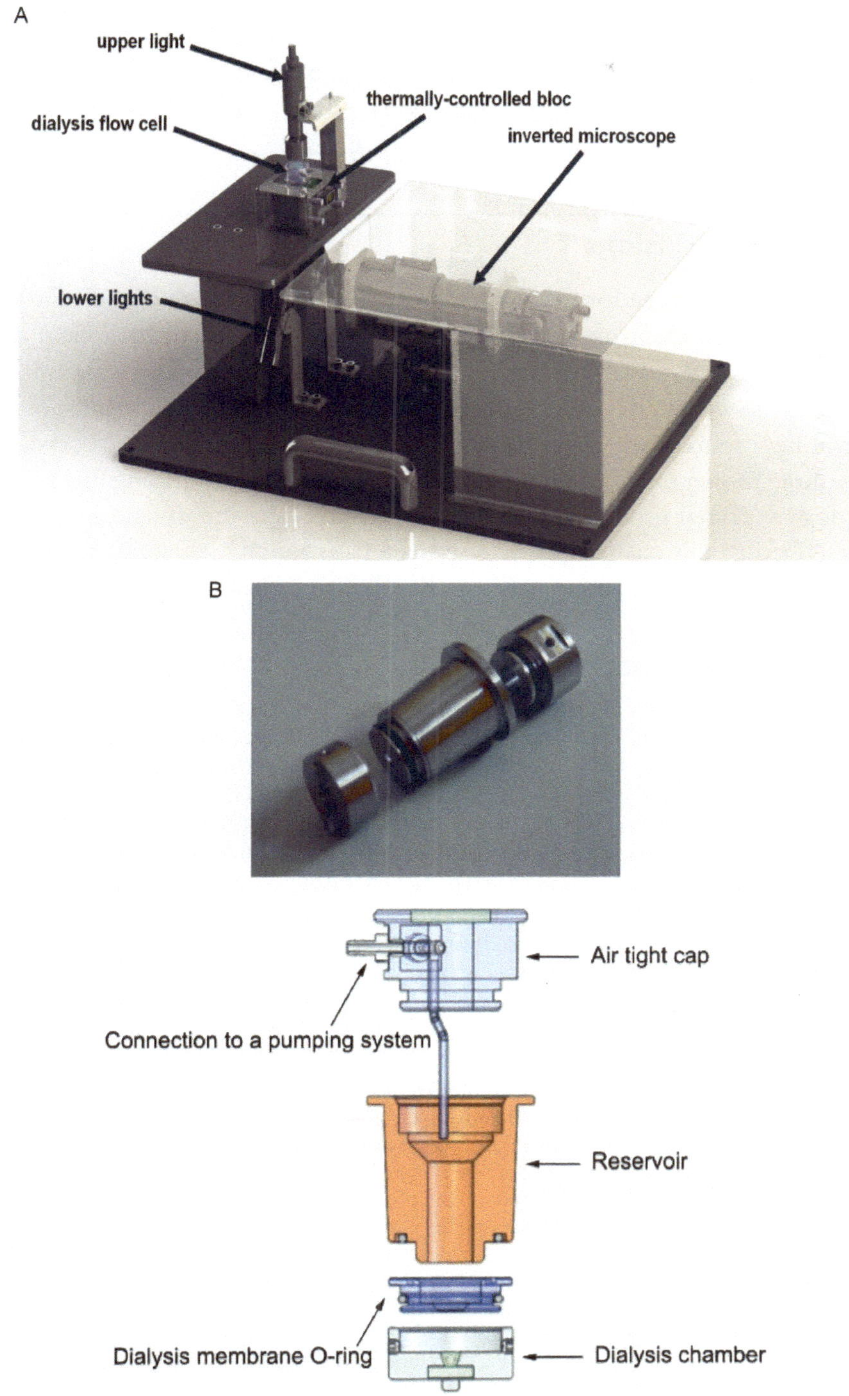

Fig. 10 (A) Simplified view of a crystallization apparatus (OptiCrys) for temperature-controlled flow cell dialysis with real-time visualization (Junius, Vahdatahar, Oksanen, Ferrer, & Budayova-Spano, 2020) and (B) photograph and a schematic view of the temperature-controlled flowing reservoir dialysis setup (Junius et al., 2016).

dehydrogenase from *Thermus thermophilus* (Coquelle, Fioravanti, Weik, Vellieux, & Madern, 2007) and YchB kinase from *Agrobacterium tumefaciens* (Junius et al., 2020; Junius, Vahdatahar, et al., 2020).

6. Conclusions

The landscape for NPX is rapidly changing, and with the advent of powerful new neutron sources, like the Spallation Neutron Source at Oak Ridge National Laboratory and the future European Spallation Source in Sweden, it is expected that this field will only expand. While powerful new sources will drive down the size requirement of protein crystals required for neutron experiments, there are still benefits to having fine control over crystal growth and diffraction quality. The theory and methods described in this chapter can be used as a basis for any structural biologist looking to expand their experimental toolkit to include NPX.

Acknowledgments

The authors would like to acknowledge SINE2020 (European Union's Horizon 2020 research and innovation program under grant agreement No. 654000) for providing partial funding for some of the methodology development described in this paper. M.B.S. acknowledges the EU under the DLAB contracts HPRI-CT-2001-50035 and RII3-CT-2003-505925, the MRCT CNRS under the contract 2010-2011, LABEX VALO GRAL under the contract 2015 as well as the European Union's Horizon 2020 Research and Innovation Program under the MARIE SKŁODOWSKA—CURIE grant agreement no. 722687 for providing funding for instrumental and methodological developments described in this paper. IBS acknowledges integration into the Interdisciplinary Research Institute of Grenoble (IRIG, CEA).

References

Aggarwal, M., Kovalevsky, A. Y., Velazquez, H., Fisher, S. Z., Smith, J. C., & McKenna, R. (2016). Neutron structure of human carbonic anhydrase II in complex with methazolamide: Mapping the solvent and hydrogen-bonding patterns of an effective clinical drug. *International Union of Crystallography Journal*, *3*, 319–325.

Asherie, N. (2004). Protein crystallization and phase diagrams. *Methods*, *34*, 266–272.

Blakeley, M. P. (2009). Neutron macromolecular crystallography. *Crystallography Reviews*, *15*, 157–218.

Blakeley, M. P., Hasnain, S. S., & Antonyuk, S. V. (2015). Sub-atomic resolution X-ray crystallography and neutron crystallography: Promise, challenges and potential. *International Union of Crystallography Journal*, *2*, 464–474.

Bryan, T., Gonzalez, J. M., Bacik, J. P., DeNunzio, N. J., Unkefer, C. J., Schrader, T. E., et al. (2013). Neutron diffraction studies towards deciphering the protonation state of catalytic residues in the bacterial KDN9P phosphatase. *Acta Crystallographica. Section F*, *69*, 1015–1019.

Budayova-Spano, M., Bonnete, F., Ferte, N., El Hajji, M., Meilleur, F., Blakeley, M. P., et al. (2006). A preliminary neutron diffraction study of rasburicase, a recombinant urate oxidase enzyme, complexed with 8-azaxanthin. *Acta Crystallographica. Section F, 62*, 306–309.

Budayova-Spano, M., Dauvergne, F., Audiffren, M., Bactivelane, T., & Cusack, S. (2007). A methodology and an instrument for the temperature-controlled optimization of crystal growth. *Acta Crystallographica Section D, 63*, 339–347.

Budayova-Spano, M., Fisher, S. Z., Dauvergne, M.-T., Agbandje-McKenna, M., Silverman, D. N., Myles, D. A. A., et al. (2006). Production and X-ray crystallographic analysis of fully deuterated human carbonic anhydrase II. *Acta Crystallographica. Section F, 62*, 6–9.

Cacioppo, E., Munson, S., & Lee Pusey, M. (1991). Protein solubilities determined by a rapid technique and modification of that technique to a micro-method. *Journal of Crystal Growth, 110*, 66–71.

Candoni, N., Grossier, R., Hammadi, Z., Morin, R., & Veesler, S. (2012). Practical physics behind growing crystals of biological macromolecules. *Protein and Peptide Letters, 19*, 714–724.

Casadei, C. M., Gumiero, A., Metcalfe, C. L., Murphy, E. J., Basran, J., Concilio, M. G., et al. (2014). Heme enzymes. Neutron cryo-crystallography captures the protonation state of ferryl heme in a peroxidase. *Science, 345*, 193–197.

Chayen, N. E. (1998). Comparative studies of protein crystallization by vapour-diffusion and microbatch techniques. *Acta Crystallographica Section D, 54*, 8–15.

Cheng, X., & Schoenborn, B. P. (1990). Hydration in protein crystals. A neutron diffraction analysis of carbonmonoxymyoglobin. *Acta Crystallographica. Section B, 46*, 195–208.

Chernov, A. A., & Komatsu, H. (1995). Principles of crystal growth in protein crystallization. In J. P. van der Eerden & O. S. L. Bruinsma (Eds.), *Science and Technology of Crystal Growth*. Dordrecht: Springer.

Coates, L., Tuan, H. F., Tomanicek, S., Kovalevsky, A., Mustyakimov, M., Erskine, P. et al. (2008). The catalytic mechanism of an aspartic proteinase explored with neutron and X-ray diffraction. *Journal of the American Chemical Society, 130*, 7235–7237.

Coquelle, N., Fioravanti, E., Weik, M., Vellieux, F., & Madern, D. (2007). Activity, stability and structural studies of lactate dehydrogenases adapted to extreme thermal environments. *Journal of Molecular Biology, 374*, 547–562.

Dajnowicz, S., Johnston, R. C., Parks, J. M., Blakeley, M. P., Keen, D. A., Weiss, K. L., et al. (2017). Direct visualization of critical hydrogen atoms in a pyridoxal 5′-phosphate enzyme. *Nature Communications, 8*, 955.

Dajnowicz, S., Seaver, S., Hanson, B. L., Fisher, S. Z., Langan, P., Kovalevsky, A. Y. et al. (2016). Visualizing the Bohr effect in hemoglobin: Neutron structure of equine cyanomethemoglobin in the R state and comparison with human deoxyhemoglobin in the T state. *Acta Crystallographica Section D, 72*, 892–903.

D'Arcy, A., Mac Sweeney, A., & Haber, A. (2003). Using natural seeding material to generate nucleation in protein crystallization experiments. *Acta Crystallographica Section D, 59*, 1343–1346.

D'Arcy, A., Villard, F., & Marsh, M. (2007). An automated microseed matrix-screening method for protein crystallization. *Acta Crystallographica Section D, 63*, 550–554.

Ducruix, A. F., & Ries-Kautt, M. M. (1990). Solubility diagram analysis and the relative effectiveness of different ions on protein crystal growth. *Methods, 1*, 25–30.

Fisher, S. Z., Aggarwal, M., Kovalevsky, A. Y., Silverman, D. N., & McKenna, R. (2012). Neutron diffraction of acetazolamide-bound human carbonic anhydrase II reveals atomic details of drug binding. *Journal of the American Chemical Society, 134*, 14726–14729.

Fisher, S. Z., Kovalevsky, A. Y., Domsic, J. F., Mustyakimov, M., McKenna, R., Silverman, D. N., et al. (2010). Neutron structure of human carbonic anhydrase II: Implications for proton transfer. *Biochemistry, 49*, 415–421.

Fisher, Z., Kovalevsky, A. Y., Mustyakimov, M., Silverman, D. N., McKenna, R., & Langan, P. (2011). Neutron structure of human carbonic anhydrase II: A hydrogen-bonded water network "switch" is observed between pH 7.8 and 10.0. *Biochemistry, 50*, 9421–9423.

Gavira, J. A. (2016). Current trends in protein crystallization. *Archives of Biochemistry and Biophysics, 602*, 3–11.

Gelin, M., Delfosse, V., Allemand, F., Hoh, F., Sallaz-Damaz, Y., Pirocchi, M., et al. (2015). Combining 'dry' co-crystallization and in situ diffraction to facilitate ligand screening by X-ray crystallography. *Acta Crystallographica Section D, 71*, 1777–1787.

Hanson, B. L., Langan, P., Katz, A. K., Li, X., Harp, J. M., Glusker, J. P., et al. (2004). A preliminary time-of-flight neutron diffraction study of Streptomyces rubiginosus D-xylose isomerase. *Acta Crystallographica Section D, 60*, 241–249.

Hazemann, I., Dauvergne, M. T., Blakeley, M. P., Meilleur, F., Haertlein, M., Van Dorsselaer, A., et al. (2005). High-resolution neutron protein crystallography with radically small crystal volumes: Application of perdeuteration to human aldose reductase. *Acta Crystallographica Section D, 61*, 1413–1417.

Jancarrik, J., & Kim, S.-H. (1991). Spare matric sampling: A screening method for crystallization of protein. *Journal of Applied Crystallography, 24*, 409–411.

Junius, N., Jaho, S., Sallaz-Damaz, Y., Borel, F., Salmon, J. B., & Budayova-Spano, M. (2020). A microfluidic device for both on-chip dialysis protein crystallization and in situ X-ray diffraction. *Lab on a Chip*. https://doi.org/10.1039/C9LC00651F (accepted 25 September 2019).

Junius, N., Oksanen, E., Terrien, M., Berzin, C., Ferrer, J. L., & Budayova-Spano, M. (2016). A crystallization apparatus for temperature-controlled flow-cell dialysis with real-time visualization. *Journal of Applied Crystallography, 49*, 806–813.

Junius, N., Vahdatahar, E., Oksanen, E., Ferrer, J. L., & Budayova-Spano, M. (2020). Optimization of crystallization using dialysis combined with temperature control. *Journal of Applied Crystallography,* (accepted).

Koruza, K., Lafumat, B., Nyblom, M., Knecht, W., & Fisher, S. Z. (2018). From initial hit to crystal optimization with microseeding of human carbonic anhydrase IX—A case study for neutron protein crystallography. *Crystals, 8*, 434–445.

Koruza, K., Mahon, B. P., Blakeley, M. P., Ostermann, A., Schrader, T. E., McKenna, R., et al. (2019). Using neutron crystallography to elucidate the basis of selective inhibition of carbonic anhydrase by saccharin and a derivative. *Journal of Structural Biology, 205*, 147–154.

Kovalevsky, A., Aggarwal, M., Velazquez, H., Cuneo, M. J., Blakeley, M. P., Weiss, K. L., et al. (2018). "To be or not to be" protonated: Atomic details of human carbonic anhydrase-clinical drug complexes by neutron crystallography and simulation. *Structure, 26*, 383–390.

Kovalevsky, A. Y., Chatake, T., Shibayama, N., Park, S.-Y., Ishikawa, T., Mustyakimov, M., et al. (2008). Preliminary time-of-flight neutron diffraction study of human deoxyhemoglobin. *Acta Crystallographica. Section F, 64*, 270–273.

Kovalevsky, A. Y., Hanson, L., Fisher, S. Z., Mustyakimov, M., Mason, S. A., Trevor Forsyth, V., et al. (2010). Metal ion roles and the movement of hydrogen during reaction catalyzed by D-xylose isomerase: A joint X-ray and Neutron diffraction study. *Structure, 18*, 688–699.

Leal, R. M., Teixeira, S. C., Blakeley, M. P., Mitchell, E. P., & Forsyth, V. T. (2009). A preliminary neutron crystallographic study of an A-DNA crystal. *Acta Crystallographica. Section F, 65*, 232–235.

Luft, J. R., & DeTitta, G. T. (1999). A method to produce microseed stock for use in the crystallization of biological macromolecules. *Acta Crystallographica Section D, 55*, 988–993.

Luft, J. R., Wolfley, J., Jurisica, I., Glasgow, J., Fortier, S., & DeTitta, G. T. (2001). Macromolecular crystallization in a high throughput laboratory—The search phase. *Journal of Crystal Growth, 232*, 591–595.
Madras, G., & McCoy, B. J. (2003). Growth and ripening kinetics of crystalline polymorphs. *Crystal Growth & Design, 3*, 981–990.
Maeda, M., Chatake, T., Tanaka, I., Ostermann, A., & Niimura, N. (2004). Crystallization of a large single crystal of cubic insulin for neutron protein crystallography. *Journal of Synchrotron Radiation, 11*, 41–44.
Maeki, M., Yamaguchi, H., Yamashita, K., Nakamura, H., Miyazaki, M., & Maeda, H. (2011). Analysis of kinetic behavior of protein crystallization in nanodroplets. *Chemistry Letters, 40*, 825–827.
Manzoni, F., Saraboji, K., Sprenger, J., Kumar, R., Noresson, A.-L., Nilsson, U. J., et al. (2016). Perdeuteration, crystallization, data collection and comparison of five neutron diffraction data sets of complexes of human galectin-3C. *Acta Crystallographica Section D, 72*, 1194–1202.
Mason, S. A., Bentley, G. A., & McIntyre, G. J. (1984). Deuterium exchange in lysozyme at 1.4-Å resolution. In B. P. Schoenborn (Ed.), *Neutrons in biology*. Boston, MA: Springer US.
Matsumura, H., Adachi, M., Sugiyama, S., Okada, S., Yamakami, M., Tamada, T., et al. (2008). Crystallization and preliminary neutron diffraction studies of HIV-1 protease cocrystallized with inhibitor KNI-272. *Acta Crystallographica. Section F, 64*, 1003–1006.
Michalczyk, R., Unkefer, C. J., Bacik, J. P., Schrader, T. E., Ostermann, A., Kovalevsky, A. Y., et al. (2015). Joint neutron crystallographic and NMR solution studies of Tyr residue ionization and hydrogen bonding: Implications for enzyme-mediated proton transfer. *Proceedings of the National Academy of Sciences of the United States of America, 112*, 5673–5678.
Ohlin, M., von Schantz, L., Schrader, T. E., Ostermann, A., Logan, D. T., & Fisher, S. Z. (2015). Crystallization, neutron data collection, initial structure refinement and analysis of a xyloglucan heptamer bound to an engineered carbohydrate-binding module from xylanase. *Acta Crystallographica. Section F, 71*, 1072–1077.
Oksanen, E., Blakeley, M. P., Bonneté, F., Dauvergne, M. T., Dauvergne, F., & Budayova-Spano, M. (2009). Large crystal growth by thermal control allows combined X-ray and neutron crystallographic studies to elucidate the protonation states in Aspergillus flavus urate oxidase. *Journal of the Royal Society Interface, 6*, 599–610.
Rayment, I. (2002). Small-scale batch crystallization of proteins revisited: An underutilized way to grow large protein crystals. *Structure, 10*, 147–151.
Saridakis, E., & Chayen, N. E. (2000). Improving protein crystal quality by decoupling nucleation and growth in vapor diffusion. *Protein Science, 9*, 755–757.
Sippel, K. H., Bacik, J., Quiocho, F. A., & Fisher, S. Z. (2014). Preliminary time-of-flight neutron diffraction studies of *Escherichia coli ABC* transport receptor phosphate-binding protein at the protein crystallography station. *Acta Crystallographica. Section F, 70*, 819–822.
Stsiapanava, A., Chaloupkova, R., Fortova, A., Brynda, J., Weiss, M. S., Damborsky, J., et al. (2011). Crystallization and preliminary X-ray diffraction analysis of the wild-type haloalkane dehalogenase DhaA and its variant DhaA13 complexed with different ligands. *Acta Crystallographica. Section F, 67*, 253–257.
Stura, E. A., & Wilson, I. A. (1991). Applications of the streak seeding technique in protein crystallization. *Journal of Crystal Growth, 110*, 270–282.
Stura, E. A., & Wilson, I. A. (1992). Seeding techniques. In A. Ducruix & R. Giege (Eds.), *Crystallization of nucleic acids and proteins: A practical approach*. Oxford: Oxford University Press.
Tanaka, I. (2019). Necessity and effectiveness of a biological diffractometer for installation in second target station as a new neutron pulsed source in Japan. Proceedings of the 3rd

international symposium of quantum beam science at Ibaraki university "quantum beam science in biology and soft materials". *Journal of the Physical Society of Japan, 29,* 0110151–0110155.

Thaller, C., Eichele, G., Weaver, L. H., Wilson, E., Karlsson, R., & Jansonius, J. N. (1985). Seed enlargement and repeated seeding. In H. W. Wycoff, C. H. W. Hirs, & S. N. Timasheff (Eds.), *Methods in enzymology*. New York: Academic Press.

Tilander, B., Strandeberg, B., & Fridborg, K. (1965). Crystal structure studies on human erythrocyte carbonic anhydrase C. (II). *Journal of Molecular Biology, 12,* 740–760.

Tomanicek, S. J., Wang, K. K., Weiss, K. L., Blakeley, M. P., Cooper, J., Chen, Y., et al. (2011). The active site protonation states of perdeuterated Toho-1 β-lactamase determined by neutron diffraction support a role for Glu166 as the general base in acylation. *FEBS Letters, 585,* 364–368.

Vekilov, P. G., & Vorontsova, M. A. (2014). Nucleation precursors in protein crystallization. *Acta Crystallographica. Section F, 70,* 271–282.

Wiedenmann, J., Ivanchenko, S., Oswald, F., Schmitt, F., Rocker, C., Salih, A., et al. (2004). EosFP, a fluorescent marker protein with UV-inducible green-to-red fluorescence conversion. *Proceedings of the National Academy of Sciences of the United States of America, 101,* 15905–15910.

Wlodawer, A., Savage, H., & Dodson, G. (1989). Structure of insulin: Results of joint neutron and X-ray refinement. *Acta Crystallographica. Section B, 45,* 99–107.

Wlodawer, A., & Sjolin, L. (1982). Hydrogen exchange in RNase A: Neutron diffraction study. *Proceedings of the National Academy of Sciences of the United States of America, 79,* 1418–1422.

Yamada, T., Kurihara, K., Ohnishi, Y., Tamada, T., Tomoyori, K., Masumi, K., et al. (2013). Neutron and X-ray crystallographic analysis of the human α-thrombin–bivalirudin complex at pD 5.0: Protonation states and hydration structure of the enzyme–product complex. *Biochimica et Biophysica Acta, 1834,* 1532–1538.

Zeppezauer, M., Eklund, H., & Zeppezauer, E. S. (1968). Micro diffusion cells for the growth of single protein crystals by means of equilibrium dialysis. *Archives of Biochemistry and Biophysics, 126,* 564–573.

Zhang, J. (2012). Protein-protein interactions in salt solutions. In W. Cai & H. Hong (Eds.), *Protein-protein interactions*. IntechOpen: Rijeka.

CHAPTER THREE

Prospects for membrane protein crystals in NMX

Samuel John Hjorth-Jensen[a], Esko Oksanen[b,c], Poul Nissen[a], Thomas Lykke-Møller Sørensen[a,*]

[a]Department of Molecular Biology and Genetics—DANDRITE, Aarhus University, Aarhus C, Denmark
[b]European Spallation Source ERIC, Lund, Sweden
[c]Department of Biochemistry and Structural Biology, Lund University, Lund, Sweden
*Corresponding author: e-mail address: tlms@mbg.au.dk

Contents

Abstract

Adding hydrogen atoms and protonation states to structures of membrane proteins requires successful implementation of neutron macromolecular crystallography (NMX). This information would significantly increase our fundamental understanding of the transport processes membrane proteins undertake. To grow the large crystals needed for NMX studies requires significant amounts of stable protein, but once that challenge is overcome there is no intrinsic property of membrane proteins preventing the growth of large crystals per se. The calcium-transporting P-type ATPase (SERCA) has been thoroughly characterized biochemically and structurally over decades. We have extended our crystallization efforts to assess the feasibility of growing SERCA crystals for NMX—exploring microdialysis and capillary counterdiffusion crystallization techniques as alternatives to the traditional vapor diffusion crystallization experiment. Both methods possess crystallization dynamics favorable for maximizing crystal size and we used them to facilitate the growth of large crystals, validating these approaches for membrane protein crystallization for NMX.

Methods in Enzymology, Volume 634
ISSN 0076-6879
https://doi.org/10.1016/bs.mie.2019.11.019

1. Introduction

Membrane proteins are cellular gatekeepers conveying environmental signals, translocating molecules, and generating electrochemical gradients across the hydrophobic cell membrane. Understanding how membrane proteins work will enable us to explore the evolutionary technology they represent as well as target them to combat disease (Misawa, Osaki, & Takeuchi, 2018; Ng, Poulsen, & Deber, 2012; Schmit & Michiels, 2018). Membrane proteins are very diverse with individual proteins having multiple domains adapted to reside in hydrophobic/hydrophilic environments, intra- and extracellularly. This multidomain structure make membrane proteins not only scarce but also notoriously unstable and very challenging targets to study both biochemically and structurally. But many impressive studies over the last decade have demonstrated that it can be done (see https://blanco.biomol.uci.edu/mpstruc/ for an updated list of membrane protein structures).

Numerous membrane protein structures have been determined using X-ray crystallography—and more recently cryo-electron microscopy (https://www.wwpdb.org). In combination with detailed biochemical characterization, these structures have increased our understanding of how transport and signaling across membranes work. But for many of these processes we are still short of a fundamental mechanistic understanding as we do not normally resolve hydrogen atoms and protonation states in these structures. We can infer hydrogen localization when circumstantial evidence provides a plausible model, although the importance of protonation for a given reaction mechanism may not be clear. Detailed structural views with "hydrogen resolution" would significantly change this and allow us to address unanswered questions. The method of choice for obtaining structures with "hydrogen resolution" is neutron macromolecular crystallography (NMX). Because neutrons scatter from nuclei, the coherent scattering lengths of hydrogen (and deuterium) are similar in magnitude to carbon, nitrogen and oxygen (Sears, 1992). Hydrogen and deuterium atoms can therefore be easily identified in the density maps derived from neutron diffraction data because of their increased contribution to diffraction. NMX is not a new tool in structural biology, but it is a tool that in the coming decade is going to experience a steep increase in performance once the European Spallation Source (ESS) in Lund, Sweden becomes operational. (https://europeanspallationsource.se/instruments/nmx).

In structural biology, ongoing technological advancements continue to push the limit for what is possible. New synchrotrons (https://lightsources.org) and free-electron lasers (Chapman, 2019) deliver beam sizes small enough and flux densities high enough to enable useful diffraction data to be extracted from micron-sized crystals. Furthermore, recent advances in detector technology has made it possible to capture the diffraction generated from the very intense and short exposures of these tiny crystals (Forster, Brandstetter, & Schulze-Briese, 2019). Also, software capable of processing the vast amount of data generated during these experiments has now been developed (Winter et al., 2018). Advanced detector technology has also spawned the resolution revolution in cryogenic electron microscopy (cryoEM) (Cheng, 2018), allowing atomic resolution models of larger biomolecules and their complexes to be generated without the requirement for growing crystals at all (Cheng, 2015).

With the tiny beam sizes and high flux densities now available, X-ray crystallographers have abandoned the pursuit of large crystals. Crystallization itself still presents a significant challenge, but if small crystals analyzed by micro-focused beams provide a high-resolution structure substantiating the relevant scientific question(s), the crystallographic afterthought is minimal. Only once crystal quality is poor or data analysis challenging do biochemists have to venture into the world of crystal optimization—a realm populated with empirical knowledge of dubious worth. This is also where one needs to venture on the quest to grow the large crystals required for NMX. To collect useful NMX data requires crystals that are hundreds of microns in all dimensions, a result that requires extensive optimization of the crystallization process. The reason that crystals need to be this large for NMX is that the diffracted intensity measured is proportional to the incoming beam intensity (O'Dell, Bodenheimer, & Meilleur, 2016). Even with the predicted impressive performance of the ESS, the neutron flux to be provided is high only relative to what is currently available and will still be orders of magnitude less than the X-ray flux experienced at synchrotron facilities.

Will it then be feasible to use neutron crystallography for membrane proteins? We recently reviewed potential membrane protein targets based on reported crystal sizes and cell parameters (Sorensen et al., 2018) and found that membrane proteins should indeed be considered for NMX studies. Over the years, we have produced many well-diffracting crystals for X-ray crystallography from a number of different membrane-protein projects. Obviously, each project has presented a range of intricate challenges and required resourceful insight and hard work for a successful outcome, but in our experience, there is nothing per se limiting the size of membrane protein crystals once an ample supply of stable protein has been organized. We have had significant success with the P-type ATPases family of membrane proteins and a key facet to this success has been a thorough understanding of the biochemical properties of this class of proteins (Moller, Olesen, Winther, & Nissen, 2010).

In this chapter, we describe two of the methods that we have applied to SERCA to increase crystal size—dialysis and counter-diffusion crystallization. These methods have been reviewed elsewhere (McPherson & Gavira, 2014; Otalora, Gavira, Ng, & Garcia-Ruiz, 2009) and are known to researchers in the neutron crystallography field. Ultimately, the goal is to produce a large perfectly well-ordered crystal, but the reality may be quite different. One intuitively considers the ability for a given crystal to grow to a large size indicative of a "healthy" crystal form with a sound lattice that supports continuous growth without imperfections. But local order does not necessarily propagate across hundreds of micrometers. As a crystal grows in size so does the likelihood of crystal imperfections originating from lattice variations between crystal domains as well as lattice variations within domains (Nave, 2014).

2. Microdialysis crystallization

The principle for microdialysis crystallization is illustrated in Fig. 1. The protein solution—either by itself or mixed with precipitant solution (but remaining below the nucleation zone of the phase diagram)—is separated from a larger volume of precipitant solution by a size-discriminating membrane allowing free diffusion of all components except the protein. Over time, precipitating agents diffuse across the membrane into the protein solution, and at the correct combination of concentrations, the phase enters the nucleation zone. From here, the formation and growth of nuclei results in the progressive reduction of protein concentration until the solubility limit is reached. At this point, growth is halted and crystal size remains constant over time. However, further crystal growth may be initiated by increasing the precipitant concentration within the reservoir. As this extra precipitant diffuses across the separating membrane, the protein solubility is further decreased, driving additional protein from the solution into the crystal. Expanding upon this principle are double-contained dialysis setups, in which an initial reservoir is itself dialyzed against a larger, secondary reservoir containing higher concentrations of precipitants (Thomas, Rob, & Rice, 1989).

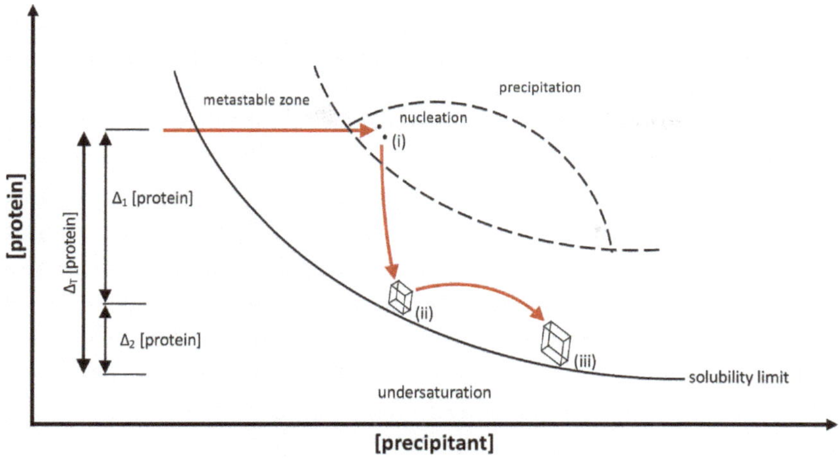

Fig. 1 Crystal phase-time pathway within microdialysis. The mother liquor starts in the undersaturation/metastable zone and slowly moves toward the nucleation zone as precipitants diffuse across the dialysis membrane (i). As a few nuclei form, the mother liquid begins its trajectory through the metastable (i.e., crystal growth) zone, stopping at the solubility line (ii). Further grow can be induced by increasing the precipitant concentration inside the reservoir, slowly lowering the solubility limit of the mother liquor due to the diffusion of additional precipitants across the membrane (iii). A maximal amount of protein is added (Δ_T [protein]) due to the contributions of two growth phases (Δ_1 [protein] and Δ_2 [protein]), resulting in very large crystals.

The key advantage of the microdialysis technique is the ability to control the phase-time pathway, which is essential for producing crystals of the size needed for NMX studies. By contrast, the batch method offers very limited ability for its saturation state to be slowly altered over time—the protein and crystallization solution are mixed together and an equilibrium is quickly reached. This most often results in uniform nucleation with many crystals of equal size that do not undergo an extended growth period. Vapor diffusion on the other hand, does offer the ability to increase or decrease the saturation state of the mother liquor over time. However, this change cannot be precisely controlled, i.e., the concentrations of each component within the mother liquor cannot be estimated at equilibrium. Also, vapor diffusion affects the concentrations of all components, with each non-volatile chemical species becoming more or less concentrated as drop volumes decrease or increase. These excessive changes in chemical composition may result in unwanted additional nucleation or osmotic shock, damaging or destroying the crystal.

Microdialysis allows each individual component within the mother liquor to be reasonably estimated and controlled over time in terms of conditions at equilibrium. For example, to increase the size of a crystal once equilibrium is reached, the polyethylene glycol (PEG) concentration of the mother liquor can be increased such that: (a) only PEG concentration significantly changes while the concentration of other components remain constant hence avoiding osmotic shock; (b) the change in PEG concentration can be accurately controlled making it feasible to reproducibly increase PEG concentration by steps of 1–2%, an increase that may extend crystals growth without inducing additional nucleation; and (c) a gradual change in PEG concentration (i.e., slow kinetics) that is conducive to the growth of an ordered lattice and less likely to result in heavy precipitation and/or nucleation.

An additional crystallization element offered by microdialysis is the ability to not just add, but to remove individual components from the mother liquor. In the case of membrane proteins, lowering the concentration of the solubilizing detergent could potentially promote crystal growth. In the case of SERCA, removing solubilizing octaethyleneglycol dodecylether ($C_{12}E_8$) may drive protein-containing micelles out of solution and into any growing crystal lattice present. We have explored this to a limited extent, and indeed crystals do appear to grow in response to the gradual removal of this detergent. However, further experiments to test the usefulness of this approach are needed to assess effects on diffraction quality, as removing detergent may affect crystal packing significantly.

A schematic presentation of the microdialysis setup we use to crystallize SERCA is shown in Fig. 2. Protein or batch solution is dispensed into the chamber of a microdialysis button, the volume of which can vary from five to several hundred microlitres. A dialysis membrane with a suitable molecular weight cut-off is placed on top, locked and sealed with an O-ring. We use PEG 6000 when crystallizing SERCA and hence use membranes with a cut-off of 12–14 kDa to allow its free diffusion across the dialysis membrane. The microdialysis button is then transferred into a reservoir chamber and fully submerged with precipitant solution. Finally, the reservoir chamber is sealed and the entire setup placed in a temperature-controlled environment.

A typical microdialysis-grown SERCA crystal is shown in Fig. 3, displaying a diamond-shaped morphology similar to the morphology observed for crystals obtained from hanging drop vapor diffusion experiments. The size of the crystal in Fig. 3 is approximately $500 \times 500 \times 100\,\mu m$, although

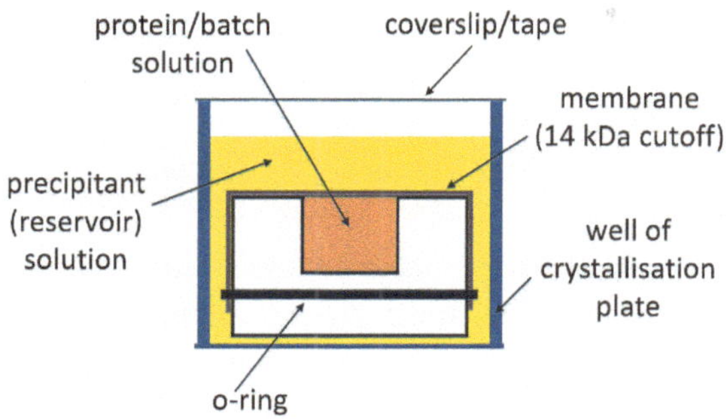

Fig. 2 Schematic of the microdialysis format. Buttons are placed inside 24-well crystallization plates, allowing for the easy setup of replicates and parallel experiments. Reservoir solutions are able to be changed over time by first removing sealing coverslips/tape, then either adding stock straight to the existing reservoir solution (to increase component concentrations) or exchanging the entire reservoir with that of another composition (to decrease component concentrations), then re-sealing.

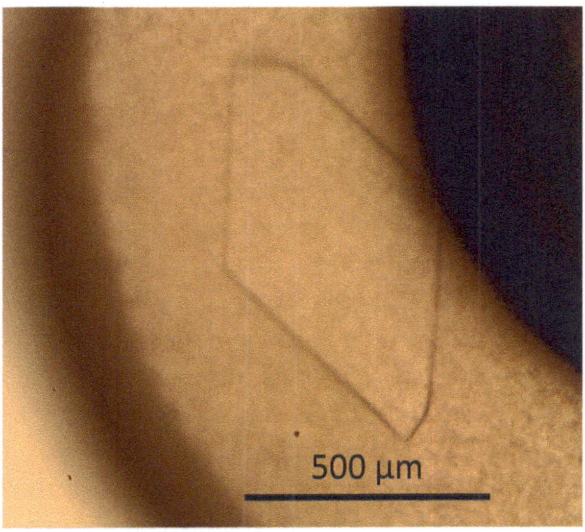

Fig. 3 SERCA crystal grown using microdialysis. Protein chamber contained 5 μL of batch solution (11.6 mg/mL solubilized SERCA, 1.33% (*w/v*) PEG 6000) and was initially dialyzed against a reservoir solution containing 4% (*w/v*) PEG 6000. Crystal is shown 15 days after setup (the PEG 6000 concentration within the reservoir was increased to 6% at day 8).

we have grown SERCA crystals up to 1 mm across the largest dimension using this method. In this experiment, 12 mg/mL solubilized SERCA was mixed in a 5:1 ratio with an 8% (*w/v*) PEG 6000 precipitant solution creating an undersaturated batch solution as the starting point for the crystallization experiment. The 5 μL batch solution was dialyzed against 1300 μL of a precipitant solution containing 4% (*w*/*v*) PEG 6000. To further increase the size of the crystal, the concentration of PEG 6000 within the reservoir solution was increased from 4% to 6% eight days after the initial experiment was setup.

When considering the use of microdialysis, there are a couple of challenges to keep in mind. First, microdialysis is much more challenging to set up than the more straightforward batch and hanging/sitting drop vapor diffusion crystallization formats. Particular care is needed when placing the membrane, as even the slightest movement can result in air being introduced at the protein-membrane interface. If the amount of air introduced is significant, diffusion across the membrane becomes greatly impeded as does the ability to visualize the contents of the dialysis button itself. Bubble formation is of particular concern when crystallizing membrane proteins, as the use of detergents make the occurrence of bubbles far more likely. Also, while this situation may normally be potentially reversed by removing the membrane, restoring the protein meniscus, and repeating the process, this again is more difficult when dealing with detergent-containing membrane proteins. The general advice is to practice using the actual solutions (omitting protein) to be able to handle detergent- and precipitant-containing solutions without introducing air, thereby conserving valuable protein.

Second, as microdialysis crystallization involves changing the saturation state of the mother liquor over time by dialyzing it against a reservoir of much larger volume, it is required that any ions, ligands, or other components present in the protein solution also be added to the reservoir solution. This includes additives used to stabilize the protein, to induce a specific functional state, or to generate a protein-ligand complex. These inclusions are necessary to prevent the additives from slowly diffusing out of the protein chamber over time and maintain protein stability and/or the desired functional state or complex. Again, this issue is more complicated for membrane proteins, as they often require more complex chemical mixtures to remain stable and soluble. Detergent is a key additional component that has to be included in the reservoir mixture when setting up a membrane protein microdialysis crystallization experiment. Often detergents are present in concentrations above the critical micelle concentration (CMC) to prevent the protein-containing micelles within the button from disintegrating and the protein precipitating. Consequently, expensive and highly purified detergents are often added to reservoirs in large quantities, meaning microdialysis consumes far more resources than other methods.

Finally, crystals grown within microdialysis buttons can be more difficult to harvest than crystals obtained using other methods, as the button wall surrounding the protein chamber restricts access to any crystals located inside. This is exacerbated by the fact that membrane protein crystals typically contain high levels of solvent and are therefore very fragile, making them particularly susceptible to damage during the crystal harvesting process. Again, it is worth emphasizing that practice is important, as this will maximize the chance that any prize crystal grown can be recovered safely.

Despite these challenges, microdialysis is a crystallization technique with a unique advantage in terms of the control that it offers over the crystal phase-time pathway. Results with SERCA so far indicate that it is feasible to grow NMX-compatible crystals using microdialysis. Future experiments will continue to explore this method, using protein/batch volumes of 100 μL and greater.

3. Capillary counter-diffusion crystallization

The principle for crystallization within the capillary counterdiffusion format is illustrated in Fig. 4. One end of the protein solution, confined within a capillary, is brought in contact with a precipitant solution, either confined within the same capillary or within a separate chamber. Immediately following setup, the two solutions begin to mix by diffusion, with the precipitating agents generally diffusing faster into the protein than vice versa. The high precipitant concentration at the liquid-liquid interface immediately results in large amounts of crystalline precipitate, and the local protein concentration begins to decline. The precipitant, which is not consumed in this process, continues to move further into the protein solution, where protein solubility is again reduced and another round of precipitation and/or nucleation occurs. As this progressively weakening process repeats, a temporally and spatially varying supersaturation gradient forms along the length of the capillary. Farthest away from the interface, low levels of supersaturation (featuring low concentrations of precipitant and high concentrations of protein) result in fewer crystals of larger volumes (Otalora et al., 2009).

As a crystallization technique, capillary counterdiffusion possesses several inherent advantages. A major advantage with respect to diffraction data collection is that crystals, once grown, can be analyzed *in situ* with no additional handling required (Ng et al., 2003). This is very useful, as membrane protein crystals, at least in our hands, only become more fragile and difficult to

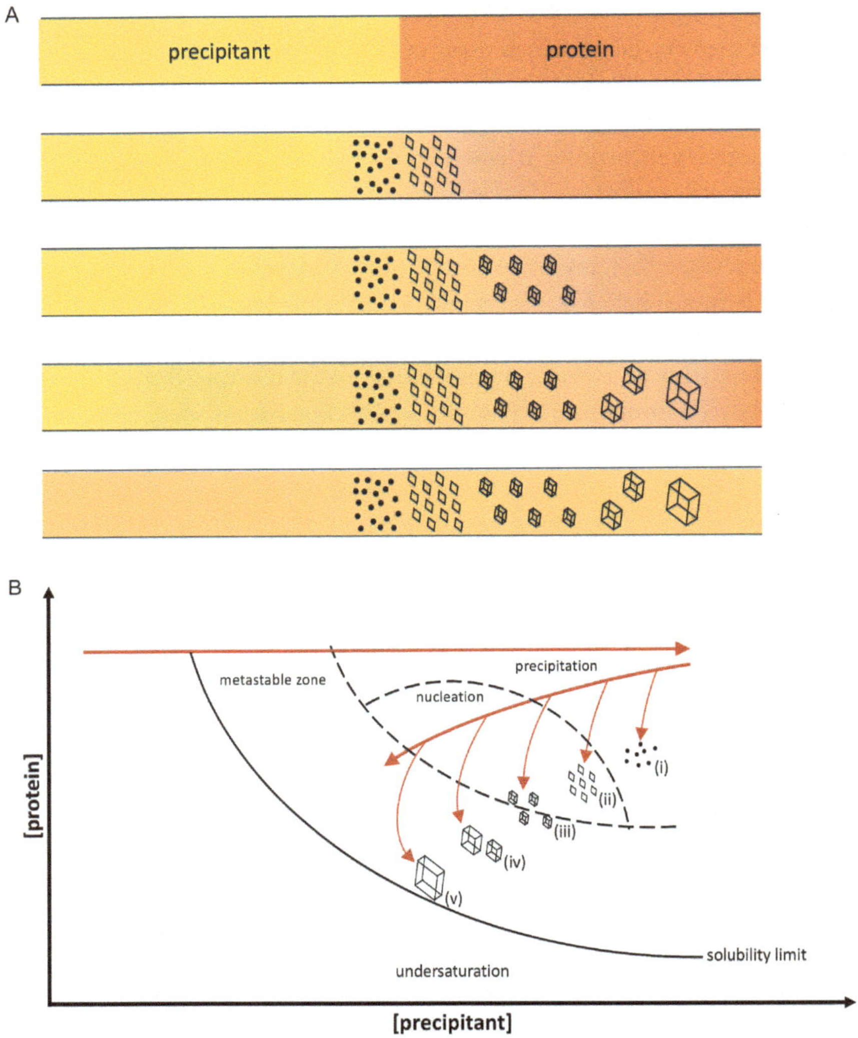

Fig. 4 Crystal phase-time pathways within capillary counterdiffusion. (A) The protein and precipitate solutions are initially fully separated within a capillary with either the two liquids in direct contact or separated by a gel barrier. The two solutions then proceed to diffuse into one another, resulting in a wide range of conditions (and outcomes) that vary along the length of capillary (Ng, Gavira, & Garcia-Ruiz, 2003). (B) In terms of phase space, conditions at/near the interface immediately following setup consist of high precipitant and high protein concentrations, resulting in precipitate formation and the associated local reduction of [protein] (i). Farther away from the interface, the precipitant continues to diffuse into the protein (both solutions are at slightly lower concentrations as the precipitant dilutions as it is extending and the protein is both extending and precipitating/nucleating) resulting in high density nucleation (ii). As the weakening precipitant front continues to advance, nuclei continue to form at a progressively reduced rate (iii and iv). Far into the protein solution, only few nuclei form from the now diminished precipitant wave. These nuclei spend an extended time within the metastable zone as precipitant continues to arrive over time, becoming large crystals (v) (Otalora et al., 2009).

handle with increasing crystal size. Another unique advantage of capillary counterdiffusion is that its dynamics are inherently favorable for generating large crystals and hence it is useful for growing crystals for NMX. Though it does not offer the precise control that microdialysis affords, the wide range of conditions and kinetics explored in each individual experiment often results in a section where only a few nuclei form, with these nuclei undergoing extensive growth to become large crystals (Otalora et al., 2009). The fact that a wide range of conditions are covered per experiment means this method in particular is able to compensate for variations between protein batches. Given the highly complex mixtures involved in membrane protein preparations, this class of proteins is particularly susceptible to batch-to-batch variations and results are often challenging to reproduce.

There are several methods for setting up capillary crystallization experiments (for a detailed review, refer to the comprehensive article by Otalora et al. (2009), with the most common setups often involving the precipitant and protein solutions being separated by a physical buffer, usually a gel such as agarose, in order to delay the onset of diffusion or avoid osmotic shock. An example is the Gel Acupuncture Method (Garcia-Ruiz, Moreno, Viedma, & Coll, 1993), where a capillary filled with protein is embedded in agarose gel. Precipitant solution is added on top of or in the gel, which then proceeds to diffuse into the protein solution and up the capillary. For SERCA, we have noted that the PEG 6000 precipitant we use to crystallize is unable to diffuse through this physical buffer at a sufficient rate to induce nucleation, and as such we have opted for a setup where both the precipitant and protein solutions are contained within the capillary and a free liquid-liquid interface exists between the two.

The setup typically used to crystallize SERCA is shown in Fig. 5. It is important to note that this particular procedure works best when using

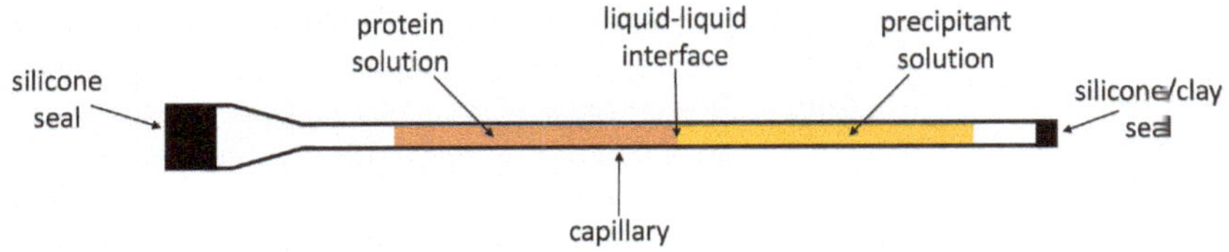

Fig. 5 Schematic of the capillary counterdiffusion format. First protein, then an equal volume of precipitant solution is carefully aspirated into a capillary (such that the two liquids are touching). Both ends of the capillary are then sealed with silicone sealant/clay and the entire setup moved into a temperature-controlled environment to incubate.

capillaries with relatively narrow diameters (≤0.5 mm), where geometries are restrictive enough to limit convective mixing of the protein and precipitant solution and preserve the supersaturation gradient. We are also performing experiments in large diameter capillaries (≥1 mm) in which solubilized SERCA or undersaturated batch solution is mixed with molten low-melting point agarose such that the mixture contains between 0.1 and 0.3% (*w/v*) agarose. This solution is then injected into the capillary and allowed to set before the precipitant solution is added adjacently.

Results from a capillary counterdiffusion experiment with SERCA are shown in Fig. 6. In this experiment, equal volumes of 12 mg/mL solubilized SERCA and a 12% (*w/v*) PEG 6000 precipitant solution were placed in liquid-liquid interface setup within a 0.5 mm diameter glass capillary. After 6 days, the supersaturation gradient characteristic of capillary counterdiffusion can be seen, with heavy precipitant becoming a few well-formed crystals as one moves along the length of the capillary.

There are a few issues to be aware of when considering capillary counterdiffusion crystallization experiments using membrane proteins. First, setting up capillary experiments require more careful handling than other crystallization methods due to the fragile nature of glass and quartz capillaries (a typical wall thickness is 0.01 mm). It is possible for capillaries to break

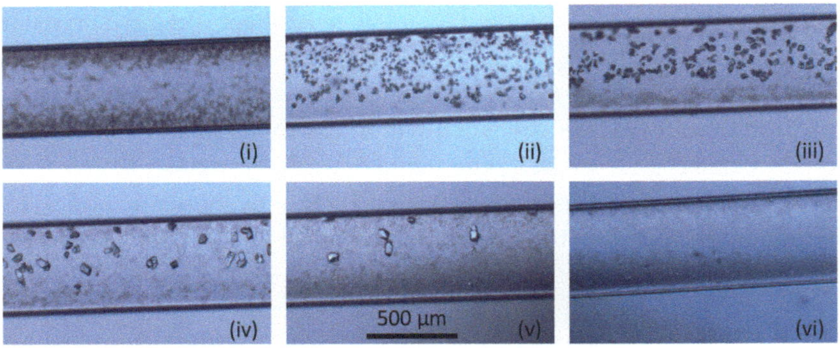

Fig. 6 SERCA crystals grown using capillary counterdiffusion. A 0.5 mm glass capillary was filled with 4.5 µL of 12.6 mg/mL solubilized SERCA and 4.5 µL of precipitant solution containing 12% (*w/v*) PEG 6000, imaged 6 days after setup. Though the crystals shown are relatively small, the advantageous behavior of capillary counterdiffusion is clearly demonstrated. Heavy precipitate is seen at the liquid-liquid interface, where concentrations of both protein and precipitate were at their peak (i). Moving along the capillary one can see the results of a supersaturation gradient, with large numbers of small crystals (ii), becoming fewer in number while simultaneously increasing in size (iii–v). The final section of the capillary contains protein-rich solution free of nuclei, which over time, can serve as a protein source to grow the few single crystals adjacent.

during and after setup, resulting in the loss of significant amounts of precious protein. Second, to prevent chaotic mixing and convection from occurring within in capillaries, the protein solution sometimes needs to be gelled (Biertumpfel, Basquin, Birkenbihl, Suck, & Sauter, 2005; Gavira, de Jesus, Camara-Artigas, Lopez-Garriga, & Garcia-Ruiz, 2006; Otalora et al., 2009). However, gelling the protein solution requires mixing and dispensing at temperatures high enough to maintain the agarose in a molten state. While SERCA is able to withstand these elevated temperatures of up to 30 °C for short periods, a significant number of membrane proteins may not, and the thermal stability of your membrane protein should be demonstrated before proceeding with this technique.

Finally, it is important to keep the diffusion properties of the precipitants being used in mind. In the case of SERCA, we use high molecular weight PEGs which require extremely long equilibration times due to their relatively low diffusion coefficients (Carter et al., 1999). As previously mentioned, we found diffusion kinetics to be insufficient for SERCA nucleation when using setups employing a physical buffer between the precipitant and protein solutions and consequently focused on setups with a free liquid-liquid interface. This is an important issue one should consider when deciding upon the experimental setup best suited for one's particular system.

4. Crystallization of SERCA by microdialysis and capillary counterdiffusion

4.1 Equipment

1. Analytical balance (VWR)
2. TLA 100.3 rotor (Beckman Coulter)
3. Optima MAX-XP Ultracentrifuge (Beckman Coulter)
4. UV/visible spectrophotometer (ThermoFischer Scientific)
5. 12–14 kDa dialysis membrane discs (Hampton Research)
6. 30 μL dialysis kit including buttons, O-rings, and applicator (Hampton Research)
7. 24 well, VDX crystallization plate without sealant (Hampton Research)
8. High vacuum grease (Dow Corning)
9. 22 mm siliconized glass coverslips (Hampton Research)
10. 0.5 mm and 1 mm glass or quartz capillaries (Hampton Research)
11. 1 mL luer syringe (Chirana)
12. Parafilm
13. 1.5 mL microtubes (Sarstedt)

14. 250 mL Schott bottle
15. 250 mL beaker
16. Mounting clay (Hampton Research)

4.2 Chemicals

1. BME (β-Mercaptoethanol; Sigma Aldrich)
2. $C_{12}E_8$ (octaethyleneglycol dodecylether; Nikko Chemicals)
3. $CaCl_2$ (Sigma Aldrich)
4. KCl (Sigma Aldrich)
5. AMPPCP (β,γ-Methyleneadenosine 5′-triphosphate disodium salt; Sigma Aldrich)
6. SR vesicles (sarcoplasmic reticulum vesicles from rabbit fast twitch skeletal muscle; prepared as per Andersen, Lassen and Moller (1985)
7. Glycerol (Sigma Aldrich)
8. $MgCl_2$ (Sigma Aldrich)
9. MOPS (3-(N-Morpholino)propanesulfonic acid; Sigma Aldrich)
10. PEG 6000 (polyethylene glycol 6000; Sigma Aldrich)
11. Sodium acetate (Fluka BioChemika)
12. *tert*-butanol (Sigma Aldrich)
13. Low-melting point (LMP) agarose powder (Invitrogen)

4.3 Protocol

4.3.1 Preparation of solubilized SERCA Protein (Ca_2E1-AMPPCP form)

1.1. Prepare the detergent and buffer solutions. Calculate the volumes of stock needed to prepare a
370 mM $C_{12}E_8$ detergent solution and a
100 mM MOPS pH 6.8, 80 mM KCl buffer.

1.2. Working on ice, add 1 mL of the 100 mM MOPS pH 6.8, 80 mM KCl buffer to the SR vesicles vial (containing 4.7 mg of protein) and suspend the vesicles by gently pipetting up and down. Transfer the mixture into a 2 mL polyallomer ultracentrifuge tube.

1.3. Wash the vial with another 1 mL of the 100 mM MOPS pH 6.8, 80 mM KCl buffer and add the wash to the ultracentrifuge tube (for a total volume of 2 mL).

1.4. Balance the ultracentrifuge tube before placing them inside a TLA100.3 rotor. Place the rotor inside a Optima Max Ultracentrifuge and spin at 50000 RPM, 4 °C for 35 min.

1.5. While the suspended SR vesicles are spinning, prepare the following 2× incubation mixture in a 1.5 mL microtube. Add the required amount of each stock to produce 350 µL of solution containing the following: 2 mM AMPPCP, 10 mM BME, 20 mM $CaCl_2$, 40% (*v*/*v*) glycerol, 160 mM KCl, 6 mM $MgCl_2$, and 200 mM MOPS pH 6.8.

Tip: An alternative SERCA state, the Ca_2E1-ADP:AlF_4^- form, can be prepared by replacing the 2 mM AMPPCP with 0.67 mM $AlCl_3$, 10 mM NaF, and 2 mM ADP.

1.6. Recover the ultracentrifuge tube now containing a pellet of SERCA microsomes. Discard the supernatant, then add 110 µL of the 100 mM MOPS pH 6.8, 80 mM KCl buffer to the pellet and resuspend by gently pipetting up and down.

Tip: To reduce the chance of introducing bubbles, reduce the pipetting volume to ~70 µL. This will prevent air from being aspirated into the tip while drawing solution up and down over the SERCA pellet.

1.7. Once the pellet has been fully resuspended, transfer 125 µL into the 2× incubation mixture along with 159 µL of H_2O.

1.8. To complete the incubation mix, add 66.2 µL of the 370 mM $C_{12}E_8$ solution (final concentration of 35 mM) and mix by lightly vortexing. The solution (final volume of 700 µL) should turn from translucent to clear.

1.9. Allow the mixture to incubate on ice for 1 h, then ultracentrifuge again at 50000 RPM, 4 °C for 35 min. Collect the supernatant and determine the protein concentration by measuring 280 nm absorbance with a spectrophotometer. This solution can be used directly for crystallization experiments.

4.3.2 Crystallization of the SERCA Ca_2E1-AMPPCP form by Microdialysis

2.1. Prepare the dialysis solution. Calculate the volumes of stock needed to prepare 1.5 mL of the following: 4% (*w*/*v*) PEG 6000, 18% (*v*/*v*) glycerol, 80 mM sodium acetate, 4% (*v*/*v*) *tert*-butanol, 100 mM MOPS pH 6.8, 80 mM KCl, 3 mM $MgCl_2$, 10 mM $CaCl_2$, 0.5 mM AMPPCP, 5 mM BME, and 15 mM $C_{12}E_8$.

2.2. Prepare a precipitant solution (to be mixed with the protein to form an unsaturated batch solution). Calculate the volumes of stock needed to prepare 1 mL of the following: 8% (*w*/*v*) PEG 6000, 8% (*v*/*v*) glycerol, 200 mM sodium acetate, 4% (*v*/*v*) *tert*-butanol, 100 mM MOPS pH 6.8, and 5 mM BME.

2.3. Mix the protein with the 8% PEG 6000 crystallization solution in a 5:1 ratio to produce an unsaturated batch. Make sure the volume prepared is in excess of the nominal button volume: e.g., aim to set aside ~35 μL of solution for a 30 μL microdialysis button to ensure a reasonable sized meniscus is able to be formed.

2.4. Half fill a 250 mL beaker with water and place a 12–14 kDa cut-off dialysis membrane inside to soak.

2.5. Take a microdialysis button and carefully dispense the unsaturated batch solution inside the protein chamber until a convex meniscus rises above the button surface.

Tip: As the protein solution contains a high concentration of detergent, take extra care when pipetting to avoid bubble formation. An easy way to do this is to aspirate a larger volume of solution then required and stop dispensing before the last of the liquid leaves the pipette tip. Although this method will not deliver an accurate volume of solution, this does not matter as a specific volume is not necessary so long as there is a clear meniscus rising above the button surface—minimizing bubble formation is more important.

2.6. Remove the dialysis membrane from the beaker with tweezers. Using gloves, hold the membrane in both hands and carefully unfold and straighten. Remove any excess water by dabbing the membrane edge with a paper towel, then gently lower the membrane until it touches the top of the protein meniscus. The protein solution should "grab" the membrane. At this point slowly lower your fingers and allow the weight of the membrane to drape itself over the button top. Carefully release the membrane.

2.7. Take the applicator and place it over the top of the button. Holding the applicator in position with one hand, take an O-ring and push it down over the applicator and onto the button (making sure it locks into the groove running along the side of the button and securely holds the membrane in place).

2.8. Cut away the excess membrane around the sides of the button using scissors, then transfer the button into the well of a 24-well crystallization plate. Dispense the dialysis solution into the well, making sure the button is fully immersed in the liquid.

2.9. Seal the well using vacuum grease and a glass coverslip, then place the tray in a 20 °C environment to equilibrate.

2.10. To increase the size of any crystals that form, increase the PEG concentration of the reservoir (i.e., dialysis solution). Increasing PEG concentration by ~2% is a reasonable place to start. This can easily be achieved by simply removing the well seal, adding PEG stock directly, then gently mixing and replacing the seal.

Tip: The volume of PEG stock needed to be added to the reservoir to achieve the desired concentration increase can easily be calculated using the following formula:

$$V_{PEG} = \left(V_R / \left(C_{i(PEG)} - C_{f(PEG)}\right)\right) / \left(C_{f(PEG)} - C_{PEG}\right).$$

where V_{PEG} = volume of PEG stock to be added, V_R = current volume of reservoir, $C_{i(PEG)}$ = current concentration of PEG in reservoir, $C_{f(PEG)}$ = desired concentration of PEG in reservoir, and C_{PEG} = concentration of PEG stock. This formula can of course be applied to any precipitant used to crystallize your particular protein.

2.11. To harvest crystals from microdialysis buttons, first cut and remove the membrane covering the protein chamber using a scalpel. Crystals can now be fished and flash-cooled in liquid nitrogen using a standard loop, however be aware that the high walls of the protein chamber restrict access somewhat. Alternatively, crystals can be mounted within a capillary for room temperature data collection. A simple and effective method, described in detail by Mac Sweeney and D'Arcy (2003), is to place clay around the base of the magnetic cap holding the loop, then inserting the loop containing the crystal into a capillary containing some reservoir solution (a seal is formed as the capillary is pressed into the clay).

4.3.3 Crystallization of the SERCA $Ca_2E1\text{-}ADP{:}AlF_4^-$ form by Capillary Counterdiffusion

3.1. Prepare the precipitant solution. Calculate the volumes of stock needed to prepare 1 mL of the following: 12% (*v*/*v*) PEG 6000, 16% (*v*/*v*) glycerol, 200 mM Sodium Acetate, 6% (*v*/*v*) *tert*-butanol, 100 mM MOPS pH 6.8, and 10 mM BME.

3.2. Take a 0.5 mm glass or quartz capillary and open it just above the narrow sealed end (opposite of the bulb end) by scoring with a glass cutting stone and then snapping. Solutions are now able to be aspirated into the capillary.

3.3. Dispense 5 μL of protein onto a siliconized glass coverslip. Holding the capillary at the bulb end, aspirate the protein into the capillary by touching the drop with the opened narrow end (through capillary action).

3.4. Dispense 5 μL of precipitant solution onto a clean section of the coverslip. Carefully aspirate the drop as per *3.3*. The precipitant solution should directly follow the protein solution, i.e., the two solutions touching with no air between them. Both solutions should now be contained within the capillary with a liquid-liquid interface existing between the two.

3.5. Seal the opened narrow end of the capillary by pressing into clay (this seal can be reinforced by applying silicone sealant on top of the clay to make an air-tight seal that is stable over time).

3.6. Seal the bulb end of the capillary with silicone sealant and place the capillary in a 20 °C environment to equilibrate (horizontal position, with the bulb end slightly raised to allow any air pockets that form to rise to the end of the protein solution away from the interface).

3.7. To prepare any grown crystals for data collection, first open the capillary at both ends (the silicone sealing the bulb end may be cut away with a scalpel and other end may be cut through). Immobilize the opened capillary and use another capillary of narrower diameter (e.g., 0.1 mm) to remove the material surrounding the crystal to be isolated (keeping some solution on either side to maintain a stable environment for the crystal). The capillary can then be resealed at both ends with clay and mounted directly onto a goniometer for X-ray/neutron analysis.

4.3.4 Alternative method for crystallizing SERCA Ca_2E1-ADP:AlF_4^- form by Capillary Counterdiffusion (with gelled batch solution)

4.1. Prepare the precipitant solution. Calculate the volumes of stock needed to prepare 1.5 mL of the following: 18% (*w/v*) PEG 6000, 15% (*v/v*) glycerol, 100 mM sodium acetate, 4% (*v/v*) *tert*-butanol, 100 mM MOPS pH 6.8, 40 mM KCl, 1.5 mM $MgCl_2$, 5 mM $CaCl_2$, 0.5 mM ADP, 0.167 mM $AlCl_3$, 2.5 mM NaF, and 7.5 mM BME.

4.2. Prepare another precipitant solution (to be mixed with the protein and molten agarose to form an unsaturated gelled batch). Calculate the volumes of stock needed to prepare 1 mL of the following: 5.7% (*w/v*) PEG 6000, 14.3% (*v/v*) glycerol, 286 mM sodium acetate, 11.4% (*v/v*) *tert*-butanol, 100 mM MOPS pH 6.8, and 10 mM BME.

4.3. Prepare a custom syringe to dispense the solutions inside the crystallization capillary. First cut a small rectangular section of parafilm (measuring approx. 10 × 50 mm) and pre-stretch it to about twice its initial length. Attach one end of the parafilm to the end of a 1 mL syringe and wrap it 1–2 times around the syringe end (leaving the remaining length of the parafilm hanging). Take a 0.5 mm capillary and cut off the narrow sealed end (removing as little length as possible). Hold the bulb of the capillary against the syringe end with one hand. Use your free hand to carefully wrap the loose end of the parafilm around the capillary bulb while applying constant tension to attach it to the syringe. To make sure the seal is airtight, pre-stretch and wrap another piece of parafilm around the joint between the capillary bulb and the syringe end.

4.4. Prepare a 1% (*w*/*v*) solution of low-melting point agarose by adding the agarose powder to a buffer of 100 mM MOPS pH 6.8, 80 mM KCl in a Schott bottle. Dissolve the agarose by repeatedly heating the solution in a microwave, stirring every 30 s until the liquid is completely clear. Allow the solution to start cooling down.

4.5. Take another capillary (of 1 mm diameter) and lay it flat on the bench next to a ruler. Using a permanent marker, make a mark 15 mm from the narrow sealed end of the capillary. From this point make another two marks along the capillary shaft, at distances of 20 and 40 mm respectively. These mark the sections where the protein and precipitant solutions will be dispensed. Immobilize the capillary in a horizontal position, with the open bulb end slightly raised, using a portion of modelling clay.

4.6. Mix 100 μL of the protein with 70 μL of the 5.7% PEG 6000 precipitant solution in a 1.5 mL microtube to create an undersaturated batch, then place inside a 30 °C incubator. Also transfer the custom syringe, a 200 μL pipette tip, and 30 μL molten agarose (within its own pre-warmed microtube) into the incubator.

4.7. Once thermally equilibrated, remove the batch solution, the molten agarose, and the pipette tip from the incubator. Using the warmed pipette tip, add the batch solution to the molten agarose and mix thoroughly by gently pipetting up and down. Remove the custom syringe from the incubator and aspirate the batch-agarose mixture. Quickly but carefully, insert the custom syringe inside the 1 mm crystallization capillary and dispense the mixture between the two marks nearest the sealed end.

4.8. Allow the batch-agarose mixture to set, then use the cleaned custom pipette to carefully dispense the 18% PEG 6000 precipitant solution inside the crystallization capillary, from the gelled batch to the mark farthest away from the sealed ended. Make sure that there is no air present at the interface between the gelled batch and the dispensed precipitant solution.

4.9. Seal the open end of the capillary with silicone and place the capillary in a 20 °C environment to incubate (in a horizontal position).

4.10. To prepare any grown crystals for data collection, first open the capillary at both ends (the silicone sealing the bulb end may be cut away with a scalpel and other end may be cut through). Immobilize the opened capillary and use a custom pipette of the variety previously mentioned to remove the material surrounding the crystal to be isolated (keeping some solution on either side to maintain a stable environment for the crystal). The capillary can then be resealed at both ends with clay and mounted directly onto a goniometer for X-ray/neutron analysis.

5. Summary

Growing large membrane protein crystals is challenging, but here we demonstrate that it is possible in the case of SERCA. Working toward producing crystals of membrane proteins for NMX, it is essential that we learn from past experience and how the first membrane protein structures were solved using MX. The early targets were available in sufficient amounts and biochemically well-described. This highlights the importance of starting with well-characterized targets to be able to take a systematic reproducible crystallographic approach. Crystallographic success is most often built on a foundation of biochemical insight, and for SERCA we have benefitted from decades of biochemical characterization, which, combined with a good deal of crystallographic understanding, has enabled a successful outcome. We suggest that a similar approach is considered for other membrane protein projects pursuing large crystals for NMX.

There are many approaches to solubilizing and purifying membrane protein for crystallization. The crystals of SERCA described here have been solubilized directly from the native membrane and hence crystallized in the presence of high concentrations of lipid and detergent (HILIDE) (Gourdon et al., 2011). The technique has been described as mimicking solubilized membranes and appears to be able to "re-form" repetitive bilayers in the crystal when stimulated by crystallization conditions. This approach has worked in numerous cases and here we have demonstrated that it also is compatible with microdialysis and capillary crystallization techniques.

Other approaches are being explored and we are working to test the applicability of other techniques for membrane protein crystallization. There are aspects of crystallization space we do not explore in any great detail during standard crystallization. For example, the effect of temperature on crystallization is explored rather coarsely by placing duplicate experiments at 4 °C and 20 °C. Recent studies (Junius et al., 2016) have shown that continuous monitoring and adjustment of temperature enables careful control of nucleation and crystal growth. Going forward, it will be attractive to further explore continuous monitoring of parameters in crystallization space to extend stable growth from MX- to NMX-sized crystals.

Acknowledgment

The authors would like to thank Anne Lillevang for expert technical assistance.

Funding information

This project has received funding from the European Union's Horizon 2020 research and innovation programme under the Marie Skłodowska-Curie grant agreement No. 722687.

References

Andersen, J. P., Lassen, K., & Moller, J. V. (1985). Changes in Ca^{2+} affinity related to conformational transitions in the phosphorylated state of soluble monomeric Ca^{2+}-ATPase from sarcoplasmic reticulum. *Journal of Biological Chemistry*, *260*(1), 371–380.

Biertumpfel, C., Basquin, J., Birkenbihl, R. P., Suck, D., & Sauter, C. (2005). Characterization of crystals of the Hjc resolvase from Archaeoglobus fulgidus grown in gel by counter-diffusion. *Acta Crystallographica. Section F, Structural Biology and Crystallization Communications*, *61*(Pt 7), 684–687. https://doi.org/10.1107/S1744309105018269.

Carter, D. C., Wright, B., Miller, T., Chapman, J., Twigg, P., Keeling, K., et al. (1999). Diffusion-controlled crystallization apparatus for microgravity (DCAM): flight and ground-based applications. *Journal of Crystal Growth*, *196*, 602–609.

Chapman, H. N. (2019). X-ray free-Electron lasers for the structure and dynamics of macromolecules. *Annual Review of Biochemistry*, *88*, 35–58. https://doi.org/10.1146/annurev-biochem-013118-110744.

Cheng, Y. (2015). Single-particle Cryo-EM at crystallographic resolution. *Cell*, *161*(3), 450–457. https://doi.org/10.1016/j.cell.2015.03.049.

Cheng, Y. (2018). Membrane protein structural biology in the era of single particle cryo-EM. *Current Opinion in Structural Biology*, *52*, 58–63. https://doi.org/10.1016/j.sbi.2018.08.008.

Forster, A., Brandstetter, S., & Schulze-Briese, C. (2019). Transforming X-ray detection with hybrid photon counting detectors. *Philosophical Transactions. Series A, Mathematical, Physical, and Engineering Sciences*, *377*(2147), 20180241. https://doi.org/10.1098/rsta.2018.0241.

Garcia-Ruiz, J. M., Moreno, A., Viedma, C., & Coll, M. (1993). Crystal quality of lysozyme single crystals grown by the gel acupuncture method. *Materials Research Bulletin*, *28*(6), 541–546.

Gavira, J. A., de Jesus, W., Camara-Artigas, A., Lopez-Garriga, J., & Garcia-Ruiz, J. M. (2006). Capillary crystallization and molecular-replacement solution of haemoglobin

II from the clam Lucina pectinata. *Acta Crystallographica. Section F, Structural Biology and Crystallization Communications*, *62*(Pt. 3), 196–199. https://doi.org/10.1107/S1744309106002648.

Gourdon, P., Andersen, J. L., Hein, K. L., Bublitz, M., Pedersen, B. P., Liu, X.-Y., et al. (2011). HiLiDe—Systematic approach to membrane protein crystallization in lipid and detergent. *Crystal Growth & Design*, *11*(6), 2098–2106. https://doi.org/10.1021/cg101360d.

Junius, N., Oksanen, E., Terrien, M., Berzin, C., Ferrer, J. L., & Budayova-Spano, M. (2016). A crystallization apparatus for temperature-controlled flow-cell dialysis with real-time visualization. *Journal of Applied Crystallography*, *49*(Pt. 3), 806–813. https://doi.org/10.1107/S1600576716004635.

Mac Sweeney, A., & D'Arcy, A. (2003). A simple and rapid method for mounting protein crystals at room temperature. *Journal of Applied Crystallography*, *36*(1), 165–166. https://doi.org/10.1107/S0021889802019623.

McPherson, A., & Gavira, J. A. (2014). Introduction to protein crystallization. *Acta Crystallographica. Section F, Structural Biology Communications*, *70*(Pt. 1), 2–20. https://doi.org/10.1107/S2053230X13033141.

Misawa, N., Osaki, T., & Takeuchi, S. (2018). Membrane protein-based biosensors. *Journal of the Royal Society, Interface*, *15*(141), 20170952. https://doi.org/10.1098/rsif.2017.0952.

Moller, J. V., Olesen, C., Winther, A. M., & Nissen, P. (2010). The sarcoplasmic Ca2+-ATPase: Design of a perfect chemi-osmotic pump. *Quarterly Reviews of Biophysics*, *43*(4), 501–566. https://doi.org/10.1017/S003358351000017X.

Nave, C. (2014). Matching X-ray beam and detector properties to protein crystals of different perfection. *Journal of Synchrotron Radiation*, *21*(Pt. 3), 537–546. https://doi.org/10.1107/S1600577514003609.

Ng, J. D., Gavira, J. A., & Garcia-Ruiz, J. M. (2003). Protein crystallization by capillary counterdiffusion for applied crystallographic structure determination. *Journal of Structural Biology*, *142*(1), 218–231.

Ng, D. P., Poulsen, B. E., & Deber, C. M. (2012). Membrane protein misassembly in disease. *Biochimica et Biophysica Acta*, *1818*(4), 1115–1122. https://doi.org/10.1016/j.bbamem.2011.07.046.

O'Dell, W. B., Bodenheimer, A. M., & Meilleur, F. (2016). Neutron protein crystallography: A complementary tool for locating hydrogens in proteins. *Archives of Biochemistry and Biophysics*, *602*, 48–60. https://doi.org/10.1016/j.abb.2015.11.033.

Otalora, F., Gavira, J. A., Ng, J. D., & Garcia-Ruiz, J. M. (2009). Counterdiffusion methods applied to protein crystallization. *Progress in Biophysics and Molecular Biology*, *101*(1–3), 26–37. https://doi.org/10.1016/j.pbiomolbio.2009.12.004.

Schmit, K., & Michiels, C. (2018). TMEM proteins in Cancer: A review. *Frontiers in Pharmacology*, *9*, 1345. https://doi.org/10.3389/fphar.2018.01345.

Sears, V. F. (1992). Neutron scattering lengths and cross sections. *Neutron News*, *3*(3), 26–37. https://doi.org/10.1080/10448639208218770.

Sorensen, T. L. M., Hjorth-Jensen, S. J., Oksanen, E., Andersen, J. L., Olesen, C., Moller, J. V., et al. (2018). Membrane-protein crystals for neutron diffraction. *Acta Crystallographica Section D, Structural Biology*, *74*(Pt. 12), 1208–1218. https://doi.org/10.1107/S2059798318012561.

Thomas, D. H., Rob, A., & Rice, D. W. (1989). A novel dialysis procedure for the crystallization of proteins. *Protein Engineering*, *2*(6), 489–491. https://doi.org/10.1093/protein/2.6.489.

Winter, G., Waterman, D. G., Parkhurst, J. M., Brewster, A. S., Gildea, R. J., Gerstel, M., et al. (2018). DIALS: Implementation and evaluation of a new integration package. *Acta Crystallographica SectionD, Structural Biology*, *74*(Pt. 2), 85–97. https://doi.org/10.1107/S2059798317017235.

CHAPTER FOUR

IMAGINE: The neutron protein crystallography beamline at the high flux isotope reactor

Flora Meilleur[a,b,*], Andrey Kovalevsky[a], Dean A.A. Myles[a]
[a]Neutron Scattering Division, Oak Ridge National Laboratory, Oak Ridge, TN, United States
[b]Department of Molecular and Structural Biochemistry, North Carolina State University, Raleigh, NC United States
*Corresponding author: e-mail address: meilleurf@ornl.gov

Contents

Abstract

IMAGINE is a high intensity, quasi-Laue neutron crystallography beamline developed at the 85 MW High Flux Isotope Reactor (HFIR) at Oak Ridge National Laboratory (ORNL). This state-of-the-art facility for neutron-diffraction enables neutron protein structures to be determined at or near atomic resolutions from crystals with volumes of $<1\ mm^{3}$ and unit cell edges of <150 Å. The beamline features include elliptical focusing mirrors that deliver neutrons into a $2.0 \times 3.2\ mm^2$ focal spot at the sample position, and variable short and long wavelength cutoff optics that provide automated exchange between multiple wavelength configurations. The beamline is equipped with a single-axis goniometer, neutron-sensitive cylindrical image plate detector and room temperature and

Methods in Enzymology, Volume 634
ISSN 0076-6879
https://doi.org/10.1016/bs.mie.2019.11.016

cryogenic sample environments. This article describes the beamline components, the diffractometer and the data collection and data analysis protocols that are used, and outlines the protein deuteration, crystallization and conventional crystallography capabilities that are available to users at ORNL's neutron facilities. We also present examples of the scientific questions being addressed at this beamline and highlight important findings in enzyme chemistry that have been made possible by IMAGINE.

1. Introduction

Neutron crystallography is an inherently flux-limited technique. Available neutron fluxes are orders of magnitude less than the X-ray fluxes available from modern in-house generators and synchrotrons. Addressing this limitation typically requires crystals much larger ($>0.1\,mm^3$) than for X-ray analysis, and is the primary motivation for enhancing the scattering power of a protein crystal through the isotopic labeling and specialized crystal growth techniques discussed in other chapters (see Budayova-Spano, Koruza, & Fisher, 2020). It also means that long exposure times are required to measure a neutron diffraction dataset to sufficient resolution and completeness. While complete X-ray datasets can be measured in seconds at high-intensity X-ray sources, neutron datasets typically require days to weeks for data collection. However, through recent advances in instrumentation and labeling it has become possible, in special cases, to obtain useful neutron datasets in a single day or less (Munshi et al., 2012).

Modern instruments can be divided into three groups based upon the neutron source and mode of operation, namely monochromatic quasi-Laue (also known as "pink-Laue") (IMAGINE, presented here) instruments at reactor sources, and time-of-flight (TOF) Laue instruments (Coates & Sullivan, 2020; Markó, Nagy, Aprigliano, & Oksanen, 2020; Tanaka et al., 2020) at spallation sources. All instrument detector systems are designed to maximize the coverage of reciprocal space so that the largest number of stimulated reflections can be recorded simultaneously, whether the instrument operates with mono- or polychromatic incident radiation. Typically, neutron protein crystallography beamlines also take advantage of longer wavelength neutrons to maximize the intensities of the diffraction peaks, which are proportional to the square of the incident wavelength λ, as shown in Eq. (1), where $\langle I \rangle$, I_0, $|F|$, V, λ, and ν_0 are the diffracted intensity, incident beam intensity, structure factor magnitude, and to reduce the spatial overlap of reflections at the detector, which is determined by Bragg's law

(2d*sin(λ) = λ). At the wavelength typically used for X-ray crystallography ($\lambda \leq 1.54\,\text{Å}$), the diffracted beams lie on cones in the forward scattering direction, and two-dimensional detectors are installed downstream of the crystal. At the longer wavelengths ($2.3\,\text{Å} < \lambda < 6.0\,\text{Å}$) typically used by cold-neutron diffractometers, the high-resolution diffraction also occurs in the backscattering direction. This requires wide angle detector arrays with approximate cylindrical or semi-spherical geometries that surround the sample in order to record both forward- and back-scattering reflections.

$$\langle I \rangle \propto \frac{I_0 \times |F|^2 \times V \times \lambda^2}{v_0^2} \tag{1}$$

The IMAGINE beamline takes advantage of the high time-integrated neutron flux provided by the steady-state operation of a reactor by using a broad-bandpass ($d\lambda/\lambda \sim 30\%$) quasi-Laue geometry to greatly increase the number of stimulated reflections at each crystal setting relative to the monochromatic mode (Meilleur et al., 2013; Schroder, O'Dell, Myles, Kovalevsky, & Meilleur, 2018). This quasi-Laue configuration avoids the need for a monochromator crystal upstream of the sample and relies instead upon paired mirror optics that can be tuned to select and transmit a broad bandpass radiation that best match the spatial resolution requirements of the sample. The choice of bandpass in quasi-Laue mode also represents a compromise between maximizing the number of reflections at each crystal setting and minimizing the incoherent neutron scattering background from hydrogen in the sample, which also scales with the band pass that is used and is thus larger than that observed with monochromatic radiation (Meilleur, Myles, & Blakeley, 2006). On balance, while monochromatic geometry may produce datasets of higher intrinsic quality compared to quasi-Laue geometry (either for X-rays or neutrons), the time required for data collection is typically much longer, and the quasi-Laue configuration provides significant advantage in allowing data to be collected much more rapidly from larger unit cell systems and from smaller, weakly scattering crystals (Helliwell, 1997).

2. Beamline overview

2.1 Neutron source

Oak Ridge National Laboratory (Oak Ridge, TN, USA) operates two powerful neutron sources for neutron scattering research, the Spallation Neutron

Source (SNS), which currently provides the most intense pulsed neutron beams in the world, and the 85 MW High Flux Isotope Reactor (HFIR), which currently provides one of the highest steady-state neutron fluxes in the world. The HFIR is equipped with a high-performance cold neutron source that has a brightness of 1.6×10^{13} n/cm^2 s·ster·Å at the peak ($\lambda = 2.6$ Å) and delivers neutrons to a suite of instruments that are optimized for structural biology and materials research (Crow et al., 2011; Heller et al., 2014; Lynn et al., 2006; Wignall et al., 2012).

2.2 Beamline optics

The IMAGINE instrument is situated on the end-station of cold guide number 4 (CG-4), which has a flux of 2×10^9 n s^{-1} cm^{-2} over the useful 2–10 Å wavelength range. The instrument receives neutrons from a 19×12 mm^2 section of CG-4. The neutron wavelength and bandpass delivered to the sample is varied by automated exchange of three neutron flat mirrors and three pairs of neutron filters that provide short wavelength ($\lambda_{min} = 2.0$, 2.8 and 3.3 Å) and long wavelength ($\lambda_{max} = 3.0$, 4.0 and 4.5 Å) cut-offs, respectively. A pair of elliptically shaped mirrors then collects and focuses the resulting beam vertically and horizontally down to 2×3.2 mm^2 at the sample position, with full width vertical and horizontal divergence of 0.5° and 0.6°, respectively. This results in a >20-fold increase in flux density at the sample position. While pairwise combination of the wavelength filters allows the bandpass to be optimized to match the resolution limits of the particular sample, in practice, most data are collected using the 2.8–4.5 Å setting. A useful feature is that removal of the long wavelength filter enables the sample to be illuminated by the full white beam, which allows candidate crystals to be rapidly screened and indexed in a full Laue mode, accelerating and facilitating crystal selection and data collection.

2.3 Diffractometer

The diffractometer uses a cylindrical neutron image plate design (Cipriani, Castagna, Wilkinson, Lehmann, & Buldt, 1996; Niimura et al., 1994). The diffractometer is available commercially from ARINAX, Grenoble, France (Fig. 1). The detector system has a broad dynamic range (>10,000), high spatial resolution (250 μm), and high detective quantum efficiency (DQE) of 47% for the standard 2.8–4.5 Å quasi-Laue configuration.

Fig. 1 IMAGINE installed on CG-4D. (A) IMAGINE is installed in a 4″-thick steel wall enclosure next to the Bio-SANS instrument in the HFIR cold guide hall. The elliptical mirror chamber sits directly upstream of the instrument. (B) The image plate detector is mounted inside a cylindrical drum with the radius of 200 mm. The detector is orientated with the erasing and readout system placed in the backscattering position. Crystals are mounted on a sample stick with automated φ-axis rotation for room temperature data collection. *Reproduced with permission of the International Union of Crystallography (Photo credit: Jason Richards, ORNL).*

2.4 Data collection and reduction

The flat mirrors and filters are remotely controlled through their respective controllers and software. Data acquisition is controlled through a graphical user interface (GUI) developed by the manufacturer of the diffractometer, ARINAX. IMAGINE is currently equipped with a single-axis goniometer. The curvature of the Ewald sphere when using longer wavelength radiation is such that the blind region is significant, and this means the crystal will need to be re-oriented at least once to ensure good data completeness. In addition, instruments that use a cylindrical detector geometry (BIODIFF, IMAGINE, and LADI-III) have an intrinsic gap in their coverage of reciprocal space along the cylindrical long axis. Reflections that are lost can be recovered by using more than one orientation. Plans to simplify this and optimize data collection strategy are discussed in Section 6 (future developments).

Cold neutrons (<50 meV) do not induce ionization of atoms or primary radiation damage in biological samples. Therefore, cryogenic sample temperatures are not necessary to protect against radiation damage in the protein crystal, and most experiments are performed at ambient temperature. For room temperature data collection, crystals are mounted in quartz capillaries. While room temperature data collection is typical, the instrument is also equipped with a closed cycle refrigerator for cryo-crystallography at temperatures down to 15 K. Data collection at low temperature can reduces thermal disorder in protein crystals and makes possible the study of trapped reaction intermediates and temperature sensitive systems (Myles, Dauvergne, Blakeley, & Meilleur, 2012). In either configuration, crystals are mounted and aligned manually on the goniometer. Given that data collection time typically extends over several days, manual centering is not a significant overhead. When data collection is complete, crystals are recovered, and corresponding X-ray dataset(s) collected for joint refinement using X-ray and neutron data (see Section 2.5).

For indexing and integration, IMAGINE uses the LAUEGEN suite of software, which was developed originally for X-ray Laue crystallography and then adapted for neutron diffraction and cylindrical detector geometries (Campbell, 1995; Campbell, Hao, Harding, Nguti, & Wilkinson, 1998). The data are then scaled and wavelength normalized using the Lscale program (Arzt, Campbell, Harding, Hao, & Helliwell, 1999; Helliwell et al., 1989). SCALA from the CCP4 suite of programs can then be used to merge the data and generate standard data collection statistics (Winn et al., 2011). A dedicated workstation is available to the users for on-site data reduction.

2.5 Structure refinement

Structure refinement is carried out against neutron data alone or simultaneously against both X-ray and neutron data. The latter protocol is referred to as joint refinement or X/N refinement (Afonine et al., 2010; Liebschner, Afonine, Urzhumtsev, & Adams, 2020). The choice of one method of refinement over the other may be influenced by the neutron dataset resolution and completeness but is ultimately the experimenter's choice. Joint X/N refinement is highly recommended for hydrogenous proteins that are partially deuterated by H/D exchange, because at neutron diffraction resolutions below ~2.0 Å, the intrinsic negative scattering length of hydrogen gives rise to negative peaks in neutron scattering density maps that can generate significant density cancelation effects for aliphatic and aromatic groups. Thus, these groups are not always well-defined, and their positions may not be accurately determined using neutron data alone. Joint refinement requires that X-ray dataset be collected at the same temperature as the neutron dataset. Collecting both datasets at the same temperature is essential as temperature alters aspects of the protein structure or of the crystal itself (Deacon et al., 1997; Fraser et al., 2011; Fukuda & Inoue, 2015; Tilton, Dewan, & Petsko, 1992; Weik & Colletier, 2010). While the X-ray and neutron dataset should ideally be collected from the same crystal for joint XN refinement, the X-ray dataset is most often measured from a smaller crystal grown in the same drop or under the same crystallization conditions as the neutron crystal. X-ray data can be collected using ORNL's in-house X-ray diffraction facilities (Section 3).

Phenix.refine is typically used to refine structures against neutron data collected on IMAGINE (Adams et al., 2011; Adams, Mustyakimov, Afonine, & Langan, 2009; Afonine et al., 2010; Liebschner et al., 2020). The phenix.refine module of the PHENIX software package is capable of refining model against both neutron data alone and joint X-ray/neutron datasets. Refinement against neutron data closely follows X-ray refinement protocols, and neutron density maps can be visualized using the program COOT (Emsley & Cowtan, 2004; Emsley, Lohkamp, Scott, & Cowtan, 2010). Model building, map interpretation and refinement are discussed further in Logan (2020).

3. Sample preparation and ancillary facilities

Replacing D for H in a protein crystal provides significant benefits to a neutron protein crystallography experiment (O'Dell, Bodenheimer, & Meilleur, 2016). The greater magnitude of the D coherent scattering length (6.671 fm) compared to that of H (−3.741 fm) results in greater average reflection intensities. Simultaneously, the decrease in the D incoherent scattering cross section compared to H dramatically reduces the isotropic background intensity. Furthermore, the D coherent scattering length is positive as is that of common atoms found in proteins while the coherent scattering length of H is negative. At the typical resolutions (1.9–2.3 Å) obtained for neutron protein crystallographic data, D substitution simplifies the analysis of neutron density maps by minimizing the number of sites where the negative coherent scattering length of H atoms leads to density cancelation.

Bulk solvent, ordered water molecules, and hydrogen atoms bound to protein oxygen, and nitrogen (i.e., titratable hydrogen atoms that are accessible) can all be replaced with deuterium by preparing all reagents in deuterium oxide (D_2O) either during or after complete crystal growth. Most frequently, crystals are grown to final size in H_2O mother liquor and then subjected to H/D exchange either by vapor equilibration in capillary or soaking in an artificial deuterated mother liquor. While in some cases, crystals can be directly grown in solution made with D_2O, in our experience many proteins do not crystallize as efficiently or grow to as large a size in heavy water.

Titratable hydrogen atoms represent approximatively 25% of the total hydrogen atoms of any proteins. In order to exchange the remaining 75% attached to carbon atoms, proteins need to be perdeuterated during synthesis. At ORNL, the Bio-Deuteration Laboratory supports the production of perdeuterated and 85% uniform D-labeled protein for users of the IMAGINE beamline with approved neutron beam time (https://www.ornl.gov/node/56459).

Users can also take advantage of the Protein Characterization & Crystallization Laboratory which is equipped with biophysical characterization instruments, including a dynamic light scattering instrument, a circular dichroism spectropolarimeter, fluorescence microscope and stereomicroscopes, a Gryphon crystallization robot and a Scorpion liquid dispensing robot, and 13 temperature-controlled chambers and other devices for crystallogenesis and large crystal growth. A Protein X-ray Crystallography Laboratory equipped with a Rigaku high intensity micro-focus rotating anode X-ray generator and an Eiger 4M detector is available to users for crystals screening and for X-ray data collection for joint X/N refinement.

4. Facility access

Access to IMAGINE is granted through the HFIR and SNS User Program (https://neutrons.ornl.gov/users). A call for proposals is issued twice a year. Beam time is allocated after proposals are scored and ranked by external scientific peer review. Access to the Bio-Deuteration Laboratory can be requested in parallel with applications for neutron beam time. Access to the protein characterization and crystallization laboratories can be coordinated with the authors and/or other scientists at ORNL.

5. Highlights

The structural biology program on IMAGINE provides an atomic-level description and understanding of the structure and function of proteins involved in diverse biological processes, including catalysis, ligand binding, light capture and conversion, cell signaling and DNA repair. Most studies are focused on understanding the physiochemical mechanisms through which enzymes recognize ligands and catalyze reactions, and to help define and understand the key chemistries that are involved. Detailed examples are presented in Kovalevsky et al. (2020), Gerlits et al. (2020), Mueser, Drago, Kovalevsky, and Dajnowicz (2020), and Kerber and Vrielink (2020). Several other examples are highlighted below.

5.1 *Neurospora Crassa* lytic polysaccharide monooxygenase 9D

Lytic polysaccharide monooxygenases (LPMOs) are copper-dependent metalloenzymes that activate molecular oxygen for the net insertion of one oxygen atom into the carbon-hydrogen bond of a glycosidic carbon in polymeric carbohydrates. LPMOs, thus, can disrupt the crystalline structure of cellulose in lignocellulosic plant biomass, and can be used in industry to enhance the efficiency of cellulose degradation by other hydrolytic enzymes for biofuels production. Therefore, detailed knowledge of the LPMO enzymatic mechanism, which requires molecular oxygen, two electrons, and two protons, is important but remains poorly understood An examination of protonation states enabled by neutron crystallography of the LPMO enzymatic resting state and density functional theory calculations identified a role for conserved histidine (H157) in promoting oxygen activation (Fig. 2) (O'Dell, Agarwal, & Meilleur, 2017; O'Dell, Swartz, Weiss, & Meilleur, 2017). The result provides a new structural description of oxygen activation by LPMOs. Fully characterizing the LPMO reaction mechanism may inform strategies for increasing the efficiency of enzymatic polysaccharide hydrolysis for applications including bioethanol production.

5.2 H-RAS GTPase

The small GTPase H-RAS is a prototype for nucleotide-binding proteins that function by cycling between bound guanosine triphosphate (GTP) and guanosine diphosphate (GDP). The neutron structure of H-RAS exemplifies the uniquely rich informational content provided by a single neutron

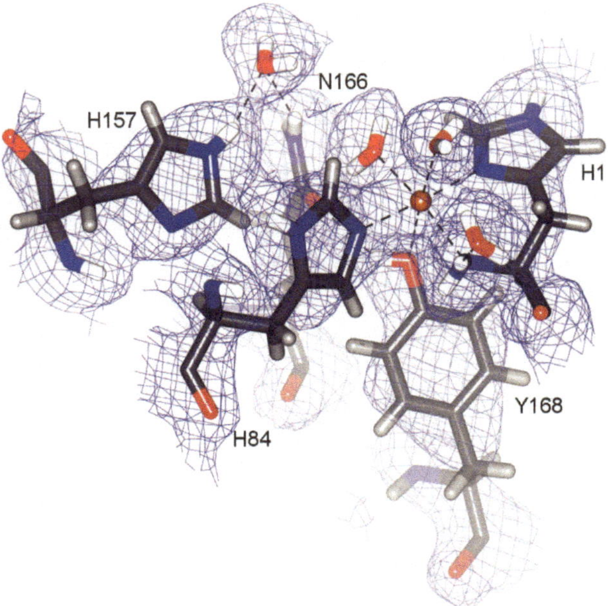

Fig. 2 Active site of *Nc*LPMO9D in the enzymatic resting state (PDB 5TKI). The conformation and tautomeric state of singly protonated H157 revealed by neutron protein crystallography was confirmed to stabilize molecular oxygen (not shown) near the LPMO active site by quantum chemical calculations. The H1 backbone amide N and carbonyl O form H-bonds with the "pocket" water molecule which has been hypothesized to participate in substrate interactions. ^{1}H atoms are shown in light gray while atoms that have undergone exchange to ^{2}H are shown in white. The copper atom is shown in brown. The blue mesh represents neutron scattering length density (SLD) $2F_O–F_C$ map contoured at $\sigma = 1.0$. *Reproduced with permission of the International Union of Crystallography.*

diffraction experiment (Knihtila, Holzapfel, Weiss, Meilleur, & Mattos, 2015; Knihtila, Volmar, Meilleur, & Mattos, 2019). Previous efforts to elucidate the GTP hydrolysis mechanism of H-RAS have relied on the assumption that the γ-phosphate group of GTP bound at the active site of the enzyme would be deprotonated as is observed in solution (Cheng, Sukal, Callender, & Leyh, 2001). This study, however, suggests that the γ-phosphate group of the GTP ligand remains protonated at physiological pH when bound at the active site of H-RAS in the ground state (Fig. 3). The neutron structure revealed the modulation of the pK_a of the gamma-phosphate group by the H-RAS active site microenvironment and opens the possibility that the pre-catalytic state of the GTP hydrolysis reaction catalyzed by H-RAS is different than previously thought.

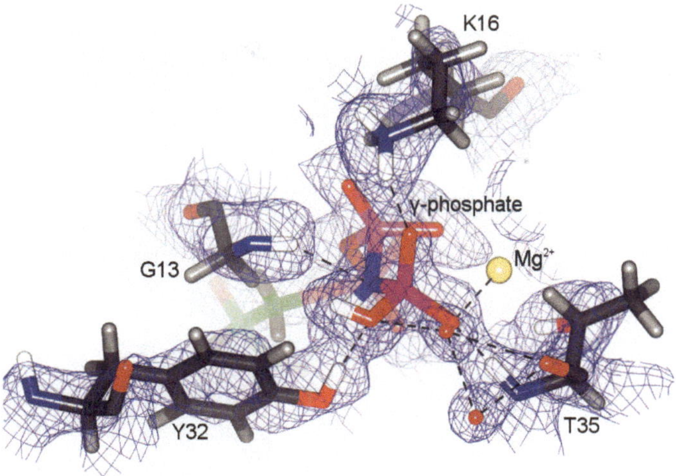

Fig. 3 H-RAS in complex with GTP analog GppNHp (PDB 4RSG). Unexpectedly, clear neutron SLD could be observed for a protonated (^{2}H) γ-phosphate in the $2F_O$–F_C map. The deuterium atoms of the nucleophilic water (hydrogen bonded to the γ-phosphate and T35) are not visible in the density map, suggesting rotational freedom. Ligand carbon atoms are shown in green. ^{1}H atoms are shown in light gray while atoms that have undergone exchange to ^{2}H are shown in white. The blue mesh represents neutron SLD $2F_O$–F_C map contoured at $\sigma = 1.0$. *Reproduced with permission of the International Union of Crystallography.*

5.3 Neutron protein crystallography at cryogenic temperature

Cold (2–10 Å) neutrons commonly used in crystallographic experiments do not cause any observable radiation damage in protein crystals and thus most experiments are performed at room temperature. However, cryo-cooling protein crystals can (in some cases) also improve data quality and reduce B-factors (Blakeley, Kalb, Helliwell, & Myles, 2004; Casadei et al., 2014), as well as enabling freeze-trapping studies of enzymatic reaction intermediates. Until quite recently, cryo-cooling had been assumed challenging for large neutron sized protein crystals, since the solvent must be rapidly (flash)-cooled to a vitreous glass state to avoid ice formation that disrupts the crystal lattice. Protocols for cryo-neutron protein crystallography have been developed to cool and maintain large protein crystals (1–5 mm^3) at cryogenic temperatures (<15 K) using a closed cycle refrigerator (Blakeley et al., 2004; Myles et al., 2012). The approach mirrors that used for off-line cryo-cooling of X-ray size samples, flash-cooling loop-mounted crystals simply by plunging directly into liquid N_2 (77K) and then transferring crystals to pre-cooled cryostat under liquid nitrogen. A closed-cycle refrigerator

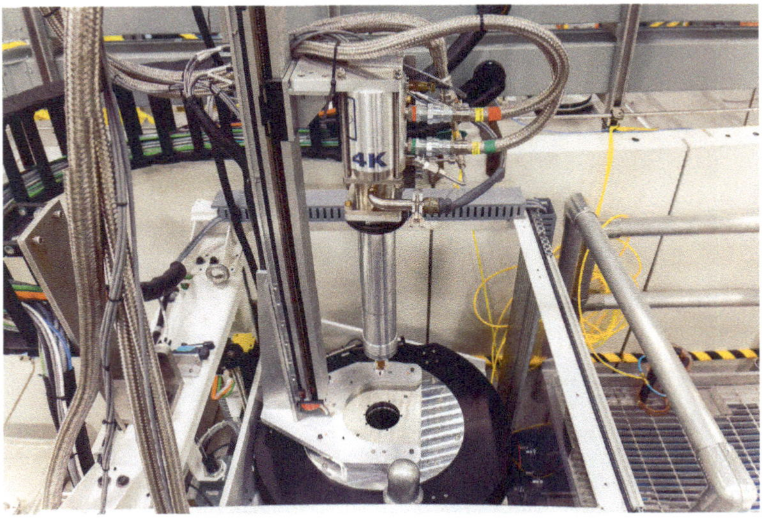

Fig. 4 An automated lift moves the CCR up and down. Crystals are mounted directly onto the CCR cold finger *via* custom magnetic bases and then sealed with a heat shield (not shown) and a dome-shaped vacuum enclosure (visible at the bottom center of the photograph). A motorized vertical lift facilitates positioning of the mounted sample into the neutron beam. *Reproduced with permission of the International Union of Crystallography (Photo credit: Genevieve Martin, ORNL).*

(CCR) that operates in the range of 4–450 K (Fig. 4) has been recently commissioned on IMAGINE, expanding the capabilities of the instrument for neutron diffraction analysis at cryogenic temperatures.

To demonstrate the use of this system for cryogenic analysis of protein crystals, a 1.8 Å dataset was collected on T4L lysozyme at 80 K (Fig. 5) (Li et al., 2017). Access to these cryo-cooling capabilities of IMAGINE should allow new types of experiments that use freeze-trapping techniques to "quench" kinetic processes and capture reaction intermediaries.

6. Future development

Collecting data with high completeness is complicated by the cylindrical detector geometry as well as a significant blind region arising from the curvature of the Ewald sphere at long wavelengths. Working with a single axis goniometer, crystals are typically reoriented to increase completeness by collecting data in a second crystal orientation. A miniaturized kappa goniometer head used at synchrotron beamlines has been modified and built for incorporation on IMAGINE. The Arinax MiniKappa MK3 provides an additional kappa and phi axis. This will enable data to be collected at

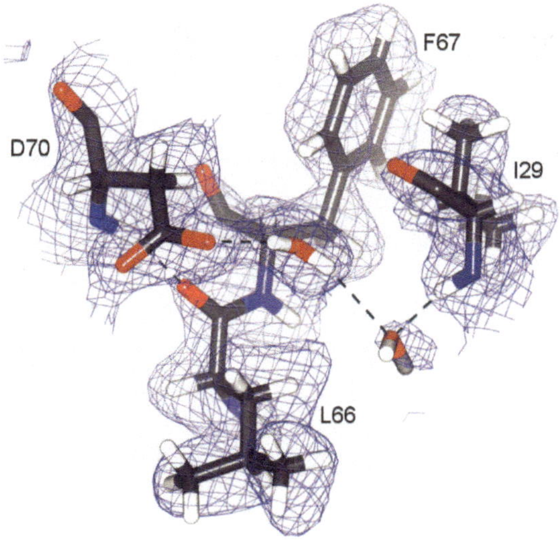

Fig. 5 T4 Lysozyme "cavity 1" as observed at 80 K (PDB 5VNQ). Hydrogen atoms that were substituted with ^{2}H by covalent labeling and exchange are shown in white. The blue mesh represents neutron SLD $2F_O$–F_C map contoured at $\sigma = 1.0$. *Reproduced with permission of the International Union of Crystallography.*

multiple crystal settings within a single data collection run, which is expected to increase the data collection efficiency of IMAGINE. The CrystalPlan software has been adapted for IMAGINE to optimize data collection strategies using the mini-Kappa goniometer (Zikovsky, Peterson, Wang, Frost, & Hoffmann, 2011). CrystalPlan uses the UB matrix determined from a short Laue geometry alignment image to predict an optimal set of crystal orientations that will maximize the completeness using a minimal number of crystal settings. These settings are then loaded and run in the instrument GUI data collection script. This will allow IMAGINE users to maximize use of available beam time per sample per experiment and increase the data collection efficiency.

Longer term, the planned 2024 replacement of the cold guide neutron delivery system at the HFIR will provide a new position for IMAGINE that is closer to the cold source, and that is expected to deliver a two- to fourfold increase in flux on the instrument.

7. Summary

IMAGINE is a world class neutron diffractometer that specializes in neutron protein crystallography. The instrument offers access to crystals with unit cells of up to 150 Å, complementing the capabilities of the

ORNL Spallation Neutron Source MaNDi time-of-flight instrument which is optimized for larger unit cells (Coates & Sullivan, 2020). Recent upgrades have added to the capabilities of IMAGINE by providing access to data collection at variable temperatures (4–450 K), which will broaden the science that can be done on this beam line. Additionally, low temperature macromolecular neutron diffraction data can now be obtained from cryo-cooled protein crystals, making experiments with freeze-trapped intermediates in proteins possible. The unique combination of scientific and technical expertise as well as facilities for biodeuteration, crystal growth and macromolecular crystallography at ORNL provides the user community with supporting capabilities for sample preparation that help drive the productivity and development of the structural biology program at IMAGINE, allowing new scientific questions to be addressed.

Acknowledgments

This research used resources at the High Flux Isotope Reactor and Spallation Neutron Source, a DOE Office of Science User Facility operated by the Oak Ridge National Laboratory. F.M. acknowledges support from the USDA National Institute of Food and Agriculture, Hatch project 1010523. The installation of the IMAGINE instrument was supported in part by the National Science Foundation award 0922719.

References

Adams, P. D., Afonine, P. V., Bunkoczi, G., Chen, V. B., Echols, N., Headd, J. J., et al. (2011). The Phenix software for automated determination of macromolecular structures. *Methods*, *55*, 94–106.

Adams, P. D., Mustyakimov, M., Afonine, P. V., & Langan, P. (2009). Generalized X-ray and neutron crystallographic analysis: More accurate and complete structures for biological macromolecules. *Acta Crystallographica. Section D, Biological Crystallography*, *65*, 567–573.

Afonine, P. V., Mustyakimov, M., Grosse-Kunstleve, R. W., Moriarty, N. W., Langan, P., & Adams, P. D. (2010). Joint X-ray and neutron refinement with phenix.refine. *Acta Crystallographica. Section D, Biological Crystallography*, *66*, 1153–1163.

Arzt, S., Campbell, J. W., Harding, M. M., Hao, Q., & Helliwell, J. R. (1999). LSCALE—The new normalization, scaling and absorption correction program in the Daresbury Laue software suite. *Journal of Applied Crystallography*, *32*, 554–562.

Blakeley, M. P., Kalb, A. J., Helliwell, J. R., & Myles, D. A. (2004). The 15-K neutron structure of saccharide-free concanavalin A. *Proceedings of the National Academy of Sciences of the United States of America*, *101*, 16405–16410.

Budayova-Spano, M., Koruza, K., & Fisher, Z. (2020). Large crystal growth for neutron protein crystallography. In P. Moody (Ed.), *Vol. 634. Methods in enzymology. Neutron crystallography in structural biology* (pp. 21–46). Elsevier.

Campbell, J. W. (1995). Lauegen, an X-windows-based program for the processing of Laue X-ray-diffraction data. *Journal of Applied Crystallography*, *28*, 228–236.

Campbell, J. W., Hao, Q., Harding, M. M., Nguti, N. D., & Wilkinson, C. (1998). LAUEGEN version 6.0 and INTLDM. *Journal of Applied Crystallography*, *31*, 496–502.

Casadei, C. M., Gumiero, A., Metcalfe, C. L., Murphy, E. J., Basran, J., Concilio, M. G., et al. (2014). Neutron cryo-crystallography captures the protonation state of ferryl heme in a peroxidase. *Science*, *345*, 193–197.

Cheng, H., Sukal, S., Callender, R., & Leyh, T. S. (2001). Gamma-phosphate protonation and pH-dependent unfolding of the Ras.GTP.Mg2+ complex: A vibrational spectroscopy study. *The Journal of Biological Chemistry*, *276*, 9931–9935.

Cipriani, F., Castagna, J. C., Wilkinson, C., Lehmann, M. S., & Buldt, G. (1996). A neutron image plate quasi-Laue diffractometer for protein crystallography. *Neutrons in Biology*, *64*, 423–431.

Coates, L., & Sullivan, B. (2020). The macromolecular neutron diffractometer at the spallation neutron source. In P. Moody (Ed.), *Vol. 634. Methods in enzymology. Neutron crystallography in structural biology* (pp. 87–99). Elsevier.

Crow, L., Robertson, L., Bilheux, H., Fleenor, M., Iverson, E., Tong, X., et al. (2011). The CG1 instrument development test station at the high flux isotope reactor. *Nuclear Instruments and Methods in Physics Research Section A: Accelerators, Spectrometers, Detectors and Associated Equipment*, *634*, S71–S74.

Deacon, A., Gleichmann, T., Kalb, A. J., Price, H., Raftery, J., Bradbrook, G., et al. (1997). The structure of concanavalin A and its bound solvent determined with small-molecule accuracy at 0.94 angstrom resolution. *Journal of the Chemical Society, Faraday Transactions*, *93*, 4305–4312.

Emsley, P., & Cowtan, K. (2004). Coot: Model-building tools for molecular graphics. *Acta Crystallographica. Section D, Biological Crystallography*, *60*, 2126–2132.

Emsley, P., Lohkamp, B., Scott, W. G., & Cowtan, K. (2010). Features and development of Coot. *Acta crystallographica. Section D, Biological crystallography*, *66*, 486–501.

Fraser, J. S., van den Bedem, H., Samelson, A. J., Lang, P. T., Holton, J. M., Echols, N., et al. (2011). Accessing protein conformational ensembles using room-temperature X-ray crystallography. *Proceedings of the National Academy of Sciences of the United States of America*, *108*, 16247–16252.

Fukuda, Y., & Inoue, T. (2015). High-temperature and high-resolution crystallography of thermostable copper nitrite reductase. *Chemical Communications*, *51*, 6532–6535.

Gerlits, O., Weiss, K. L., Blakeley, M. P., Veglia, G., Taylor, S. S., & Kovalevsky, A. (2020). Protein kinase A in the neutron beam: Insights for catalysis from directly observing protons. In P. Moody (Ed.), *Vol. 634. Methods in enzymology. Neutron crystallography in structural biology* (pp. 311–331). Elsevier.

Heller, W. T., Urban, V. S., Lynn, G. W., Weiss, K. L., O'Neill, H. M., Pingali, S. V., et al. (2014). The bio-SANS instrument at the high flux isotope reactor of Oak Ridge National Laboratory. *Journal of Applied Crystallography*, *47*, 1238–1246.

Helliwell, J. R. (1997). Neutron Laue diffraction does it faster. *Nature Structural Biology*, *4*, 874–876.

Helliwell, J. R., Habash, J., Cruickshank, D. W. J., Harding, M. M., Greenhough T. J., Campbell, J. W., et al. (1989). The recording and analysis of synchrotron X-radiation Laue diffraction photographs. *Journal of Applied Crystallography*, *22*, 483–497.

Kerber, T., & Vrielink, A. (2020). The role of hydrogen atoms in redox catalysis by the flavoenzyme cholesterol oxidase. In P. Moody (Ed.), *Vol. 634. Methods in enzymology. Neutron crystallography in structural biology* (pp. 361–377). Elsevier.

Knihtila, R., Holzapfel, G., Weiss, K., Meilleur, F., & Mattos, C. (2015). Neutron crystal structure of RAS GTPase puts in question the protonation state of the GTP gamma-phosphate. *The Journal of Biological Chemistry*, *290*, 31025–31036.

Knihtila, R., Volmar, A. Y., Meilleur, F., & Mattos, C. (2019). Titration of ionizable groups in proteins using multiple neutron data sets from a single crystal: Application to the small GTPase Ras. *Acta Crystallographica. Section F, Structural Biology Communications*, *75*, 111–115.

Kovalevsky, A., Gerlits, O., Beltran, K., Weiss, K., Keen, D. A., Blakeley, M. P., et al. (2020). Proton transfer and drug binding details revealed in neutron diffraction studies of wild-type and drug resistant HIV-1 protease. In P. Moody (Ed.), *Vol. 634. Methods in enzymology. Neutron crystallography in structural biology* (pp. 257–279). Elsevier.

Li, L., Shukla, S., Meilleur, F., Standaert, R. F., Pierce, J., Myles, D. A. A., et al. (2017). Neutron crystallographic studies of T4 lysozyme at cryogenic temperature. *Protein Science*, *26*, 2098–2104.

Liebschner, D., Afonine, P. V., Urzhumtsev, A. G., & Adams, P. D. (2020). Implementation of the riding hydrogen model in *CCTBX* to support the next generation of X-ray and neutron joint refinement in *Phenix*. In P. Moody (Ed.), *Vol. 634. Methods in enzymology. Neutron crystallography in structural biology* (pp. 177–199). Elsevier.

Logan, D. T. (2020). Interactive model building in neutron macromolecular crystallography. In P. Moody (Ed.), *Vol. 634. Methods in enzymology. Neutron crystallography in structural biology* (pp. 201–224). Elsevier.

Lynn, G. W., Heller, W., Urban, V., Wignall, G. D., Weiss, K., & Myles, D. A. A. (2006). Bio-SANS—A dedicated facility for neutron structural biology at Oak Ridge National Laboratory. *Physica B: Condensed Matter*, *385–86*, 880–882.

Markó, M., Nagy, G., Aprigliano, G., & Oksanen, E. (2020). Neutron macromolecular crystallography at the European spallation source. In P. Moody (Ed.), *Vol. 634. Methods in enzymology. Neutron crystallography in structural biology* (pp. 125–151). Elsevier.

Meilleur, F., Munshi, P., Robertson, L., Stoica, A. D., Crow, L., Kovalevsky, A., et al. (2013). The IMAGINE instrument: First neutron protein structure and new capabilities for neutron macromolecular crystallography. *Acta Crystallographica. Section D, Biological Crystallography*, *69*, 2157–2160.

Meilleur, F., Myles, D. A., & Blakeley, M. P. (2006). Neutron Laue macromolecular crystallography. *European Biophysics Journal*, *35*, 611–620.

Mueser, T. C., Drago, V., Kovalevsky, A., & Dajnowicz, S. (2020). Pyridoxal 5′ phosphate dependent reactions: Analyzing the mechanism of aspartate aminotransferase. In P. Moody (Ed.), *Vol. 634. Methods in enzymology. Neutron crystallography in structural biology* (pp. 333–359). Elsevier.

Munshi, P., Chung, S. L., Blakeley, M. P., Weiss, K. L., Myles, D. A., & Meilleur, F. (2012). Rapid visualization of hydrogen positions in protein neutron crystallographic structures. *Acta Crystallographica. Section D, Biological Crystallography*, *68*, 35–41.

Myles, D. A. A., Dauvergne, F., Blakeley, M. P., & Meilleur, F. (2012). Neutron protein crystallography at ultra-low (<15 K) temperatures. *Journal of Applied Crystallography*, *45*, 686–692.

Niimura, N., Karasawa, Y., Tanaka, I., Miyahara, J., Takahashi, K., Saito, H., et al. (1994). An imagining plate neutron detector. *Nuclear Instruments and Methods in Physics Research Section A: Accelerators, Spectrometers, Detectors and Associated Equipment*, *349*, 521–525.

O'Dell, W. B., Agarwal, P. K., & Meilleur, F. (2017). Oxygen activation at the active site of a fungal lytic polysaccharide monooxygenase. *Angewandte Chemie*, *56*, 767–770.

O'Dell, W. B., Bodenheimer, A. M., & Meilleur, F. (2016). Neutron protein crystallography: A complementary tool for locating hydrogens in proteins. *Archives of Biochemistry and Biophysics*, *602*, 48–60.

O'Dell, W. B., Swartz, P. D., Weiss, K. L., & Meilleur, F. (2017). Crystallization of a fungal lytic polysaccharide monooxygenase expressed from glycoengineered Pichia pastoris for X-ray and neutron diffraction. *Acta Crystallographica. Section F, Structural Biology Communications*, *73*, 70–78.

Schroder, G. C., O'Dell, W. B., Myles, D. A. A., Kovalevsky, A., & Meilleur, F. (2018). IMAGINE: Neutrons reveal enzyme chemistry. *Acta Crystallographica. Section D, Structural Biology*, *74*, 778–786.

Tanaka, I., Chatake, T., Fujiwara, S., Hosoya, T., Kusaka, K., Niimura, N., et al. (2020). Current status and near future plan of neutron protein crystallography at J-PARC. In P. Moody (Ed.), *Vol. 634. Methods in enzymology. Neutron crystallography in structural biology* (pp. 101–123). Elsevier.

Tilton, R. F., Jr., Dewan, J. C., & Petsko, G. A. (1992). Effects of temperature on protein structure and dynamics: X-ray crystallographic studies of the protein ribonuclease-A at nine different temperatures from 98 to 320 K. *Biochemistry*, *31*, 2469–2481.

Weik, M., & Colletier, J. P. (2010). Temperature-dependent macromolecular X-ray crystallography. *Acta Crystallographica. Section D, Biological Crystallography*, *66*, 437–446

Wignall, G. D., Littrell, K. C., Heller, W. T., Melnichenko, Y. B., Bailey, K. M., Lynn, G. W., et al. (2012). The 40 m general purpose small-angle neutron scattering instrument at Oak Ridge National Laboratory. *Journal of Applied Crystallography*, *45*, 990–998.

Winn, M. D., Ballard, C. C., Cowtan, K. D., Dodson, E. J., Emsley, P., Evans, P. R., et al. (2011). Overview of the CCP4 suite and current developments. *Acta Crystallographica. Section D, Biological Crystallography*, *67*, 235–242.

Zikovsky, J., Peterson, P. F., Wang, X. P. P., Frost, M., & Hoffmann, C. (2011). CrystalPlan: An experiment-planning tool for crystallography. *Journal of Applied Crystallography*, *44*, 418–423.

CHAPTER FIVE

The macromolecular neutron diffractometer at the spallation neutron source

Leighton Coates*, Brendan Sullivan
Neutron Scattering Division, Oak Ridge National Laboratory, Oak Ridge, TN, United States
*Corresponding author: e-mail address: coatesl@ornl.gov

Contents

Abstract

The Macromolecular Neutron Diffractometer known as MaNDi is located on beamline 11B of the Spallation Neutron Source at Oak Ridge National Laboratory. Neutrons are produced in discrete pulses at the Spallation Neutron Source, which enables Laue diffraction data through measuring the time-of-flight and thus wavelength of each diffracted neutron. The MaNDi diffractometer is optimized to collect diffraction data from protein crystals with unit cell axes between 30 and 300 Å. The instrument is designed to provide flexibility in several instrumental parameters such as wavelength bandwidth and beam divergence to allow data collection from a range systems. Data collection is possible at room temperature or 100 K using a standard nitrogen gas stream cooler combined with standard X-ray style mounting pins and loops

1. Introduction

The Macromolecular Neutron Diffractometer (MaNDi) is a wavelength-resolved Laue instrument optimized to collect neutron diffraction data on large unit cells up to a maximum cell axis length of 300 Å.

Methods in Enzymology, Volume 634
ISSN 0076-6879
https://doi.org/10.1016/bs.mie.2019.11.020

The instrument operates in a Laue mode in which a range of neutron wavelengths are used for data collection, typically neutrons from 2 to 4.16 Å are the most commonly used range as they are optimal for nearly all macromolecular experiments. The ability to resolve the wavelength of any diffracted neutron from within the range of wavelengths used for data collection is possible because of the pulsed nature of the Spallation Neutron Source (SNS), which produces 60 discrete neutron pulses per second, and the high time resolution of the SNS Anger camera detectors used on MaNDi (Coates, Stoica, Hoffmann, Richards, & Cooper, 2010). As the time of the generation of each neutron pulse and the distance from the neutron source to the neutron detector is well known, the time-of-flight (TOF) (Langan et al., 2008) of each detected neutron can be used to determine the wavelength of each detected neutron using the de Broglie formula. Wavelength resolution allows separation of harmonic and spatially overlapped peaks, which would overlap in a typical reactor-based neutron or X-ray Laue experiment. This critical advantage enables MaNDi to collect data on large unit cells and to perform experiments that are not possible on reactor-based instruments. Additionally, the background of observed peaks is also reduced when the Laue diffraction data are wavelength resolved, resulting in a higher signal-to-noise ratio for the integrated Bragg reflections. Detailed explanations of the instrument design, commissioning, and performance of MaNDi are given in a series of articles (Coates et al., 2010, 2015; Schultz et al., 2005) that are recommended for interested readers. However, the instrument end station of MaNDi is where users typically mount samples, conduct screening experiments, and collect datasets. At the heart of the instrument is a detector array frame (DAF). This hollow aluminum sphere houses 40 SNS Anger cameras around the sample position to give a detector coverage of 4.1 sr (Fig. 1). Two video cameras for sample alignment are also housed in the DAF. The significant area detector coverage, coupled with the range of wavelengths used in the diffraction experiment enables a large amount of data to be collected in a single orientation of the crystal sample. Typically, the number of orientations required for a complete dataset depends on the space group of the crystal. For high symmetry space groups such as I4 high completeness datasets can be collected from just four orientations while for the $P2_12_12_1$ space group, which is the most common space group for protein crystals, around nine orientations would be needed for a high completeness dataset.

To rotate the sample between orientations and to hold the sample in the center of the DAF, a custom goniometer is used (Fig. 1B). This goniometer

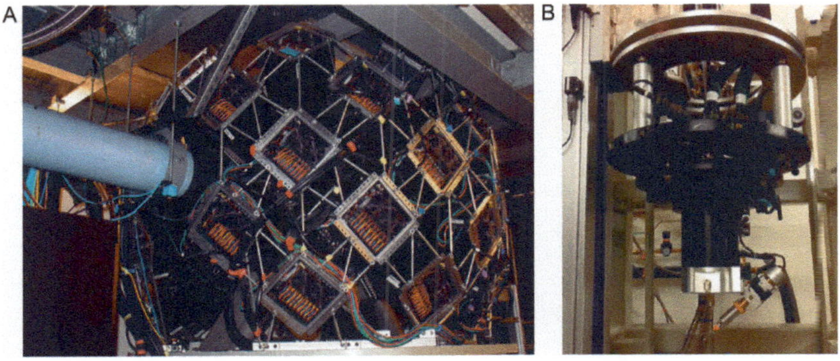

Fig. 1 (A) The detector array frame (DAF) of the MaNDi instrument contains 40 SNS Anger cameras which surround the sample position in a spherical array. The DAF also contains two cameras that view the sample position, a goniometer which rests on the top of the DAF is used to align the sample in the neutron beam. (B) The MaNDi goniometer translates vertically by 2 m between a sample loading position, which is used to mount the sample at the start of the experiment and a data collection position that places the goniometer entirely into the DAF allowing the crystal to be centered and rotated in the neutron beam.

translates perpendicular to the sample position with the DAF; in the upward position the sample is loaded onto the goniometer, the goniometer is then lowered down to rest on the top of the DAF (Fig. 1A) which positions the sample into the neutron beam that passes through the center of the DAF. During the collection from each orientation, the crystal remains static; once data collection has completed at each orientation, the crystal is then rotated before the collection of another orientation begins. Samples for MaNDi can be mounted in two different ways. For room temperature data collection, the crystal is typically mounted into a sealed fused quartz capillary tube along with a slug of mother liquor (Fig. 2). If data will be collected at 100 K, then the crystal is mounted using standard crystallographic loops and pins (Coates et al., 2014). The click-to-center sample alignment procedure on MaNDi places the crystal into the neutron beam using a series of mouse clicks on the crystal location within an image of the sample position generated from the DAF sample alignment cameras. As there are limits of travel on the goniometer motors that are used to align the sample, there are specific dimensions required for mounted crystals within capillaries for experiments at MaNDi (Fig. 2). These guidelines are also helpful when collecting X-ray data at room temperature using newer X-ray systems that utilize a triple-axis goniometer coupled with a small pixel detector, as the smaller length of the capillary helps prevent collisions between the capillary, beamstop, and

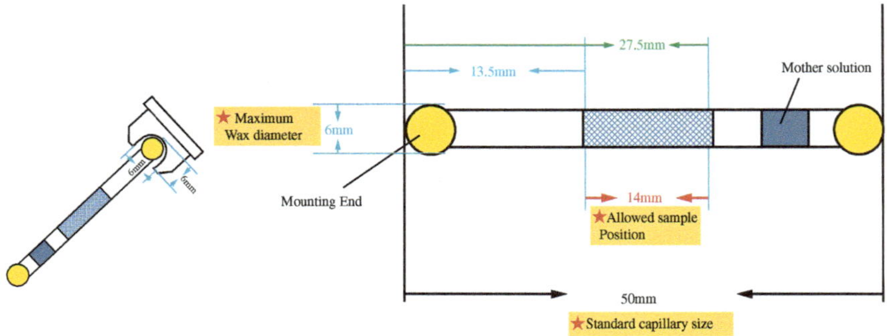

Fig. 2 Crystal mounting guidelines for room temperature data collection on the MaNDi instrument. Thick-walled fuzed quartz capillaries sealed with a high melting temperature wax are recommended as they provide an airtight seal, are resistant to handling damage and are easy to transport. As MaNDi utilizes a motorized goniometer translation of the capillary is limited by the motor travel limits, the capillary is mounted onto an adjustable crystal mount using modeling clay.

detector. These requirements are shown in Fig. 2. IMAGINE is a quasi-Laue protein crystallography beamline optimized to collect data on unit cell axes up to 150 Å which is located the High Flux Isotope Reactor (HFIR) at Oak Ridge National Laboratory (Schroder, O'Dell, Myles, Kovalevsky, & Meilleur, 2018), this instrument is optimized to collect data on smaller unit cell axes complementing the capabilities of MaNDi at the SNS and is described in the chapter "IMAGINE: The neutron protein crystallography beamline at the high flux isotope reactor" by Meilleur et al., in this volume.

Once a crystal has been centered in the neutron beam the shutter is opened and the experiment begins. Since the position-sensitive SNS Anger cameras used on MaNDi detect neutrons as soon as they arrive, it is possible to view a live diffraction image while it is being collected using the MaNDi instrument control software. At the start of the experiment, the low number of counts means there is insufficient difference between signal and background to see peaks. As time progresses, the lower angle Bragg reflections start to become visible, with the full Laue diffraction image becoming visible after several hours. An example of the time evolution of peaks in a single Anger camera is shown in Fig. 3. For crystals that diffract neutrons strongly a full data collection can be set up for automated data collection in which all the orientations required for a complete dataset are cued for collection. The time taken to collect each orientation varies based on several factors all of which influence the scattering power of the crystal these include the size of the crystal, the volume of the unit cell, the quality of the crystal and the amount of hydrogen within the sample (these factors are discussed by Fisher in this volume).

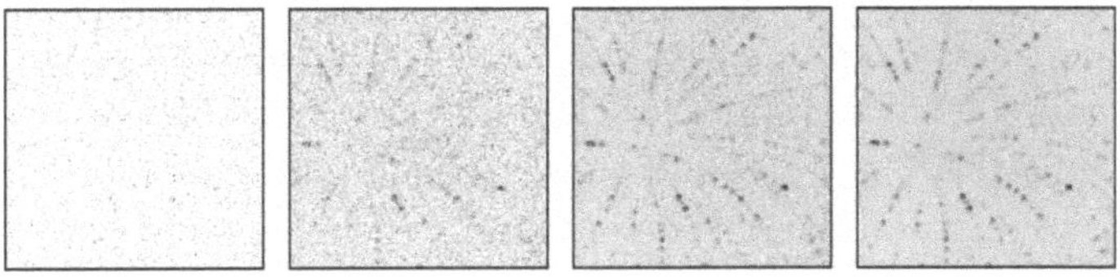

Fig. 3 Live Laue diffraction images are taken from a Toho-1 β-lactamase shown after a period of 5 min, 1, 4, and 12 h. After 5 min, the stronger low angle Bragg reflections start to become visible over the background noise while after 1 and 4 h exposures more Bragg reflections at higher scattering angles become visible.

The number of orientations required for a complete dataset can be estimated by the use of the data collection strategy program CrystalPlan (Zikovsky, Peterson, Wang, Frost, & Hoffmann, 2011), which generates an optimal list of orientations that are required for the collection of a complete dataset.

2. Auto reduction of diffraction data

The pulsed timing structure of neutrons at MaNDi allows for a straightforward deconvolution of harmonic peaks. This is evident when viewing data in reciprocal space. Following the crystallography convention in the sample frame, we define the scattering vector of each detected neutron as

$$\overrightarrow{q_s} = \lambda\left(\overrightarrow{k_f} - \overrightarrow{k_i}\right) = 2\pi UB\overrightarrow{h}$$

where $\overrightarrow{k_f}$ and $\overrightarrow{k_i}$ are the final and initial wavevectors and λ is the wavelength, all three which can be calculated from the instrument geometry. UB is the standard crystallography orientation matrix. In this formalism, Bragg peaks occur at integer $\overrightarrow{h} = (h, k, l)$, which elucidates the mathematical basis for wavelength-resolved deconvolution of peaks. In order to obtain the highest possible data quality, the UB matrix must be accurately determined which can be done using a handful of easily identifiable peaks. With the orientation matrix known, it is standard practice in macromolecular crystallography to predict peak locations for integration in order to produce a set of intensities. After this, appropriate normalizations are applied before scaling, merging, and model refinement against the structure factors, whose magnitudes are the square root of the normalized intensities. At the MaNDi beamline, this entire process has been implemented as an autoreduction routine. This section will describe how this process is done on a step-by-step basis.

The autoreduction routine is built on top of the Mantid framework (Arnold et al., 2014). For the interested reader, a detailed description of each algorithm is available through Mantid's documentation which is available at https://www.mantidproject.org. After the collection of a diffraction image is completed and written to disk, autoreduction generates a reciprocal space representation of the data using Mantid's ConvertToMD algorithm. Using Mantid's FindPeaksMD, the algorithm determines the location (in $\vec{q}$) of the brightest N peaks where N is a user-defined parameter (typically between 50 and 150). FindPeaksMD works by iterating over a sorted list of intensities and requiring that peaks be sufficiently far apart based on the estimated unit cell; as a result, it is helpful if users know the approximate unit cell of their crystal before starting an experiment at MaNDi. With the brightest N peaks found, the UB matrix can be determined. This is done using Mantid's FindUBUsingFFT algorithm. If FindUBUsingFFT fails, often a result of poor peak finding, autoreduction will try to determine a UB matrix from the estimated unit cell using Mantid's FindUBUsingLatticeParameters algorithm. With the UBMatrix known, the location of peaks is predicted out to a user-defined resolution and integrated using spherical integration. Spherically integrated intensities, which are obtained using a simple peak-minus-background routine, are used primarily to determine which peaks are strong and which are weak later in the autoreduction routine. Presently, strong peaks are taken to be those with a spherically determined intensity above a user-defined threshold (a default value of 200 counts has typically been successful for this).

At this point, the peaks in the diffraction image are integrated using 3D profile fitting, which is a recently developed integration technique that has yielded considerable improvements in data quality (Sullivan et al., 2018). Briefly, rather than counting the number of neutrons near where we expect peaks to be, a model peak is fitted to each peak and the model is integrated. The peak model is three dimensional and is easiest interpreted in spherical coordinates in reciprocal space. The TOF profile is known from moderator emission simulations and, since $TOF \sim 1/||\vec{q}||$ describes the radial direction. The remaining angular directions correspond to the view of the detector's face and are described by a bivariate Gaussian. Moreover, these bivariate Gaussian's vary slowly with their position on the detector sphere.

As a result, the peak shapes of weak peaks can be inferred by fitting nearby strong peaks. Profile fitting first fits all of the strong peaks and then forces these learned profiles onto weak peaks (with strong and weak being defined by spherical integration intensities). At the conclusion of profile fitting integration, which can take up to several hours in its current implementation, four files are saved. Two files are, the integrated intensities from both spherical and profile fitted integration. In addition, the parameters resulting from fitting each peak are saved along with the determined UB matrix.

Finally, the integrated runs must be filtered, normalized, scaled, merged, and put into a format compatible with community standard protein crystallography software programs. Filtering is necessary to remove peaks that were not correctly fitted. Typically, peaks with TOF fits having $\chi^2 < 50$ or whole 3D fits with $\chi^2 < 10$ are removed from further analysis. In standard data sets, this makes up <1% of all peaks. It is also possible to omit peaks with $\frac{I}{\sigma}$ below a user-defined value. Traditionally for MaNDi data, peaks with $\frac{I}{\sigma} < 1$ have been omitted. Normalizations for the Lorentz correction, wavelength dependence in the incident spectrum, and incident flux must then be applied. Note that high order corrections, such as sample absorption and detector solid angle are not currently implemented. The Lorentz corrected intensity is:

$$I_{corr} = I_{raw}\frac{\sin^2\theta}{\lambda^4}$$

Currently, wavelength corrections are done using Lauenorm from the Lauegen suite (Campbell, 1995). In addition to applying a scaling correction that accounts for differences in exposure time), this program splits nominally equivalent peaks into different wavelength bins and applies the wavelength-dependent scaling factor to each bin that minimizes residuals between equivalent peaks. Once these corrections have been applied, two reflection mtz files are created—one for unmerged intensities and one for merged intensities. These mtz files can be used with standard protein crystallography software. In particular, PHENIX (Adams et al., 2010) provides a convenient way to see merging statistics and phenix.xtriage can be used for quick diagnostic information as the run progresses (assuming sufficient completeness). Finally, phenix.refine, which is easily configured to handle neutron data, can be used for model building as the experiment progresses (Afonine et al., 2010, 2012). Such data can be analyzed by the user during the beamtime to make informed decisions on how to best utilize remaining beamtime.

From the perspective of a user, the entire reduction process happens automatically. To facilitate autoreduction, users must input some sample-specific parameters (e.g., approximate unit cell, space group, etc.) at the start of the experiment.

3. Early science highlights

MaNDi entered the general user program at the SNS in the Fall of 2014 and, as of the time of writing (December 2019), has produced 45 publications in the four and a half years that it has operating. Some highlights from the early stages of the user program are described in this section, with emphasis placed on experiments that were only feasible on MaNDi.

4. Human manganese superoxide dismutase

Human Manganese Superoxide Dismutase (MnSOD) is a metalloprotein that prevents mitochondrial dysfunction by mitigating reactive oxygen species (ROS) in the mitochondrial matrix. MnSOD performs its bioprotective role by converting highly reactive superoxide into oxygen and hydrogen peroxide. As the neutrons used on the MaNDi instrument do not damage the sample or alter the oxidation state of the metal ions within the sample it is possible to collect neutron datasets with different oxidation states from the same crystal. This makes it possible to study the changes in amino acid protonation states that accompany changes in the oxidation state of the manganese ion (Fig. 4) within the active site of the enzyme (Azadmanesh, Lutz, Weiss, Coates, & Borgstahl, 2018).

The unit cell of MnSOD is 81.4, 81.4, 242.3 Å; this is one of the largest unit cells studied with neutrons to date. The TOF resolution of MaNDi allows the harmonic Bragg reflections generated from the Laue diffraction pattern to be separated and integrated separately. This deconvolution allows collection of datasets with completeness values over 93% complete. As the unit cell of the enzyme is large and the available crystal volumes are limited to around 0.3 mm^3 (Azadmanesh, Trickel, Weiss, Coates, & Borgstahl, 2017) the replacement of all hydrogen atoms in the sample with deuterium (perdeuteration) is useful for further reducing experimental background to maximize the resolution of the data that can be collected. Several neutron datasets have been collected on MnSOD in various oxidation states that are currently being refined and interpreted to resolve the proton movements that take place during the reaction mechanism of the enzyme.

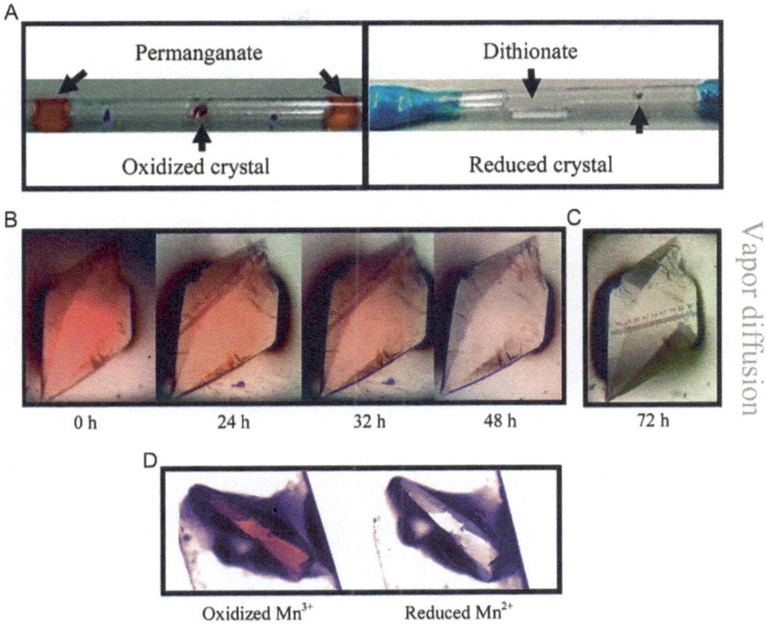

Fig. 4 Redox manipulation of perdeuterated human MnSOD crystals via vapor diffusion within capillaries. Cracks within the crystal were present prior to manipulation. (A) Images of crystals within capillaries along with deuterated reservoir supplemented with either potassium permanganate for oxidation or dithionite for reduction. (B) Time course of crystal reduction using reservoir solutions supplemented with 1% hydrogen peroxide. The purple/pink background is from the microscope polarizer with its setting maintained between images, while the lighting was consistent between the images in this panel. (C) Complete reduction by hydrogen peroxide is observed by the absence of pink color that is achieved 72h after supplementation with hydrogen peroxide. (D) The same crystal was oxidized by supplementation with potassium permanganate and reduced with sodium dithionite. The purple background is from the microscope polarizer, with its setting maintained between images. The lighting was kept consistent between the pictures. *Reproduced with permission of the International Union of Crystallography from Azadmanesh, J., Lutz, W. E., Weiss, K. L., Coates, L. & Borgstahl, G. E. O. (2018). Redox manipulation of the manganese metal in human manganese superoxide dismutase for neutron diffraction.* Acta Crystallographica, Section F: Structural Biology and Crystallization Communications *74, 677–687.*

5. Chlorite dismutase

The enzyme chlorite dismutase (CD) catalyzes the degradation of chlorite to chloride and dioxygen. During this reaction, a covalent oxygen–oxygen bond is formed, a biochemical reaction that is rare in nature and was thought to be unique to the water-splitting oxygen-evolving complex (Mn_4CaO_5) of photosystem II. The ability to degrade chlorite has been shown for several functional CDs, all of which possess a putatively flexible arginine (Arg 127) in a hydrophobic distal heme pocket. Available X-ray crystal structures reveal two distinct conformations, either pointing toward the heme iron (inward) or toward the substrate entry channel (outward). Neutron protein crystallography studies conducted on MaNDi obtained comprehensive structural information on CD in its active ferric state, including the protonation states of important amino acid residues. The protonation state of this Arg (127) has been the subject of ongoing debate as it is believed that it possesses an untypically low pKa value (6.5) that dictates the activity profiles and inactivation behavior of CDs. Neutron studies on MaNDi demonstrated that the distal Arg127 cannot act as proton acceptor as it is fully ionized even at pH 9.0 ruling out its proposed role in dictating the pH dependence of chlorite degradation (Fig. 5).

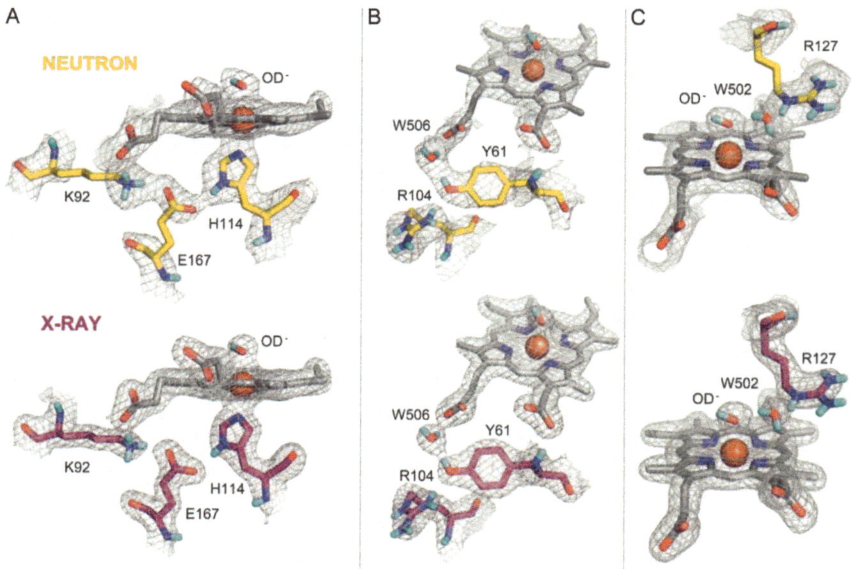

Fig. 5 Comparison of neutron (protein carbon atoms in yellow, $\sigma = 1.4$) and X-ray (protein carbon atoms in purple, $\sigma = 1.5$) $2F_O - F_c$ density maps in the active site of H/D exchanged CD at pH 9.0. (A) Proximal H bonding players His114, Glu167, and Lys92. (B) H-bonding network between Tyr61, a conserved deuterium oxide (W506) and the heme propionate. (C) Catalytically important Arg127 with the side chain oriented toward the substrate entrance channel (out-conformation). The distal heme ligand OD^- is depicted as sticks, the heme iron is shown as an orange sphere. Nitrogen atoms are shown in blue, oxygen in red, and deuterium in turquoise. *Reprinted with permission from Schaffner, I., Mlynek, G., Flego, N., Puhringer, D., Libiseller-Egger, J., Coates, L., et al. (2017). Molecular mechanism of enzymatic chlorite detoxification: INSIGHTS from structural and kinetic studies.* ACS Catalysis 7, *7962–7976 Copyright 2018 American Chemical Society.*

The spacegroup of this system is P1, which is particularly difficult for neutron diffraction studies as the only symmetry-related reflections present are the Freidel mates. Consequently, a large number of unique orientations must be collected even at moderate resolutions to collect a complete dataset. Typically, collecting this many unique orientations would require months of neutron beamtime. However, the large area detector coverage coupled with the Laue neutron bandwidth enabled MaNDi to collect a full dataset to be collected using 18 crystal orientations collected in 18 days representing the first successful neutron single crystal data collection on a system in this spacegroup.

6. Toho-1 β-lactamase

The growing resistance of bacteria to antibiotics is a growing problem with severe implications for human health (Drawz & Bonomo, 2010). The β-lactam ring found within all β-lactamase based antibiotics is a highly reactive structure that inhibits the synthesis of the bacterial cell wall leading to cell death. β-lactam antibiotics are widely prescribed because of their high specificity and low toxicity. The leading mechanism by which bacteria become resistant to β-lactam antibiotics is by the expression of β-lactamase enzymes that bind to and break down β-lactam antibiotics via a complex acylation deacylation mechanism. The acylation part of the reaction involves proton transfer between the conserved amino acids Ser 70, Lys 73, Glu166, and a catalytic water molecule within the active of the enzyme. Recently, using an approach combing neutron crystallography, X-ray crystallography and quantum mechanical and molecular mechanical (QM/MM) calculations we were able to determine how the binding of the clinical antibiotic Cefotaxime to Toho-1 β-lactamase alters the conformation and protonation states of the conserved amino acids within the active site which prime it for catalysis to occur (Langan et al., 2018; Vandavasi et al., 2016, 2017). Proton locations within the active site were determined using a series of neutron studies of different active site mutants, in one of these studies the protonation state of Lys 73 was determined for a ligand-free Ser70Ala mutant (Fig. 6). This structure served as a starting point for QM/MM simulations of the proton movements that occur just before acylation takes place. This result along with others indicated that Ser70, Lys73 and Glu166 are all neutral just before acylation of a takes place as originally suggested by Mobashery and co-workers (Meroueh, Fisher, Schlegel, & Mobashery, 2005).

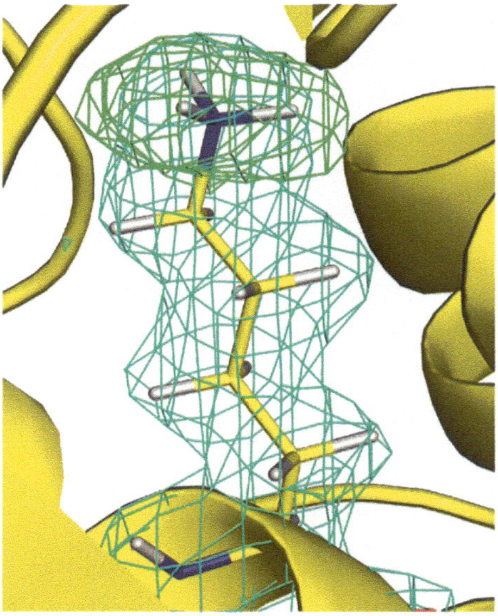

Fig. 6 A room temperature ligand-free neutron structure of a Ser70Ala Toho-1 β-lactamase mutant shows that Lys 73 in the ND_3^+ protonation state. Deuterium omit difference density map (green) is contoured at +2.5 σ with $2F_o$-F_c nuclear density contoured at 1.5 σ shown in cyan. Reprinted with permission from (Langan, 2018) Copyright 2018 American Chemical Society.

Acknowledgments

Research at ORNL's Spallation Neutron Source was sponsored by the Scientific User Facilities Division, Office of Basic Energy Sciences, U.S. Department of Energy.

References

Adams, P. D., Afonine, P. V., Bunkóczi, G., Chen, V. B., Davis, I. W., Echols, N., et al. (2010). PHENIX: A comprehensive python-based system for macromolecular structure solution. *Acta Crystallographica Section D, Biological Crystallography*, *66*, 213–221.

Afonine, P. V., Grosse-Kunstleve, R. W., Echols, N., Headd, J. J., Moriarty, N. W., Mustyakimov, M., et al. (2012). Towards automated crystallographic structure refinement with phenix.refine. *Acta Crystallographica Section D, Biological Crystallography*, *68*, 352–367.

Afonine, P. V., Mustyakimov, M., Grosse-Kunstleve, R. W., Moriarty, N. W., Langan, P., & Adams, P. D. (2010). Joint X-ray and neutron refinement with phenix.refine. *Acta Crystallographica Section D, Biological Crystallography*, *66*, 1153–1163.

Arnold, O., Bilheux, J.-C., Borreguero, J., Buts, A., Campbell, S. I., Chapon, L., et al. (2014). Mantid—data analysis and visualization package for neutron scattering and μ SR experiments. *Nuclear Instruments and Methods in Physics Research Section A: Accelerators, Spectrometers, Detectors and Associated Equipment*, *764*, 156–166.

Azadmanesh, J., Lutz, W. E., Weiss, K. L., Coates, L., & Borgstahl, G. E. O. (2018). Redox manipulation of the manganese metal in human manganese superoxide dismutase for neutron diffraction. *Acta Crystallographica, Section F: Structural Biology and Crystallization Communications*, *74*, 677–687.

Azadmanesh, J., Trickel, S. R., Weiss, K. L., Coates, L., & Borgstahl, G. E. O. (2017). Preliminary neutron diffraction analysis of challenging human manganese superoxide dismutase crystals. *Acta Crystallographica Section F, Structural Biology Communications*, *73*, 235–240.

Campbell, J. (1995). LAUEGEN, an X-windows-based program for the processing of Laue X-ray diffraction data. *Journal of Applied Crystallography*, *28*, 228–236.

Coates, L., Cuneo, M. J., Frost, M. J., He, J., Weiss, K. L., McFeeters, H., et al. (2015). The macromolecular neutron diffractometer MaNDi at the spallation neutron source. *Journal of Applied Crystallography*, *48*, 1302–1306.

Coates, L., Stoica, A. D., Hoffmann, C., Richards, J., & Cooper, R. (2010). The macromolecular neutron diffractometer (MaNDi) at the spallation neutron source, Oak Ridge: Enhanced optics design, high-resolution neutron detectors and simulated diffraction. *Journal of Applied Crystallography*, *43*, 570–577.

Coates, L., Tomanicek, S., Schrader, T. E., Weiss, K. L., Ng, J. D., Juttner, P., et al. (2014). Cryogenic neutron protein crystallography: Routine methods and potential benefits. *Journal of Applied Crystallography*, *47*, 1431–1434.

Drawz, S. M., & Bonomo, R. A. (2010). Three decades of beta-lactamase inhibitors. *Clinical Microbiology Reviews*, *23*, 160–201.

Langan, P., Fisher, Z., Kovalevsky, A., Mustyakimov, M., Valone, A. S., Unkefer, C., et al. (2008). Protein structures by spallation neutron crystallography. *Journal of Synchrotron Radiation*, *15*, 215–218.

Langan, P. S., Vandavasi, V. G., Cooper, S. J., Weiss, K. L., Ginell, S. L., Parks, J. M., et al. (2018). Substrate binding induces conformational changes in a class A beta-lactamase that prime it for catalysis. *ACS Catalysis*, *8*, 2428–2437.

Meroueh, S. O., Fisher, J. F., Schlegel, H. B., & Mobashery, S. (2005). Ab initio QM/MM study of class A beta-lactamase acylation: Dual participation of Glu166 and Lys73 in a concerted base promotion of Ser70. *Journal of the American Chemical Society*, *127*, 15397–15407.

Schroder, G. C., O'Dell, W. B., Myles, D. A. A., Kovalevsky, A., & Meilleur, F. (2018). IMAGINE: Neutrons reveal enzyme chemistry. *Acta Crystallographica Section D, Biological Crystallography*, *74*, 778–786.

Schultz, A. J., Thiyagarajan, P., Hodges, J. P., Rehm, C., Myles, D. A. A., Langan, P., et al. (2005). Conceptual design of a macromolecular neutron diffractometer (MaNDi) for the SNS. *Journal of Applied Crystallography*, *38*, 964–974.

Sullivan, B., Archibald, R., Langan, P. S., Dobbek, H., Bommer, M., McFeeters, R. L., et al. (2018). Improving the accuracy and resolution of neutron crystallographic data by three-dimensional profile fitting of Bragg peaks in reciprocal space. *Acta Crystallographica Section D Structural Biology*, *74*, 1085–1095.

Vandavasi, V. G., Langan, P. S., Weiss, K. L., Parks, J. M., Cooper, J. B., Ginell, S. L., et al. (2017). Active-site protonation states in an acyl-enzyme intermediate of a class A β-lactamase with a monobactam substrate. *Antimicrobial Agents and Chemotherapy*, *61*, e01636.

Vandavasi, V. G., Weiss, K. L., Cooper, J. B., Erskine, P. T., Tomanicek, S. J., Ostermann, A., et al. (2016). Exploring the mechanism of β-lactam ring protonation in the class A β-lactamase acylation mechanism using neutron and X-ray crystallography. *Journal of Medicinal Chemistry*, *59*, 474–479.

Zikovsky, J., Peterson, P. F., Wang, X. P. P., Frost, M., & Hoffmann, C. (2011). CrystalPlan: An experiment-planning tool for crystallography. *Journal of Applied Crystallography*, *44*, 418–423.

CHAPTER SIX

Current status and near future plan of neutron protein crystallography at J-PARC

Ichiro Tanaka[a,b,*], Toshiyuki Chatake[c], Satoru Fujiwara[d], Takaaki Hosoya[a,b], Katsuhiro Kusaka[b], Nobuo Niimura[b], Taro Yamada[b], Naomine Yano[b]

[a]Domain of Materials Science and Engineering, Graduate School of Science and Engineering, Ibaraki University, Hitachi, Ibaraki, Japan
[b]Divisions of Neutron Beamline and Research, Frontier Research Center for Applied Atomic Sciences, Tokai, Ibaraki, Japan
[c]Institute for Integrated Radiation and Nuclear Science, Kyoto University, Kumatori, Osaka, Japan
[d]Institute for Quantum Life Science, National Institutes for Quantum and Radiological Science and Technology, Tokai, Ibaraki, Japan
*Corresponding author: e-mail address: ichiro.tanaka.h27@vc.ibaraki.ac.jp

Contents

Methods in Enzymology, Volume 634
ISSN 0076-6879
https://doi.org/10.1016/bs.mie.2020.01.002

Abstract

The IBARAKI Biological Crystal Diffractometer (iBIX) has been available for use at MLF (Material and Life Science Facility) in J-PARC (Japan Proton Accelerator Research Complex) since 2008. The development in state-of-the-art detector systems could enable iBIX to become one of the highest-performance neutron single-crystal diffractometers in the world. Here, together with other various developments, such as data reduction software, crystal growth, and new techniques in measurement coupled analysis, we provided new hydrogen and water structural data of several proteins and macromolecules. Although the proton power at MLF has not yet reached its planned maximum (1 MW), a more powerful neutron source will be soon needed for neutron protein crystallography. A future idea is also proposed and discussed in this article.

1. Introduction

Neutron protein crystallography (NX) was first used in Japan in the 1990s with the appearance of Biological Crystal Diffractometer-1 (BIX-1), a research reactor, at Japan Research Reactor-3 (JRR-3) (Niimura et al., 1995). Since then, the use of NX has increased across the world thanks to the invention of the neutron imaging plate (Niimura et al., 1994). After the installation of 4 diffractometers, including BIX-1, at JRR-3 (Fujiwara et al., 1998; Kurihara, Tanaka, Refai-Muslih, Ostermann, & Niimura, 2004; Tanaka, Kurihara, Chatake, & Niimura, 2002), a new time-of-flight diffractometer, IBARAKI Biological Crystal Diffractometer (iBIX), has been constructed at J-PARC, using a pulsed neutron source as a next generation neutron source (Tanaka et al., 2010). As of September 2019 (the last run before summer), the proton power was about 500 kW, which is half of the planned power, and the beam stability was very good (over 90%). Together with other diffractometers developments abroad, the number of PDB (Protein Data Bank) entries with neutrons is increasing, with a current total of 159 entries with increasing maximum unit cells being analyzed, where 6KK8 is the latest deposition from iBIX with cell lengths of 133.4 Å (Yamada, Yano, Hosoya, & Kusaka, 2019). Including the above catalase and organic compounds, 7 papers with coordinates have been published (Kawasaki et al., 2012; Nakamura et al., 2015; Ogo et al., 2013; Unno et al., 2015; Yamada et al., 2019; Yokoyama et al., 2012, 2015). Nonetheless, expansion of capabilities from neutron experiments is requiring increasing amounts of neutrons, such that new, more intense sources are expected to be developed for NX, such as a second target station in SNS (Spallation Neutron Source) in the USA and J-PARC in Japan, using short pulse sources, as well as ESS (European Spallation Source), using a long pulse source.

In this article, iBIX, a state-of-the-art NX diffractometer at J-PARC, is described with regards to its instrumentation, its detector, the measurement mechanisms, including a discussion of crystallization and the data processing software used. An example of a typical analysis is also provided, as well as recent improvements in finding hydrogen information and the preparation of sample. In addition, the prospects of more powerful neutron sources are discussed.

2. Ibaraki biological crystal diffractometer, iBIX

Ibaraki biological crystal diffractometer, iBIX, is a high-performance neutron single crystal diffractometer developed at the 1 MW pulsed neutron source (MLF) of the Japan Accelerator Research Complex called J-PARC (Ikeda, 2009) in order to meet the increasing demand for neutron structural biology (Kusaka et al., 2013; Tanaka et al., 2010). The iBIX is designed to measure protein samples with unit cell volumes up to around $135^3 Å^3$ with a resolution up to 1.2 Å. The measurement efficiency of the diffractometer to be achieved is more than 50 times better than the diffractometer BIX-3 (JRR-3, JAEA, Japan).

2.1 Instrument design of iBIX

The iBIX station has been installed on a coupled moderator beam line (BL03) which provides a more intense peak and integrated intensity of neutron flux but a wider pulse shape than a decoupled one in order to obtain accurate integrated intensities of weak Bragg reflections around the resolution limit from a single crystal sample of proteins. The optics parameters (the beam divergence and the distance between a sample and sensitive area of detectors: L_2) were determined by taking account of overlapping of Bragg spots (Kusaka et al., 2006). To retain the time resolution, iBIX has a flight-path length L_1 (the distance between the moderator surface and the sample position) of 40 m. The super mirror guide tube, which has a length of 25 m, was installed with a vacuum chamber in order to reduce transmission loss of neutrons. For the horizontal plane, a curved guide tube was selected to protect samples from radiation damage from high energy gamma rays without requiring a T0 chopper. In the vertical plane, the tapered guide tube can serve to reduce the reflection number of neutrons on the surface of the super mirror so that more neutrons arrive at the sample position. The specification for iBIX are given in Table 1. To achieve high measurement efficiency, iBIX required a completely new detector system. Two-dimensional position sensitive detectors using a scintillator with

Table 1 Current specification of iBIX.

Moderator	Coupled H_2 (para) $100 \times 100\,mm^2$
Wavelength of incident neutron	0.7–4.0 Å (1st frame)
	4.0–8.0 Å (2nd frame)
Neutron intensity at the sample position (@1 MW)	$0.7 \times 10^6\,n/s/mm^2$
L_1	40 m
L_2	491.1 mm
No. of detectors	30
Solid angle of detectors	19.5% for 4π
Detector covered region	+26 ~ +167 deg
Standard sample size	1–2 mm^2
Maximum cell dimension	135 × 135 × 135 $Å^3$
Minimum resolution	1.1 Å
Average measurement time	8 days

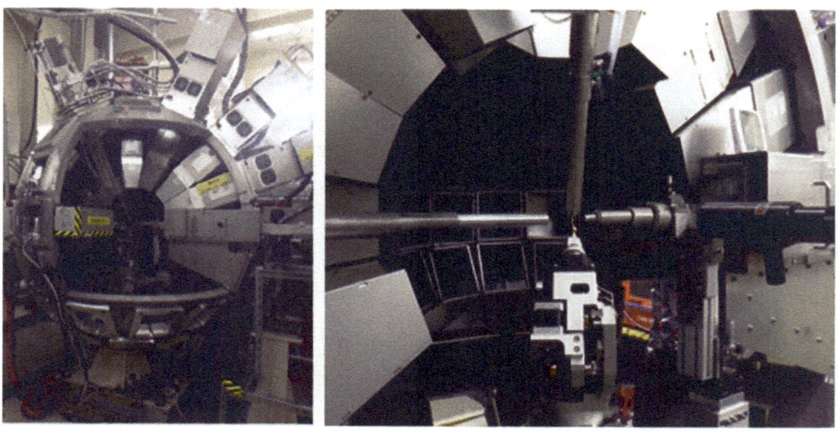

Fig. 1 Diffractometer overview (left) and detector configuration (right).

wavelength shift fiber (WLSF) were developed with a high spatial resolution, a smaller dead area, a high counting rate, and high efficiency. The details of the detector are described in Section 3. 30 detector units are arranged spherically with a distance between the sample and the detector surfaces L_2 of 491.1 mm determined by the detector support frame. The diffractometer and detector configuration is shown in Fig. 1. A gas flow-type cooling system for the sample environment is also available for low temperature experiments. N_2 gas is generated from air by pressure swing adsorption equipment, while He gas is supplied by a He gas cylinder. Attainable temperatures are 90 K and 20 K for N_2 and He gas respectively.

2.2 Current status of iBIX

Since the end of 2008, the iBIX has been available to users' experiments supported by Ibaraki University. Since August 2012, we have carried out upgrading of the 14 existing detectors and installed the 16 new detectors in iBIX. The total measurement efficiency of the present diffractometer has been improved by one order of magnitude from the previous one, and the station is further improved by the increasing accelerator power. In 2017, the accelerator power of J-PARC became 500 kW. An example of the collection of a full data set from a biological macromolecule crystal for neutron structure analysis using iBIX is as follows under the 500 MW operation. The cubic cell dimension was $133.4 \times 133.4 \times 133.4 \text{Å}^3$. The average sample volume was about 1–2 mm^3. The average measurement time was about 8 days for the protein crystal. If we extend measurement time (for example, to a few weeks), we can collect the data from samples of <1 mm^3. The resolution limit determined by the instrument design is 1.1 Å. We project that the iBIX will become one of the world's best performing diffractometers under 1 MW operation. The current specifications of the iBIX detectors are shown in Table 2.

Table 2 Specification of the iBIX detector.

Scintillator	ZnS (Ag)/$^{10}B_2O_3$
Wavelength shift fiber (WLSB)	0.5 mm φ in diameter
No. of X-axis and Y-axis of WLSB	256 and 256, respectively
Photomultiplier (PM)	Hamamatsu- photonics: 64 channels
Model number of PM	H8804MOD3-200
Scintillator sheet thickness	Front 0.2 mm, rear 0.3 mm
One detector sensitive area	133 mm × 133 mm
Dead area between two detectors	140 mm × 7 mm
Positional resolution	<1 mm
Positional homogeneity	10%
Count efficiency	58%

2.3 Measurement method with pulsed neutron source by iBIX

Spallation neutron source produce pulses of polychromatic neutrons. The repetition rate of the pulsed neutron at MLF is 25 Hz. In other words, neutron pulses are produced every 40 ms. The wavelength of the neutrons produced from pulsed neutron source can be obtained from the measurement of their flight times and the flight path length using the de Broglie relationship as follows.

$$\lambda = h/mv = ht/mL = ht/m\,(L_1 + L_2)$$

λ: Wavelength of neutron, m: Mass of neutron, v: Velocity of neutron, t: Flight time, L: Flight path length, h: Planck constant

This technique is called the time of flight (TOF) method. This TOF method of neutron diffraction is the energy dispersive Laue method. In the case of the Laue method, a large part of the reciprocal space can be recorded by using a wide range of wavelength in one exposure. Therefore, we can observe many more Bragg reflections than by using monochromatic methods. This feature is the great advantage of the Laue neutron diffraction method for improving measurement efficiency. However, in the case of the Laue method, higher order Bragg reflections are observed at overlapping positions for first order reflections on a 2-dimensional detector. Nevertheless we can separately observe these reflections by the using TOF method. Thus, TOF diffraction data has three dimensions (x, y, and TOF). It requires completely different recording techniques to the monochromatic method. In order to carry out the data reduction of TOF diffraction data measured by iBIX, we developed a data reduction program called STARGazer. The details of the program are described in the Section 4.

3. The iBIX detector

The iBIX detector system is an encoded scintillator-based position sensitive detector (PSD). The detector system consists of scintillator sheets, wavelength-shift fibers (WLSF), light-guide (optical) fibers and photomultiplier tubes (PMTs). The detector module uses the crossed-fiber read out method, in which WLSFs are arrayed in a grid pattern and sandwiched between two ceramic scintillator sheets. The scintillation photons released by the neutrons subsequently enter the adjacent X-axis and Y-axis WLSFs. The wavelength-shift fiber is a light guide fiber that contains a photo-fluorescent material. This material absorbs the higher frequency photons created by the scintillator, emitting lower frequency photons that then travel along the fiber. In the horizontal direction at one edge of the sheet, the end of each WLSF is bent at a right angle, down toward the rear. There, the WLSF is attached to an ordinary light-guide fiber that transmits the wavelength-shifted light to a PMT. The other end of the WLSF, however, does not lead to a PMT but is rather attached to a light reflector. The light reflector takes up less space on the detector surface than bending the fibers to the back does, such that using a light reflector on one end reduces the so called "dead space" of the detector, i.e. any area not sensitive to light. The WLSFs in the vertical direction are installed in a similar manner. The schematic layout of scintillator package is shown in Fig. 2. Its specification and efficiency, as well as several results, are summarized in Table 2 and Fig. 3, respectively (Hosoya et al., 2009).

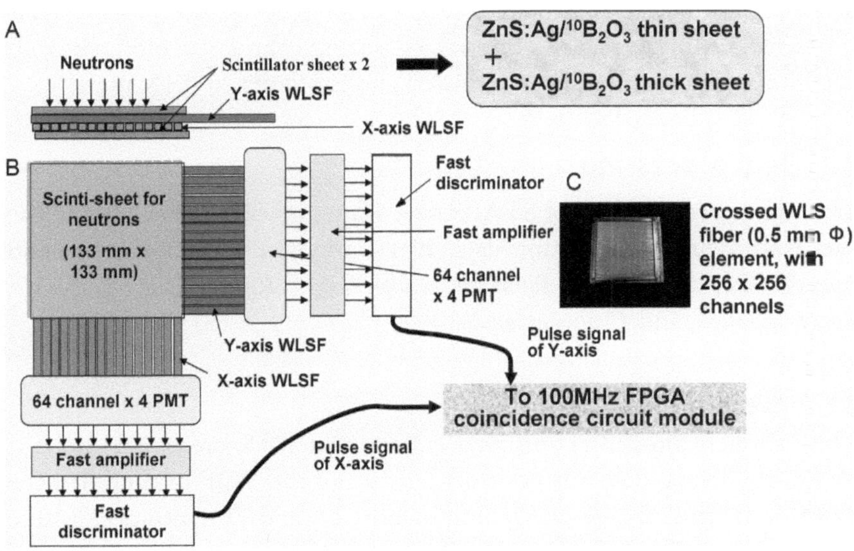

Fig. 2 Schematic diagram of scintillator package: (A) side view of scintillator package, consisting of scintillator sheets and wavelength-shift fibers (WLSF); (B) front view of scintillator package and flow of scintillation light; (C) photograph of the scintillator package.

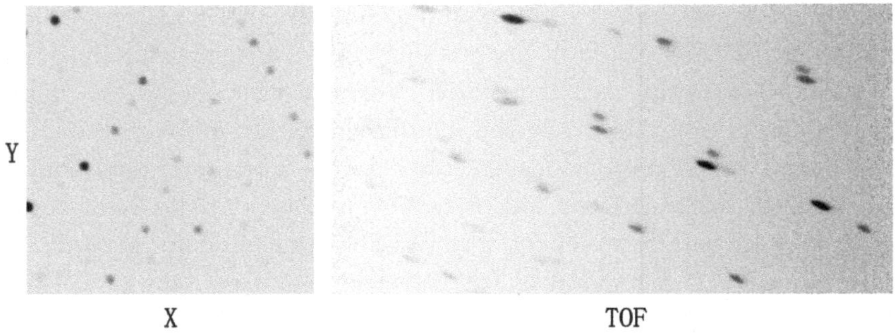

Fig. 3 Example of the TOF diffraction data on the detector (X-Y map (left) and Y-TOF map (right)).

4. Data processing of time-of-flight diffraction data

The iBIX group developed a data processing software STARGazer (Ohhara et al., 2009; Yano et al., 2018) by rewriting the ISAW (Mikkelson, Schultz, Mikkelson, & Worlton, 2005) algorithms for protein single crystals. STARGazer mainly consists of two parts: data processing and data visualization. These are described in detail in Sections 4.1 and 4.2. The software STARGazer and its manual, consisting of an installation part and a data processing part, has been prepared and is offered to iBIX users.

4.1 Data processing

The data processing part of STARGazer is used to obtain the hkl intensity data from neutron diffraction data. It is composed of eight components: *EventToHist*, *FindPeaks*, *FindCell*, *IndexPeaks*, *ReducedCell*, *LsUBMat*, *PeakIntegration*, and *EvaluateConv*.

4.1.1 EventToHist

The raw neutron diffraction data collected at iBIX are event data that record the spatial detector position at x, y and the TOF of each detected neutron. Because the neutron count in most x, y and TOF positions is zero, the event data allows for a smaller data size. For data processing, it is preferable to convert the event data into histogram data that records the neutron count at each x, y and TOF. The pixel counts are binned in the TOF direction to form channel t, and histogram data are used as the x, y and t data. Because the neutron time of flight method utilizes pulsed neutrons with a spread of wavelengths, when creating sample crystal histogram data, the variance in the detection efficiency of pixels within one detector, the difference in neutron beam intensities by wavelength, and the difference in detection efficiency by wavelength are corrected using calibration data. The event data from incoherent scattering of a vanadium sphere are used for the correction data. Additionally, the difference in the total number of neutrons irradiating the sample crystal at each sample orientation is corrected for.

4.1.2 FindPeaks

This component finds the peak position of the reflections from the histogram data. The main target of measurement by iBIX is the protein single crystal. Because neutron beam intensity is much lower than synchrotron X-ray beam intensity, and the quality of protein single crystals is lower than those of organic or inorganic compounds, the peak intensities of neutron diffraction from protein single crystals are weaker and broader. Thus, it is difficult to find the peak positions of the reflections using simple peak-search algorithms. In order to solve this problem, a rebinning and smoothing method was developed for, and implemented in, the *FindPeaks* component. Additionally, the reciprocal lattice coordinates of each peak are calculated.

4.1.3 FindCell

The UB matrix is determined after the peak search; this is a 3×3 matrix that represents the reciprocal lattice vectors a*, b*, and c* at a goniometer angle of $\omega = \chi = \varphi = 0°$. An FFT-based indexing algorithm (Steller, Bolotovsky, & Rossmann, 1997) is implemented in this component.

4.1.4 IndexPeaks

This component calculates the Miller indices of each peak from the UB matrix, the reciprocal-lattice coordinates of each peak, and the goniometer angle. If all of the absolute values of the differences obtained by rounding the calculated Miller indices, and the calculated Miller indices themselves, are less than the threshold, then these peaks are indexed. The default thresholds of h, k, and l are set to 0.2. Because the detector parameters (the distances between each detector face center and the sample, the detector position angles, and the detector tilting angles), the flight path length from the CM to the sample and the three-axis goniometer offset angles ω and χ are accurately calibrated by the beamline staff, so we can index most peaks using the UB matrix determined from all detector peaks at a single crystal orientation. Additionally, the indexed peak rate against all observed peaks is calculated to determine the accuracy of the UB matrix.

4.1.5 ReducedCell

This component is only used for complex lattice types. The candidates for a 3×3 transformation matrix from the primitive lattice to the complex lattice are calculated based on the conditions of the reduced cell.

4.1.6 LsUBMat

This component refines the UB matrix using indexed peaks. The UB matrix determined in the *FindCell* component is used as an initial value. When the lattice type is complex, the unit cells are transformed and a new UB matrix is calculated using the transformation matrix determined in the *ReducedCell* component. A least squares minimization is carried out to reduce the summation of the squares of the distances from the observed peak positions determined in the *FindPeaks* component to the peak positions predicted using the refined UB matrix on the detector. The accuracy of the refined UB matrix can be evaluated by the differences in Miller indices between the indexed values and calculated values.

4.1.7 PeakIntegration

This component predicts the peak positions on detectors and integrates reflection intensities. The intensities and their errors with Lorentz factor correction are calculated.

4.1.8 EvaluateConv

This component merges equivalent reflections and calculates data statistics, such as the number of observed reflections, the number of independent reflections, the average multiplicity, the completeness (%), R_{merge}, $R_{p.i.m}$, CC1/2, and $I/\sigma(I)$. The differences in the detection efficiency between the 30 detectors are corrected by using the total neutron count of the correction data at each detector in this component. The hkl intensity data can be output in the SCALEPACK (Otwinowski & Minor, 1997), SHELX (Sheldrick, 2008), and GSAS (Larson & Von Dreele, 2004) formats. After neutron data collection, X-ray diffraction data are collected from the crystal irradiated by neutrons or a crystal from a similar crystallization condition as the crystal used for neutron measurements for joint refinement of X-ray and neutron data.

4.2 Data visualization

The data visualization part of STARGazer is used to visualize the histogram data with searched or predicted peak positions, and to determine and confirm the integrated regions. Users can visualize the histogram data as two-dimensional X–Y, X–TOF, or Y–TOF slice maps and as a one-dimensional TOF profile. The resolution, wavelength, position on the detector, and neutron count of each reflection can be displayed using the moving cursor.

5. Large protein crystal growth and structure refinements for neutron protein crystallography

5.1 Large protein crystal growth

Neutron protein crystallography requires large crystalline samples since the intensity of the neutron beams are much smaller than that of X-ray sources. Fig. 4 shows a flowchart of the steps performed typically to obtain large crystals for neutron crystallographic analysis. Neutron protein crystallographic analysis requires the initial atomic coordinates of the protein to have been determined by X-ray crystallography. The user can judge if the sample is suited for neutron crystallography based on the unit cell dimensions, diffraction quality, and crystal shapes (O'Dell, Bodenheimer, & Meilleur, 2016).

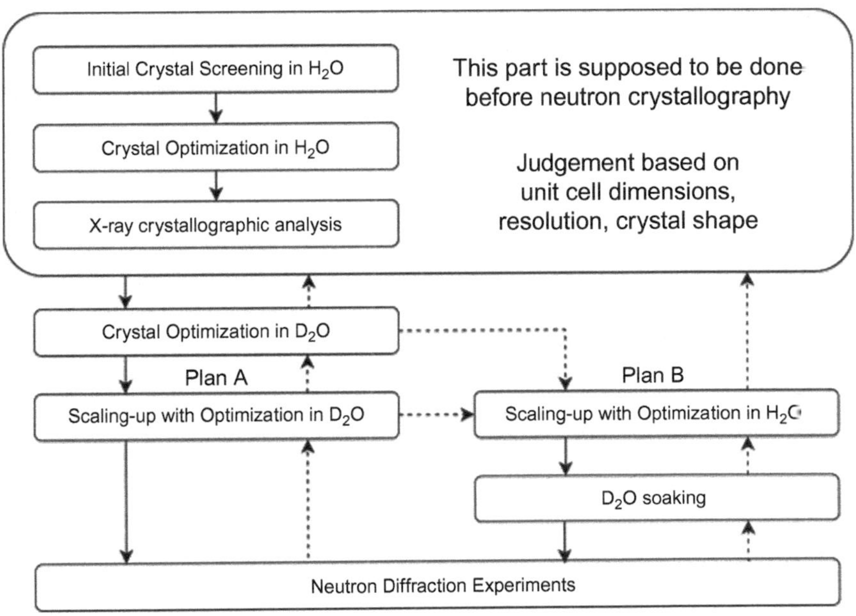

Fig. 4 Flowchart of process to obtain large crystals for neutron crystallographic analysis.

They may also have to choose crystal shapes rather than X-ray diffraction quality to obtain large crystal volumes. We recommend carrying out the crystallization in D_2O solution from the start, if possible. Crystals prepared in H_2O solutions should be soaked in D_2O solutions to reduce the background noise produced by the incoherent scattering of H atoms. Soaking methods may require the careful handling of samples and involve certain risks, including the breaking of valuable crystals. Furthermore, imperfect H/D exchange could make both atoms invisible, since the neutron scattering length of H and D have opposite signs and cancel each other.

The principle of obtaining large crystals is rather straight forward, that is, gathering more protein molecules to a fewer number of crystals. Obtaining 1 mm^3 crystals is a good initial objective for non-perdeuterated protein. An online drop volume calculator is available at https://neutrons.ornl.gov/imagine (O'Dell et al., 2016). Crystallization set-ups using ~1 mg of protein solution is a feasible start. The sitting drop vapor diffusion method is a useful way to obtain large crystals. Table 3 provides a summary of commercially available crystallization vessels and their applicable drop volumes. Fluorinert FC-70 (3M) liquid can be used as crystallization supports to prevent the adhesion of crystals. Fine tuning the crystallization conditions is necessary for scaling-up. Dialysis methods are also effective to obtain large crystals and worth a try if skin formation on the drop surface causes problems. The observation of crystals growth is also important, in order to carry out crystallization effectively by monitoring the crystallization process and speed. For reference, we built a simple crystal observation system, which can take crystal

Table 3 Vessels used for sitting drop vapor diffusion methods.

Drop volume	Reservoir volume	Vessels can be used
~50 μL	~1 mL	Cryschem plate, Micro-bridges with VDX plate (H.R.)[a]
~350 μL	~5 mL	Falcon 60 mm center well organ culture dish (Corning) with a 18 mm siliconized circle cover slide (~150 μL, H.R.) or with an order-made glass vessel,[b] or with a[c] dialysis button (H.R.)
~800 μL	~50 mL	9 Well glass plate and sandwich box setup (H.R.)[d], 150 mm culture dish (Corning) with a glass blood reaction plate (Toshin Riko, Co. Ltd.)

[a]H.R. Hampton Research, Corp.
[b]Yamada et al. (2019).
[c]The author referred Shibazaki et al. (2018).
[d]Kovalevsky et al. (2011).

photos in every 12h, using built-in camera type stereomicroscopes (EZ4 E, Leica microsystems) set in incubators. See the chapter "Large crystal growth for neutron protein crystallography" by Budayova-Spano et al. in this volume for further discussion of the means of obtaining large crystals.

5.2 Structure refinements

In neutron protein crystallography, as well as the non-hydrogen atoms, many deuterium and hydrogen atoms are included in the protein structures. Structure refinement using X-ray and neutron diffraction data (XN-joint refinement) is necessary to determine the precise structure of the crystals. X-ray diffraction data should be collected immediately after the neutron diffraction experiments to avoid radiation damage from X-rays, if the same crystal is used. To minimize the extent of the damage, it is advisable to use attenuated beams, a short exposure time, and different irradiation positions, if possible. XN-joint refinements can be carried out using the *PHENIX.refine* in the *PHENIX* software suits (Adams et al., 2010) linked with the molecular graphic program *Coot* (Emsley, Lohkamp, Scott, & Cowtan, 2010). The *PHENIX* system has simple, user-friendly GUIs and detailed online manuals and tutorials (https://www.phenix-online.org). *PHENIX* also has a useful project manager that can be used to restore all inputs and results at any stage of refinements. Using the *PHENIX* features, an atomic coordinate and four maps ($2mF_o$–DF_c and mF_o–DF_c maps for X-ray and neutron data sets) obtained in any run can be displayed in the *Coot* program. All possible hydrogen and deuterium atoms in proteins and solvent molecules can be generated by *PHENIX.readyset*. Non-H/D atoms and H/D atoms should be modeled by referring to the electron density maps and neutron scattering length maps, respectively. One should consider the possibility of disordered structures of H/D atoms and the canceling effect of mixed H/D atoms before concluding the protonation states of polar H/D sites. *PHENIX.refine* generates mmcif files for structure factors and atomic coordinates, which are PDB submission-ready.

6. D/H contrast neutron crystallography at iBIX

6.1 D/H contrast neutron crystallography

Neutron research facilities including iBIX operate neutron diffractometers, with the aim of determining the positions of hydrogen atoms in protein crystals. However, detection of hydrogen atoms in neutron Fourier maps by neutron protein crystallography (NX) requires diffraction data at a resolution better than at least ~2.0 Å, and higher resolution data are necessary to elucidate precisely the hydrogen and hydration structures of proteins. This limitation is a major issue in NX. Replacement of hydrogen with deuterium is the preferred technique to improve the quality of neutron Fourier maps without the need to improve the resolution of the diffraction, but this exchange can be exploited further. The large difference in the scattering length between deuterium and hydrogen (D and H) atoms ($b_D = 6.67$ fm and $b_H = -3.74$ fm) leads to a high density contrast in neutron Fourier maps. By replacing the H_2O solvent with D_2O in crystals, H atoms of amino and hydroxyl groups in proteins are also replaced with D atoms. This replacement of deuterium/hydrogen (D/H) exchangeable atoms in the protein and the solvent water significantly enhances the density contrast of the replaced atoms in the difference Fourier map between H_2O-solvent and D_2O-solvent crystals (*h*- and *d*- crystals). This D/H contrast technique was first applied to NX by Kossiakoff and coworkers to observe the hydration structure of trypsin (Kossiakoff, Sintchak, Shpungin, & Presta, 1992; Shpungin & Kossiakoff, 1986). In their study, the D/H contrast map was calculated in "reciprocal space," where the structure factors of the *h*-crystal were subtracted from those of the *d*-crystal (Shpungin & Kossiakoff, 1986). We developed an alternative protocol, where the difference is obtained between the densities of the neutron Fourier maps of the *d*- and *h*- crystals in "real space." The D/H contrast map in our protocol is suitable for standard structure determination, which involves the iteration of model building and atomic parameter refinement (Chatake & Fujiwara, 2016). These studies using the D/H contrast techniques were first performed using monochromatic neutrons from a reactor source. With the establishment of new spallation neutron sources, we have updated the D/H contrast technique. Here, we describe the first report of the neutron crystallographic analysis at a spallation neutron source using the D/H contrast technique.

6.2 Materials and methods

We carried out measurements on the crystals of hen egg white lysozyme (HEWL). Crystallization of HEWL was carried out using the concentration gradient method reported by Ataka and Katsura (1992). *H*-crystals were grown at 293 K in H_2O solution containing 15–25 mg/mL HEWL, 50 mM sodium acetate (pH 4.5), and 1 g/mL $NiCl_2$. The *h*-crystals reached their maximum size after 3–6 months. Replacement of H_2O solvent in *h*-crystals with D_2O to obtain *d*-crystals was carried out by the vapor equilibrium method using D_2O solution at pD 4.1 containing the same other components as the H_2O solution. Incubation was continued for 3 months. The initial model for the D/H contrast technique, the non-hydrogen structure of HEWL, was determined by X-ray diffraction measurements of the *d*-crystal at BL5A in Photon Factory in KEK. Measurements of the X-ray diffraction of the *h*-crystal were also carried out to compare the crystal structures of the *d*- and *h*- crystals, since a high degree of isomorphism between the *h*- and *d*- crystals is essential for the D/H contrast technique. Neutron diffraction measurements of the *h*- and *d*-crystals were carried out using iBIX diffractometer (Tanaka et al., 2010) installed at BL03 of MLF Materials and Life Science Experimental Facility in J-PARC.

Prior to the neutron analysis using D/H contrast maps, non-exchangeable H-atoms in C—H bonds were included using the program *PHENIX* (Adams et al., 2010) in the *d*-crystal structure determined by X-rays. D/H contrast maps were calculated from the two of $2|Fo|-|Fc|$ maps of the *d*- and *h*- crystals using an in-house program developed using Python3. Peaks higher than 1.5 σ were assigned as D/H exchangeable atoms in the D/H contrast $2|Fo|-|Fc|$ map. The atomic parameters were optimized by N-X joint refinement using X-ray and neutron data sets of *d*-crystal using the program *PHENIX* (Adams et al., 2010).

6.3 Results and discussion

The statistics of the neutron and X-ray data measurements and the structure determination are provided in Table 4. The degree of isomorphism between the *d*- and *h*- crystals was assessed by comparing the X-ray data sets of the two crystals. Very similar cell dimensions of the two crystals (a=b=79.14 Å, c=37.78 Å, and a=b=79.22 Å, c=37.82 Å for the *d*- and *h*- crystals, respectively) indicate the high level of isomorphism. The cell dimensions of the *d*-crystal obtained by X-ray measurements were used for the present analysis. Both of the neutron diffraction images of the *d*- and *h*- crystals were processed to 1.80 Å resolution. Due to the data statistics in the outer shell, the resolution limit in the refinements of the atomic parameters was limited to 2.0 Å resolution, although the whole set of intensities were used in the Fourier map calculations.

Fig. 5 shows an example of the density maps obtained in the present analysis. X-ray densities were found at the positions of non-hydrogen atoms in

Table 4 Statistics of measurements and structure determination.

	X-ray		Neutron	
Space group	$P4_32_12$			
Cell dimensions (Å, °)	$a = b = 79.14$ Å, $c = 37.38$ Å, $\alpha = \beta = \gamma = 90°$			
d_{min}	1.42 Å	1.43 Å	1.80 Å	1.80 Å
Crystal	d-crystal	h-crystal	d-crystal	h-crystal
R_{pim}	0.015	0.017	0.071	0.076
Completeness	1.00	1.00	0.98	0.97
d_{min} for refinement	1.42 Å		2.00 Å	
R_{work}	0.151		0.169 (0.216)	
R_{free}	0.180		0.216	

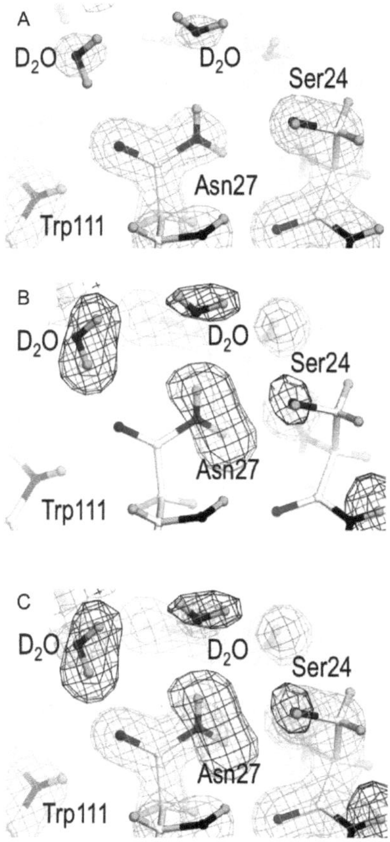

Fig. 5 Various Fourier maps near Asn27 in hen egg white lysozyme: (A) 1.5 σ X-ray 2|Fo|−|Fc| map; (B) 2.5 σ neutron contrast 2|Fo|−|Fc| map; (C) superimposed X-ray and neutron contrast 2|Fo|−|Fc| maps.

the X-ray map (Fig. 5A). On the other hand, the individual densities in the D/H contrast map (Fig. 5B), corresponding to D/H exchangeable atoms in the protein (the amino group of the Asn27 side chain, the hydroxyl group of the Ser24 side chain, and the D atoms of the D_2O molecules) were observed. Each density formed a discrete and distinct peak to facilitate the model building of these D/H atoms. The resolution of the present analysis was 2.0 Å for the refinement and 1.8 Å for map calculation, respectively. It is worth noting that these are not high resolutions in NX. Nevertheless, the D/H contrast map provided us clear and interpretable densities. This result is in agreement with our previous D/H contrast NX (1.8 Å analysis of ribonuclease A by monochromatic neutron (Chatake & Fujiwara, 2016)). Since the *d*-crystal was prepared by D_2O-solvent exchange, some of the amino groups in the main chain remained as H atoms. Therefore, the D/H exchange ratios of the main chain were optimized in the model structure. Interestingly, the same situation was observed in several side chains. There is no significant density corresponding to the D/H atom of the side chain of Trp111 (Fig. 5B and C). However, these sites in the side chains are very few, and they were not observed in the previous analysis (Chatake & Fujiwara, 2016). The D/H contrast technique using a D_2O-exchange crystal is a powerful tool to visualize the protonation state of each amino acid in proteins, and to elucidate the directions of N-D(H) and O-D(H) bonds of proteins and water molecules. Moreover, when all H atoms, including those of C—H bonds, in a protein are replaced with D atoms, they are all observable in the D/H contrast map. The present analysis demonstrates the high performance of the D/H contrast technique in NX for use in spallation source neutron experiments, as well as other experiments, at reactor sources.

7. New techniques for the detection of hydrogen at a higher sensitivity and future perspectives for neutron sources

7.1 New techniques for the detection of hydrogen at a higher sensitivity

7.1.1 High-pressure freezing

In principle, it is not necessary to freeze crystals for NX to prevent radiation damage since the neutrons used have low energy, close to that of hydrogen bonds. However, there are some important reasons for the freezing crystals in NX: (1) to keep crystals fresh, as the neutron measurement time remains long and a neutron experiment is often left on hold while the sample crystals are prepared; (2) to obtain an enzymatic reaction intermediate; (3) to observe hydrogen at a higher sensitivity by modifying neutron scattering lengths with dynamic nuclear polarization (DNP) technique.

Despite using the optimal conditions for (3), the presently available neutron intensity is still low, such that a relatively large crystal of good quality is needed. Generally, cryo-vitrification of crystals is better in smaller volume crystals as this avoids stress damage on contraction, but still requires rapid cooling and cryoprotectant(s) to avoid ice formation. However, because the cryoprotectant(s) may affect the active sites or otherwise perturb the structure, the precise/intact structural analysis becomes sometimes difficult. High pressure cryo-vitrification is expected to avoid these problems. Here, a machine used to pressurize the crystal up to 200 MPa and cool it quickly while maintaining the high pressure, an HPC-201 (manufactured by ADC Inc. in the USA) can be used. Another machine for the handling frozen crystals under a low temperature atmosphere (Leica EM CPC) can also be used. Crystals frozen using 6.35- and 2.11-mm diameter pressure tubes were found to diffract to nearly the same resolution as crystals cooled normally (Tanaka, Komatsuzaki, et al., 2018; Tanaka, Kusaka, & Niimura, 2018). Pressurizing up to 200 MPa takes about 3 and 8 min when using 2.11- and 6.35-mm diameter pressure tubes, respectively.

7.1.2 Dynamic nuclear polarization technique

The application of the dynamic nuclear polarization (DNP) technique to NPC was proposed at the beginning of the 2010s (Niimura & Podjarny, 2011), which could be used to increase the contrast of hydrogen and reduce the background in NPC. However, to achieve this, a low temperature around 1 K, a high magnetic field around 3 T, and the suitable paramagnetic centers are necessary. Here, we describe experiments that explored the use of the paramagnetic TEMPOL (1-oxyl-2,2,6,6-tetramethyl-4-hydroxypiperidine), which was used as it dissolves well in water and does not appear to disturb the protein crystal structures.

For the initial tests with a protein, a polycrystalline preparation of hen egg white lysozyme was used, made by co-crystalizing with 50 mM TEMPOL. Due to the sensitivity of the NMR coil used, about 0.3 g of protein crystal was necessary. For a single crystal with a density of 1 g/cm^3, the volume becomes about 300 mm^3, corresponding to ~6.7 mm in cube (which is very difficult to prepare), and so a polycrystalline preparation was required. A preliminary DNP experiment was carried out using a ^{3}He refrigerator, which can cool the sample to a temperature of 0.5 K, with a 2.5 T normal conducting magnetic field under which about 70 GHz microwave were applied. As a result, a 22.3% polarization rate was observed within 2 h (Tanaka, Komatsuzaki, et al., 2018). A higher polarization rate of nearly 100% can be developed by selecting better paramagnetics and increasing their concentration, other improvements would arise from degassing the water to be used in crystallization to remove oxygen molecules, and increasing the magnetic field up to around 7 T (Noda & Koizumi, 2019). A more sensitive NMR coil could be also developed for a single crystal of 1 mm^3 in volume, which is the preferred size for current NX.

7.2 Future perspectives for upgrading neutron sources

As pulsed neutron sources, such as MLF in J-PARC, have a much higher peak flux than conventional reactor neutron sources, they can provide higher resolution NX data and allow for the use of smaller crystals. In Fig. 6, the relationship between diffracted resolution and effective crystal volume both in JRR-3 (20 MW reactor in Japan) and MLF of J-PARC (proton beam power from 120 to 400 kW) is presented. Here, the effective volume is defined as the ratio of the volume of a crystal to the volume of the unit cell. It is clearly shown that pulsed source is superior to reactor (Tanaka, 2019).

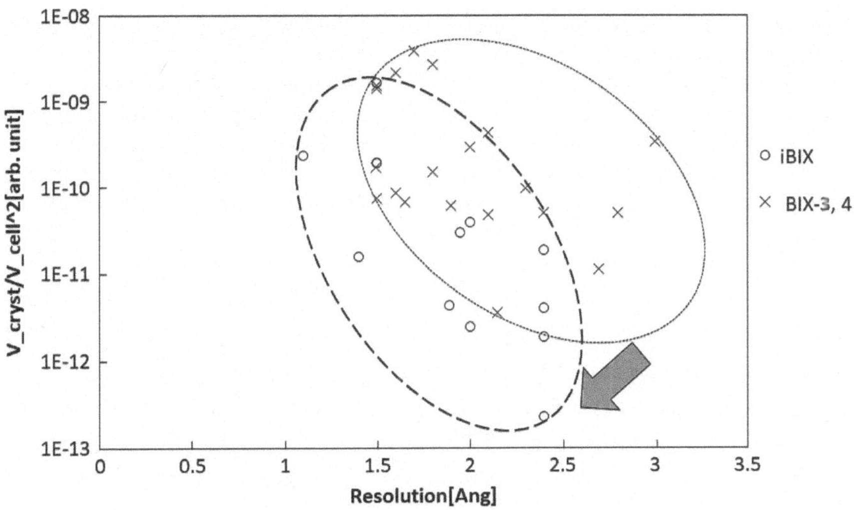

Fig. 6 Comparison between pulsed neutron source (iBIX in MLF, J-PARC) and reactor source (BIX-3 and 4 in JRR3), resolution vs. effective crystal volume. Crosses denote data obtained at BIX-3 or BIX-4, whereas circles denote data obtained at iBIX. The dotted and dashed lines represent each group of diffractometers, respectively.

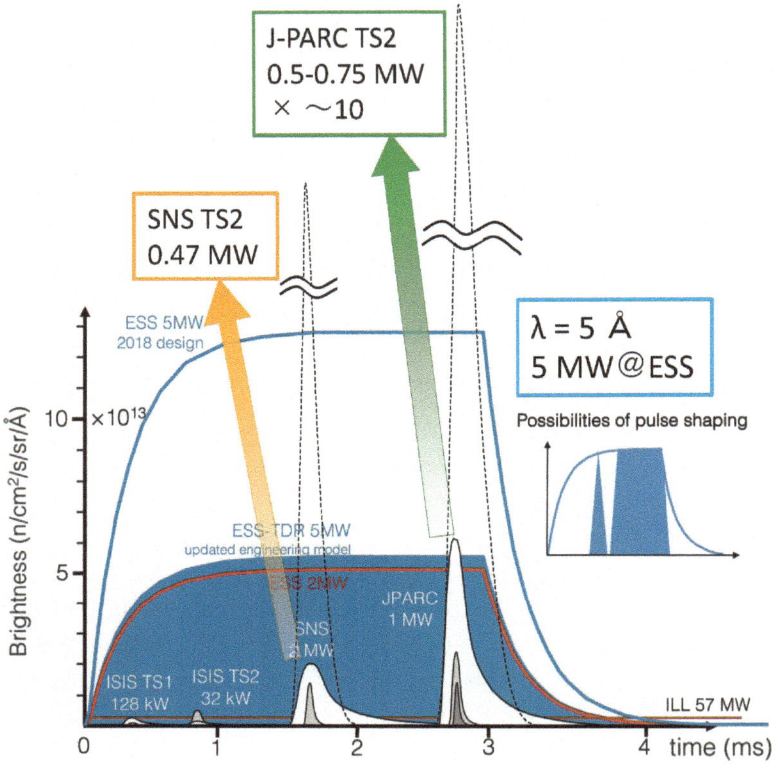

Fig. 7 Intensities both of each pulse shape at different neutron sources, including the planned sources and the most powerful reactor at present. The horizontal axis indicates the emission time in milliseconds, while the vertical axis is the brightness or the intensity. For the pulsed neutrons, the curves indicate the intensity of the time dependences per pulse. ISIS is a pulsed neutron and muon facility based in the Rutherford Appleton Laboratory in the UK, and ILL (Grenoble, France) is the most powerful reactor neutron source. J-PARC TS1 (1st target station) corresponds a 1 MW pulse shape.

In MLF in J-PARC, as well as SNS in the USA, the current first target station are to be upgraded to a more powerful ones, with a higher pulse peak, by increasing the proton beam current, as well as optimizing the moderators and beam transport. As shown in Fig. 7 (modified from ESS (European Spallation Source) home page (ESS, 2018) characterizing the second target station (TS2) projects both of SNS (Oak Ridge National Laboratory, 2015) and MLF at J-PARC (Harada & Seto, 2018), the intensity of the current brightest constant reactor source, ILL (Institute of Laue-Langevin in France), is shown at the horizontal axis, whereas the intensities of the current pulsed sources give much greater peak intensities. Furthermore, the peak intensities of the planned second targets in Japan and the USA show an increase of over 10-fold compared to the current targets.

8. Conclusion

NX is a very attractive method, however, it is very difficult to carry it out. Several authors have made various efforts to promote NX, and the number of neutron PDB entries have increased gradually, most likely as a result of this. Since 2014, there has been a trend of over 10 entries submitted to PDB per year. The combination of a pulsed neutron source and a time-of-flight diffractometer seems powerful, and the use of NX is expected to grow as a result.

Acknowledgments

We would like to thank all the users of BIX-1,2,3,4 in JRR-3 and iBIX in J-PARC, and especially Dr. Katagiri (former JAEA) for his effort on detector development. For future developments, Prof. Iwata and his lab members (Yamagata Univ.), Prof. Seto (KEK), and Drs. Harada and Nakajima (JAEA) were very helpful. These projects have been supported mainly by Ibaraki Pref. Government.

References

Adams, P. D., Afonine, P. V., Bunkóczi, G., Chen, V. B., Davis, I. W., Echols, N., et al. (2010). PHENIX: A comprehensive python-based system for macromolecular structure solution. *Acta Crystallographica, Section D: Biological Crystallography, 66*, 213–221.

Ataka, M., & Katsura, T. (1992). *A large single crystal of the tetragonal form of lysozyme can be grown in a concentration gradient of* $NiCl_2$. JAERI-M (Japan Atomic Energy Research Institute-Memos). (92–199), 61.

Chatake, T., & Fujiwara, S. (2016). A technique for determining the deuterium/hydrogen contrast map in neutron macromolecular crystallography. *Acta Crystallographica Section D, 72*, 71–82.

Emsley, P., Lohkamp, B., Scott, W. G., & Cowtan, K. (2010). Features and development of coot. *Acta Crystallographica Section D, 66*, 486–501.

ESS. (2018). *The ESS mandate.* https://europeanspallationsource.se/ess-mandate/. Accessed 22 Sep. 2019.

Fujiwara, S., Karasawa, Y., Tanaka, I., Minezaki, Y., Yonezawa, Y., & Niimura, N. (1998). A neutron diffractometer for biological macromolecules using neutron imaging plates. *Physica B, 241–243*, 207–209.

Harada, M., & Seto, H. (2018). Private communication.

Hosoya, T., Nakamura, T., Katagiri, M., Birumachi, A., Ebine, M., & Soyama, K. (2009). Development of a new detector and DAQ systems for iBIX. *Nuclear Instruments and Methods in Physics Research A, 600*, 217–219.

Ikeda, Y. (2009). J-PARC status update. *Nuclear Instruments and Methods in Physics Research A, 600*, 1–4.

Kawasaki, T., Takahashi, M., Ohhara, T., Tanaka, I., Kusaka, K., Hosoya, T., et al. (2012). Structure of morpholinium tribromoplumbate $C_4H_8ONH_2PbBr_3$ studied using single-crystal neutron diffraction. *Journal of the Physical Society of Japan, 81*, 094602, 1–6.

Kossiakoff, A. A., Sintchak, M. D., Shpungin, J., & Presta, L. G. (1992). Analysis of solvent structure in proteins using neutron $D_2O - H_2O$ solvent maps: Pattern of primary and secondary hydration of trypsin. *Proteins, 12*, 223–236.

Kovalevsky, A. Y., Hanson, B. L., Seaver, S., Fisher, S. Z., Mustyakimov, M., & Langan, P. (2011). Preliminary joint X-ray and neutron protein crystallographic studies of endoxylanase II from the fungus *Trichoderma longibrachiatum*. *Acta Crystallographica Section F*, 7, 283–286.

Kurihara, K., Tanaka, I., Refai-Muslih, M., Ostermann, A., & Niimura, N. (2004). A new neutron single crystal diffractometer dedicated for biological macromolecules (BIX-4). *Journal of Synchrotron Radiation, 11*, 68–71.

Kusaka, K., Hosoya, T., Yamada, T., Tomoyori, K., Ohhara, T., Katagiri, M., et al. (2013). Evaluation of performance for IBARAKI biological crystal diffractometer iBIX with new detectors. *Journal of Synchrotron Radiation, 20*, 994–998.

Kusaka, K., Ohhara, T., Tanaka, I., Niimura, N., Ozeki, T., Kurihara, K., et al. (2006). Peak overlapping and its de-convolution in TOF diffraction data from neutron biological diffractometer in J-PARC. *Physica B, 385–386*, 1062–1065.

Larson, A. C., & Von Dreele, R. B. (2004). *GSAS. Los Alamos National Laboratory report LAUR* (pp. 86–748).

Mikkelson, D. J., Schultz, A. J., Mikkelson, R., & Worlton, T. G. (2005). Visualization and analysis of single crystal time-of-flight neutron scattering data using ISAW. In *IUCr Commission on crystallographic computing newsletter No. 5* (p. 32).

Nakamura, A., Ishida, T., Kusaka, K., Yamada, T., Fushinobu, S., Tanaka, I., et al. (2015). "Newton's cradle" proton relay with amide-imidic acid tautomerization in inverting cellulase visualized by neutron crystallography. *Science Advances, 1,* e1500263, [1–7].

Niimura, N., Karasawa, Y., Tanaka, I., Miyahara, J., Takahashi, K., Saito, et al. (1994). An imaging plate neutron detector. *Nuclear Instruments and Methods in Physics Research Section A, 349*, 521–525.

Niimura, N., & Podjarny, A. (2011). *Neutron protein crystallography. Hydrogen, protons, and hydration in bio-macromolecules*. Oxford: Oxford University Press.

Niimura, N., Tanaka, I., Minezaki, Y., Karasawa, Y., Tanaka, I., Miki, K., et al. (1995). Diffractometer for neutron crystallography in biology. *Physica B, 213–214*, 786–789.

Noda, Y., & Koizumi, S. (2019). Dynamic nuclear polarization apparatus for contrast variation neutron scattering experiments on iMATERIA spectrometer at J-PARC. *Nuclear Instruments and Methods in Physics Research A, 923*, 127–133.

Oak Ridge National Laboratory. (2015). *Technical design report, ORNL/TM-2015/24.* https://neutrons.ornl.gov/sts/documents. Accessed 24 Sep. 2019.

O'Dell, W. B., Bodenheimer, A. M., & Meilleur, F. (2016). Neutron protein crystallography: A complementary tool for locating hydrogens in proteins. *Archives of Biochemistry and Biophysics, 602*, 48–60.

Ogo, S., Ichikawa, K., Kishima, T., Matsumoto, T., Nakai, H., Kusaka, K., et al. (2013). A functional [NiFe]hydrogenase mimic that catalyzes electron and hydride transfer from H2. *Science, 339*, 682–684.

Ohhara, T., Kusaka, K., Hosoya, T., Kurihara, K., Tomoyori, K., Niimura, N., et al. (2009). Development of data processing software for a new TOF single crystal neutron diffractometer at J-PARC. *Nuclear Instruments and Methods in Physics Research Section A, 600*, 195–197.

Otwinowski, Z., & Minor, W. (1997). Processing of X-ray diffraction data collected in oscillation mode. *Methods in Enzymology, 276*, 307–326.

Sheldrick, G. M. (2008). A short history of SHELX. *Acta Crystallographica, Section A: Foundations of Crystallography, 64*, 112–122.

Shibazaki, C., Arai, S., Shimizu, R., Saeki, M., Kinoshita, T., Ostermann, A., et al. (2018). Hydration structures of the human protein kinase CK2α clarified by joint neutron and X-ray crystallography. *Journal of Molecular Biology, 430*, 5094–5104.

Shpungin, J., & Kossiakoff, A. A. (1986). A method of solvent structure analysis for proteins using D_2O-H_2O neutron difference maps. *Methods in Enzymology, 127*, 329–342.

Steller, I., Bolotovsky, R., & Rossmann, M. G. (1997). An algorithm for automatic indexing of oscillation images using Fourier analysis. *Journal of Applied Crystallography, 30*, 1036–1040.

Tanaka, I. (2019). Necessity and effectiveness of a biological diffractometer for installation in second target station as a new neutron puled source in Japan. In *JPS Conference Proceedings*. 25, 011015 (1–5).

Tanaka, I., Komatsuzaki, N., Yue, W.-X., Chatake, T., Kusaka, K., Niimura, N., et al. (2018). Cryoprotectant-free high-pressure freezing and dynamic nuclear polarization for more sensitive detection of hydrogen in neutron protein crystallography. *Acta Crystallographica Section D, 74*, 787–791.

Tanaka, I., Kurihara, K., Chatake, T., & Niimura, N. (2002). A high performance neutron diffractometer for biological crystallography-3 (BIX-3). *Journal of Applied Crystallography, 35*, 34–40.

Tanaka, I., Kusaka, K., Hosoya, T., Niimura, N., Ohhara, T., Kurihara, K., et al. (2010). Neutron structure analysis by IBARAKI biological crystal diffractometer (iBIX) in J-PARC. *Acta Crystallographica Section D, 66*, 1194–1197.

Tanaka, I., Kusaka, K., & Niimura, N. (2018). Protonation/deprotonation of proteins by neutron diffraction structure analysis. In M. Suzuki (Ed.), *Role of water in ATP hydrolysis energy transduction by protein machinery*, (pp. 123–140). Singapore: Springer Nature Singapore Pte Ltd. Chapter 9.

Unno, M., Ishikawa-Suto, K., Kusaka, K., Tamada, T., Hagiwara, Y., Sugishima, M., et al. (2015). Insights into the proton transfer mechanism of a bilin reductase PcyA following neutron crystallography. *Journal of the American Chemical Society, 137*, 5452–5460.

Yamada, T., Yano, N., Hosoya, T., & Kusaka, K. (2019). Single-crystal TOF neutron Laue methods: Application to manganese catalase from *Thermus thermophilus* HB27. *Journal of Applied Crystallography, 52*, 972–983.

Yano, N., Yamada, T., Hosoya, T., Ohhara, T., Tanaka, I., Niimura, N., et al. (2018). Status of the neutron time-of-flight single-crystal diffraction data-processing software STARGazer. *Acta Crystallographica Section D, 74*, 1041–1052.

Yokoyama, T., Mizuguchi, M., Nabeshima, Y., Kusaka, K., Yamada, T., Hosoya, T., et al. (2012). Hydrogen-bond network and pH sensitivity in transthyretin: Neutron crystal structure of human transthyretin. *Journal of Structural Biology, 177*, 283–290.

Yokoyama, T., Mizuguchi, M., Ostermann, A., Kusaka, K., Niimura, N., Schrader, T. E., et al. (2015). Protonation state and hydration of bisphosphonate bound to farnesyl pyrophosphate synthase. *Journal of Medicinal Chemistry, 58*, 7549–7556.

CHAPTER SEVEN

Neutron macromolecular crystallography at the European spallation source

Márton Markó[a], Gergely Nagy[a,b], Giuseppe Aprigliano[b], Esko Oksanen[b,*]

[a]Institute for Solid State Physics and Optics, Wigner Research Centre for Physics, Budapest, Hungary
[b]European Spallation Source ERIC, Lund, Sweden
*Corresponding author: e-mail address: esko.oksanen@esss.se

Contents

Abstract

The long-pulse spallation source of the European Spallation Source—a facility under construction in Lund, Sweden—is well suited for macromolecular crystallography experiments. We review briefly the particular properties of the long-pulse source and the associated high-brilliance moderators from the point of view of instrument design. We then outline the design philosophy and current design of the NMX macromolecular diffractometer. We also briefly describe the supporting facilities available for users and finish by an evaluation of the expected performance.

Methods in Enzymology, Volume 634
ISSN 0076-6879
https://doi.org/10.1016/bs.mie.2020.01.005

1. The European spallation source—Long pulse and high-brilliance moderators

The various types of neutron sources available for macromolecular crystallography have been discussed in the chapters "IMAGINE: The neutron protein crystallography beamline at the high flux isotope reactor" by Meilleur et al. (in this volume), "The macromolecular neutron diffractometer at the spallation neutron source" by Coates and Sullivan (in this volume), "Current status and near future plan of neutron protein crystallography at J-PARC" by Tanaka et al. (in this volume), "Neutron macromolecular crystallography at the European spallation source" by Markó et al. (in this volume), "Dynamic nuclear polarization enhanced neutron crystallography: Amplifying hydrogen in biological crystals" by Pierce et al. (in this volume), "Implementation of the riding hydrogen model in *CCTBX* to support the next generation of X-ray and neutron joint refinement in *Phenix*" by Liebschner et al. (in this volume), "Interactive model building in neutron macromolecular crystallography" by Logan (in this volume), and "What are the current limits on determination of protonation state using neutron macromolecular crystallography?" by Liebschner et al. (in this volume). The European Spallation Source (ESS, Peggs, 2013) is a European project to build a novel type of spallation source in Lund, Sweden. In contrast to the existing high power spallation sources, the ESS will not have a synchrotron ring to compress the proton pulse that hits the spallation target (Garoby et al., 2017). While the proton pulse length at a short pulse source is typically in the range of a microsecond, at ESS—which is a long pulse spallation source (Mezei, 1997a)—it is 2.86 ms. The rationale behind this is that the moderation times for thermal and, particularly, cold neutrons are tens or hundreds of microseconds, so a ~1 microsecond proton pulse does not allow the moderator to reach its maximum peak brilliance. The disadvantage is that when the pulse of neutrons becomes non-negligible compared to the typical times-of-flight of the neutrons the wavelength resolution is degraded unless the neutron pulse is shaped by a chopper (Mezei & Russina, 2002).

As the maximum current that can be injected to the compressing synchrotron ring typically limits the total neutron current from the spallation target (Mezei, 1997b), the ESS solution also allows a significantly higher time-averaged neutron current. This makes ESS qualitatively different from the existing short-pulse spallation sources (Peggs, 2013), affecting also instrument design. The main difference is that ESS will have a significantly higher time-averaged neutron production—exceeding even reactor sources—but at the cost of significantly lower intrinsic time-of-flight—and hence wavelength resolution at a given instrument length.

The total current of neutrons from the spallation target is, however, not a very useful metric for the actual usefulness of a neutron source for a particular purpose, because the design of the moderator—the source that the scattering instrument sees—dramatically affects how many of the neutrons produced in the target can be productively used in the experiment. The quantity that best captures the source characteristics desired for scattering experiments is brilliance—neutrons emitted per unit area, time, solid angle and wavelength. Brilliance is also a quantity that can only be reduced—due to losses—in an optical system, whereas the number of neutrons per unit area (intensity) can be traded with beam divergence by optical devices. As neutron moderators emit neutrons in a much wider cone than what can be used for protein crystallography, the practical usefulness of such trade-offs is limited.

Another feature that distinguishes the ESS from currently existing neutron sources is the moderator design, which focuses on maximizing the brilliance at the cost of the total neutron current (Andersen et al., 2018). Moderators are typically surrounded by neutron-reflecting material such as beryllium with an opening for the beam to exit. Moderators tend to be brighter close to the surface and the reflector, as some of the back-reflected neutrons will also end up in the beam. The ESS low-dimensional moderator design makes use of this effect by a flat shape of the moderator vessel (Andersen et al., 2018), which changes the surface-to-volume ratio in such a way that the majority of the moderator surface benefits from being close to the reflector. This in turn allows a higher brilliance than a conventional volume moderator would provide.

The efficiency of the low dimensional moderator is highly increased with the use of para-hydrogen (Kai, Harada, Teshigawara, Watanabe, & Ikeda, 2004)—the spin isomer of H_2 where the proton spins are opposite. Liquid para-hydrogen moderates the thermal neutrons efficiently, while its scattering cross section is more than an order of magnitude lower for the cold neutrons. Therefore the thermal and epithermal neutrons become moderated close to the flat surface and can exit almost parallel to the surface—in the direction of the instruments—with low loss.

These two features of the ESS source—long pulse and brilliant moderators—are *a priori* unrelated, but define the types of instrument designs that can make optimal use of the source characteristics. Of course this does not make other types of neutron instruments unfeasible, it simply affects the gain factors that can be realized. It turns out that a macromolecular diffractometer is one of the instrument types that can make best use of such a source.

2. Implications of the long pulse to instrument design

From the perspective of neutron instrument design, a long-pulse spallation source presents different challenges and opportunities than either a reactor or a short-pulse spallation source (Aizawa et al., 1997; Mezei, 1997b). The usefulness of pulsed sources for a neutron crystallographic experiment stems from illuminating the crystal in a single experimental setup with a broad wavelength band from the source spectrum (Jauch, 1997; Schoenborn & Langan, 2004). This allows measuring diffracted intensities from different wavelengths sequentially at their time of arrival at the detector, which depends on the speed of the neutrons, hence also on their wavelength. The use of multiple wavelengths allows more neutrons from the source to produce signal—this advantage is identical to polychromatic (Laue) crystallography at a continuous reactor source. The disadvantage of polychromatic crystallography at a continuous source is that while the signal for any one reflection only arrives from a fairly narrow wavelength band, background will arise from all the incident wavelengths. This is, however, not the case at pulsed sources, where the various wavelengths are separated by their times-of-flight and hence are detected at different arrival times at the detector. This means that the background is also distributed in the time-of-flight dimension. Therefore time-of-flight Laue (TOF-Laue) crystallography combines the advantages of both the monochromatic and the Laue methods at a reactor source. A critical design parameter in both types of reactor source instruments is the bandpass $\Delta\lambda/\lambda$. For a monochromatic instrument this defines the monochromaticity that in turn determines what fraction of the available neutrons are used for the measurement. For a Laue instrument $\Delta\lambda/\lambda$ essentially defines the signal-to-background ratio. For a pulsed source instrument the question is rather about the uncertainty in λ, which stems from the finite length of the neutron pulse. The wavelength is calculated from the time-of-flight, which in turn is measured at the detector based on the time that the neutron pulse was generated. The finite length of the neutron pulse leads to an unavoidable uncertainty in the time-of-flight measurement, which is also propagated to the wavelength determination.

Since the arrival time of a neutron with a given wavelength is a linear function of the wavelength and the instrument length, the wavelength resolution of a TOF instrument is proportional to the pulse duration, and inversely proportional to the instrument length.

The uncertainty of the time-of-flight δTOF is given by the neutron pulse length—a longer flight path will lead to a smaller $\delta\lambda/\lambda$ since δTOF/TOF is smaller. As a consequence the *relative* wavelength uncertainty $\delta\lambda/\lambda$ depends on the wavelength itself—longer wavelengths will have better relative precision. At a long-pulse spallation source it is also possible to use a mechanical selection device—known as a chopper—to artificially reduce the length of the apparent pulse. This of course also leads to a loss of neutrons, but it makes the length of the pulse a variable parameter instead of a moderator design parameter like at short-pulse sources.

The advantages that are inherent in a short pulse spallation source—a large simultaneous wavelength band and high intrinsic wavelength resolution—leading to a low background are not obvious at a long-pulse source. At a reactor source, on the other hand, there is a trade-off between monochromatizing the beam at the cost of brilliance or accepting a higher background from a polychromatic beam. At a long-pulse source it would be difficult to justify monochromatizing the beam with a crystal monochromator, but neutron choppers can of course be used to select certain velocities and hence wavelengths (Lechner, 1992). The real questions for a macromolecular diffractometer at a long-pulse source are the desired value of $\delta\lambda/\lambda$ and the range of wavelengths to be used simultaneously. These together will determine the length of the flight path and the need for pulse shaping. Typical flight path lengths of ~40 m at short pulse spallation sources are often driven by cost considerations, both for the neutron optics and possible additional buildings, but the ESS buildings allow flight paths up to ~160 m without additional buildings.

One consequence of a longer flight path is that the range of wavelengths that can unambiguously be resolved—known as the wavelength frame—gets narrower. This is a function of the source repetition rate as well, since the reason for the ambiguity is that neutrons (of a given wavelength) from one pulse arrive at the detector at the same time as shorter wavelength (faster) neutrons from the *next* source pulse. This phenomenon is known as frame overlap. The unambiguously resolvable wavelength frame is given by Eq. (1). One advantage of a short source pulse is that it allows a shorter flight path for a given $\delta\lambda/\lambda$, leading to a wider wavelength frame and hence more neutrons arriving at the sample simultaneously. In practice the wavelength spectrum of the moderator will not be uniform, and widening the wavelength frame around the spectral maximum results in diminishing returns.

$$\Delta\lambda = \frac{C}{Lf} \tag{1}$$

Maximal unambiguous wavelength range for a time-of-flight neutron instrument, where C is the speed of 1 Å neutrons (3996 Å ms^{-1}), L is the instrument length and f is the source frequency.

3. The NMX instrument—Design philosophy

It was recognized early on that the ESS long pulse would be well suited for a macromolecular diffractometer but it was not obvious what the basic design of the instrument should be. The ESS used a systems engineering approach to instrument design, which starts with the identification of high-level scientific requirements (Table 1). These high-level requirements are not rigorous in the system engineering sense (e.g. they are not necessarily easily verifiable), but they were used as a basis for generating an instrument conceptual design.

$$I_{pk} = t\Phi(\lambda)\varepsilon(\lambda)\kappa \frac{V_{cryst}\lambda^4}{2V_{cell}\sin^2\theta} T_{DW}\langle |F_{hkl}|^2 \rangle \tag{2}$$

Table 1 High-level scientific requirements with their identification numbers.

ID	Statement
1	The instrument shall allow data collection from crystals with unit cell repeats >300 Å
2	The instrument shall allow data to be collected to a d_{min} of 1.5 Å
3	The instrument shall match the size of the neutron beam to the size of the sample
4	The instrument shall match the divergence of the neutron beam to the mosaicity of the sample
5	The instrument should maximize the signal-to-background (S/B) ratio of the Bragg reflections
6	The instrument should allow data collection from crystals of <0.01 mm^3 volume

The integrated intensity for an average Bragg peak (I_{pk}), where t is the duration of the measurement (s), $\Phi(\lambda)$ is the incident spectral flux at the sample (neutrons cm^{-2} s^{-1} $Å^{-1}$), $\varepsilon(\lambda)$ is the detector efficiency, κ is a conversion factor (1×10^{24} (cm^2 bn^{-1}), V_{cryst} is the crystal volume ($Å^3$), V_{cell} is the unit cell volume ($Å^3$), 2θ is the Bragg angle, T_{DW} is the temperature factor and $|F_{hkl}|^2$ is the squared structure factor modulus of the reflection *hkl* (bn) from Jauch (1997).

The requirements reflect a scientific case that is focused on neutron macromolecular crystallography, where it is usually not feasible or necessary to collect data at atomic resolution (<1.2 Å, Sheldrick, 1990). This implies that the instrument is not intended for chemical crystallography studies, which typically require atomic resolution. This choice is important because the achievable crystallographic resolution d_{min} ultimately depends on the shortest wavelength available when 2θ reaches its maximum at 180°. From Bragg's law d_{min} is then $\lambda/2$. This is practically important because the two available moderator types—thermal and cold—have different wavelength spectra. The decision not to strive for atomic resolution drives this choice in favor of the cold moderator and this choice of required wavelengths has heavy implications for the design of the neutron optical system as well as the detectors. The rationale is to use as long wavelengths as allowed by the d_{min} constraint above. There are two reasons for this. Firstly, the scattered signal in a generic reflection given by Eq. (2) depends on the fourth power of λ, which is a strong driver in itself. The second reason is more practical: the angle for the total reflectivity range of neutron optical elements also depends on λ (with increasing critical angles for total reflectivity for increasing λ). This means that it is much more feasible and affordable to deliver to the sample a neutron beam with longer wavelength neutrons. Additionally ^{157}Gd, which is the best neutron absorber (Sears, 1992) by unit volume, only absorbs strongly at $\lambda > 1.8$ Å, which affects the collimation system and possible detector technologies.

Several instrument concepts were studied, including the use of wavelength frame multiplication and pulse shaping choppers (Mezei & Russina, 2002; Strobl, Bulat, & Habicht, 2013). The key question about pulse shaping was the ability to resolve Bragg reflections sufficiently as to obtain reliable integrated intensities. At continuous source instruments it is unavoidable to resolve the Bragg peaks on the detector spatial dimension (Niimura et al., 1997), but at a time-of-flight instrument reflections can also be resolved in the TOF dimension (Schoenborn & Langan, 2004). This can make up for the lack of time-of-flight detectors with high spatial resolution, but the requirement for the pulse length for resolving two reflections by TOF quickly becomes very stringent with increasing unit cell dimensions.

This led to a concept relying on resolving the reflections on the detector spatial dimension, necessitating the development of a new type of neutron detector (Pfeiffer et al., 2015, 2016). As the distance of the detector from the sample affects the ability to resolve the reflections, it was decided to make the detectors moveable. However, since with long wavelengths the solid angle coverage needed to collect complete data is very large, there is an incentive to surround the crystal with detectors. This is not trivial to combine with moveable detectors since the detector units will have to be moved in three dimensions. For this reason, six-axis robots were considered for positioning the detectors at NMX in a three-dimensional space around the sample.

In order to quantify the requirements 3 and 4 in Table 1 some assumptions have to be made about the size and mosaicity of crystals to be measured. Based on experience at other macromolecular diffractometers it was concluded that while the crystal size is generally the parameter limiting the experiments, crystal dimensions >5 mm are highly uncommon. Therefore the maximum beam size of 5×5 mm was chosen as a starting point for the optics design. Similarly, the mosaicity of the crystals is difficult to generalize, but would not be expected to exceed $\pm 0.2°$, which then defines the maximal beam divergence. A decision was also made that the smallest beam size would be 200 μm.

All of these considerations are input for selecting the wavelength range for which the beam transport system is optimized. The wavelength spectrum of the cold moderator decreases fairly steeply at <1.8 Å wavelength, while the performance of the neutron optics for the divergence range of $\pm 0.2°$ decreases—despite increased cost. Also, the efficiency of a Gd-based detector decreases sharply at <1.8 Å wavelength, so it was natural to limit the shortest wavelength to 1.8 Å, which at $2\theta = 180°$ corresponds to a d_{min} of 0.9 Å, significantly higher than the requirement (ID2, Table 1) of 1.5 Å d_{min}. With a 1.75 Å wide wavelength frame (Eq. 1) the most typical simultaneous wavelength range used would be 1.8–3.55 Å, but it will also be possible to use longer wavelengths up to 10 Å.

4. The NMX design

Based on the conceptual instrument design the instrument was broken down into subsystems according to the ESS systems engineering design process. Rigorous—quantified, verifiable and traceable—requirements were then identified for each subsystem, resulting in 61 distinct requirements. An example of the beam transport system requirements is shown in Table 2. The requirements are traced back to either the high-level scientific requirements (Table 1) or the Concept of Operations (ConOps) document that describes how the NMX instrument will be operated.

The NMX instrument will be located in the West sector of ESS at beam port W1 (Fig. 1). The neutron beam is delivered from the target monolith to the sample position 157 m away by a beam transport system that includes a neutron guide for delivering the beam and choppers for selecting the desired wavelengths. With the full 2.9 ms pulse and a moderator-to-sample distance of 157 m, the total width of the wavelength frame is 1.75 Å.

4.1 Beam transport system

The design of the neutron guide is a compromise between many factors—it was realized that the physical properties of the guide, such as the macroscopic waviness as well as realistic alignment tolerances, place strong constraints on the guide design. Therefore the guide first expands vertically in the target monolith, curves in the common shielding bunker for radiation shielding reasons and then also expands in the horizontal direction to guarantee efficient beam transport (Fig. 2).

It was recognized that, even though the moderator surface that the optical system needs to view is approx. 3 × 3 cm, it is not practical to build a neutron guide with this cross-section for a distance of 150 m due to the significant losses predicted. The reason for this is that there are limits to the precision with which the guide elements can be aligned and these misalignments lead to some losses at each junction. The neutron transport losses depend on the relative magnitude of the misalignment compared to the guide cross-section, but the absolute misalignment is *a priori* independent of the cross-section. This means that the guide system can be more robust by simply defocussing slightly to increase the cross-section and hence decrease the relative misalignments. The optimized value for the cross-section is 4.5 × 4.5 cm.

There is, however, a contrasting requirement in the horizontal direction, since the guide system should curve the neutron beam away from a direct line-of-sight with the moderator surface. The reason for this is the high background of high-energy neutrons and other particles emanating from the moderator; not closing the line-of-sight would result in significantly

Table 2 System requirements for the NMX beam transport system (BTS).

Id	Text	Rationale	Verification
13.6.4.1(1) Wavelength resolution	The BTS shall transport from the moderator a *beam of neutrons* to the *sample* at a distance that leads to a maximal wavelength uncertainty of 5% ($\Delta\lambda/\lambda$) for the detected neutrons using the full ESS pulse	A moderate wavelength resolution allows the full pulse to be used while conserving the advantage of TOF for the S/B (see 13.6.4 (5))	Measurement of the pulse length at sample
13.6.4.1(2) Beam size	The BTS shall transport from the moderator to the sample a beam of neutrons with maximum size (full width half maximum) of 5 ± 0.1 mm and minimum size of 0.2 ± 0.02 mm	Matching the beam size to the sample size maximizes the S/B (see 13.6.4 (3,5-6))	Measurement of the beam intensity profile at sample
13.6.4.1(3) Beam divergence	The BTS shall transport from the moderator to the sample a beam of neutrons with maximum *divergence* of $\pm 0.2°$ and minimum divergence of $\pm 0.02°$	Matching the beam divergence to the sample mosaicity maximizes the S/B (see 13.6.4 (4,5-6))	Measurement of the beam divergence at sample
13.6.4.1(4) Beam size selection	The BTS shall transport from the moderator to the sample a beam of neutrons with a size *selectable* in 0.05 ± 0.02 mm steps	Matching the beam size to the sample size maximizes the S/B (see 13.6.4 (3,5-6))	Measurement of the *beam intensity profile* at sample
13.6.4.1(5) Beam divergence selection	The BTS shall transport from the moderator to the sample a beam of neutrons with a divergence selectable in 0.02° steps	Matching the beam divergence to the sample mosaicity maximizes the S/B (see 13.6.4 (4,5-6))	Measurement of the beam divergence at sample

13.6.4.1(6) Bandwidth selection	The BTS shall transport from the moderator to the sample a beam of neutrons with a wavelength *bandwidth* selectable between 0.1 and 1.8 Å with a *resolution* better than ±0.08 Å	The width of the wavelength band changes the number of reflections observed and the resolvable unit cell size (see 13.6.4 (1–2))	Measurement of the neutron spectrum at sample
13.6.4.1(7) Useful wavelength range	The BTS shall transport from the moderator to the sample a beam of neutrons with a bandwidth that lies within the wavelength range 1.8 ± 0.08 Å and 10.0 ± 0.08 Å	The width of the wavelength band changes the number of reflections observed and the resolvable unit cell size (see 13.6.4 (1–2))	Measurement of the neutron spectrum at sample
13.6.4.1(8) Shutter	When the experimental or optical cave is not *interlocked* the BTS shall block radiation from the target to a level required by safety regulations	The user must be able to change a sample (see ConOps)	Measurement of the dose rate in the experimental cave
13.6.4.1(9) Brilliance transfer	The BTS should transport from the moderator to the sample a beam of neutrons with 100% *brilliance transfer* within the wavelength and divergence range	The more neutrons are incident on the sample the higher the signal (see 13.6.4 (5))	Measurement of the flux at sample vs. flux at moderator surface
13.6.4.1(10) Divergence profile	The BTS should transport from the moderator to the sample a beam of neutrons with a smooth divergence profile	The beam divergence profile is convoluted to the reflection profiles, which have to be smooth to allow integration by profile fitting (see ConOps)	Measurement of the beam divergence profile at sample

Continued

Table 2 System requirements for the NMX beam transport system (BTS).—cont'd

Id	Text	Rationale	Verification
13.6.4.1(11) Wavelength selection	The BTS shall transport from the moderator to the sample a beam of neutrons where the total *brilliance* outside the selected wavelength band is <1%	The neutrons outside the used wavelength band add to the background (see 13.6.4 (5))	Measurement of the neutron spectrum at sample
13.6.4.1(12) Accessibility	The BTS should be accessible for repairs while the *proton beam is on target*	The instrument will be more fault tolerant hence increasing throughput of experiments (see ConOps)	Measurement of the dose rate at the relevant technical component
13.6.4.1(13) Beamstop	The BTS shall absorb >99.9% of the neutrons not scattered by the sample	The neutrons not scattered by the sample contribute to background and should not be incident on the detectors (see 13.6.4.(5) and ConOps)	Measurement with and without sample
13.6.4.1(14) Wavelength selection goal	The BTS should transport from the moderator to the sample a beam of neutrons containing only neutron within the selected wavelength band	The neutrons outside the used wavelength band add to the background (see 13.6.4 (5))	Measurement of the neutron spectrum at sample
13.6.4.1(15) Biological shielding	The BTS shall be shielded against the radiation produced by the neutron beam to safe levels according to ESS radiation safety regulations	The surroundings of the instrument must be safe for personnel	Dose rate measurement

Fig. 1 Overview of the NMX instrument from the target station (bottom left) to the experimental cave (top right) with the roof of the common shielding bunker (left) omitted.

higher radiation shielding requirements closer to the sample. At an identical guide curvature, the larger the guide cross-section, the larger guide length is required to get out or line-of-sight and the smaller the brilliance transfer is for the low wavelength neutrons. However, only the first elements of all ESS instruments—up to 24m from the moderator in the West sector where NMX is—are enclosed in a common shielding bunker (Fig. 3). This bunker provides effective shielding against the high-energy background, but its dimensions impose that the line-of-sight should be lost before 24m from the source. The geometrical condition for avoiding the line-of-sight (Eq. 3) contains the curvature radius, cross-section and length of the curve. With the length fixed by the bunker dimensions, the required curvature radius is given by the cross-section, but smaller curvature radii require neutron mirrors with higher m-values (i.e. that they are reflective to a higher angle). This not only increases cost but also leads to losses, as the reflectivities of high m-value mirrors are lower. Therefore the horizontal cross-section remains 3cm for the curving part of the neutron guide system. After the loss of the line-of-sight at ~23m from the moderator the guide expands also horizontally to 4.5cm width. Before reaching the sample position and the beam collimation the neutron guide is contracted back to 3 × 3cm cross-section over a stretch of 10m with parabolically arranged focusing mirrors.

$$\mathrm{LOS} = \sqrt{8dR} \tag{3}$$

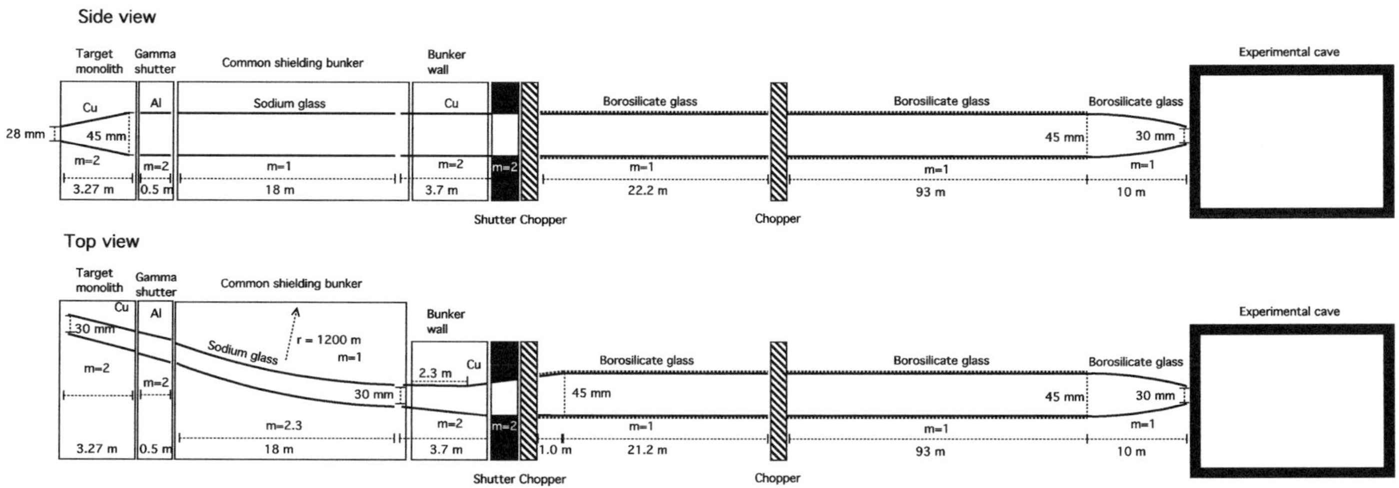

Fig. 2 Overview of the NMX Beam Transport and Conditioning system.

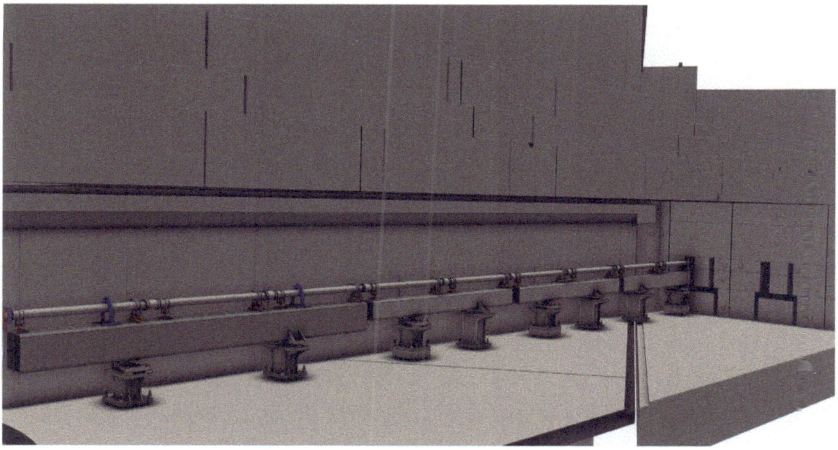

Fig. 3 Three-dimensional view of the NMX neutron optical system in the common shielding bunker.

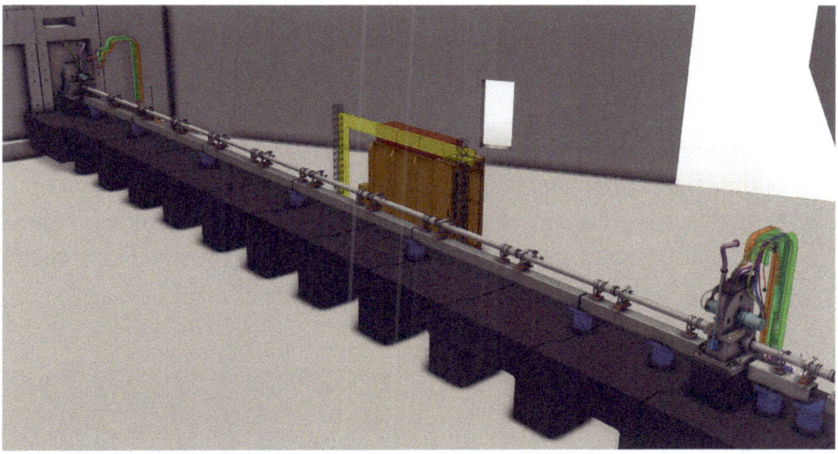

Fig. 4 The NMX neutron beamline showing the positions of the choppers: first chopper unit with a single disc top left and the second chopper unit with a dual disc bottom right.

The geometrical condition for line-of-sight avoidance, where LOS is the distance at which line-of-sight is lost, d is the width of the guide and R is the curvature radius.

The issue of frame overlap and wavelength selection is addressed by two chopper units (Fig. 4), the first one immediately after the bunker at 29 m and the second 51 m from the source. The purpose of these choppers is not to shape the pulse of neutrons, but merely to select only the desired neutron wavelengths. The purpose of the first chopper is mostly to stop the frame overlap from the neighboring neutron pulses, while the range of wavelengths that arrive at the sample is mostly defined by the second chopper. Therefore the second chopper actually has two separate discs with openings that are much larger than the effective opening needed for the full wavelength frame. By phasing these choppers accordingly so that one provides the leading edge and the other the trailing edge of an effective opening, the wavelength range that is allowed to pass can be easily adjusted by just changing the chopper phasing. This flexibility in wavelength selection can be used to reduce the inelastic scattering background in the TOF dimension at the expense of the time-averaged neutron brilliance.

Since both the guides and the choppers produce gamma radiation—harmful for both the operating personnel and the quality of data collection—when neutrons are absorbed, they have to be encased in a concrete shielding tunnel even outside the common shielding bunker mentioned above. A shutter at 28 m—just upstream of the first chopper—allows access to the instrument components outside the shielding bunker, whereas components inside the bunker are only accessible when the neutron source is not operating.

The neutron guide system ends before the beam enters the experimental area, which is the only area of the instrument where users will have access (Fig. 5). It is housed in a thick concrete cave to shield against the neutron and gamma radiation from the beam and an interlock system will ensure that no person can be inside the cave when the shutter is open.

Fig. 5 Overview of the NMX experimental cave and the adjoining control hutch and sample preparation area.

The first component in the cave is the collimation system that defines the beam using slits and apertures. The first slit immediately after the end of the guide allows principally the control of beam divergence, while the size of the beam is adjusted with a collimating aperture. In order to reduce the gamma radiation background in the cave, the neutrons not accepted by the collimation system are absorbed as far away from the sample position as possible; this is achieved by absorbing baffles in a flight tube after the first slit and a second slit set 70 cm upstream from the sample. All the collimating elements up to this point will be enclosed in a metal shielding box to minimize the gamma background.

Matching the beam size with the crystal size with pinhole collimation is not trivial when the divergence is $\pm 0.2°$. The reason is that to avoid an excessive halo around the crystal it has to be quite close to the pinhole, but for the smallest beam sizes (200 μm) the required distance is smaller than the sample diameter for the larger (3–5 mm) samples. Since the long wavelengths also impose collecting reflections at high 2θ angles (close to backscattering conditions), the shadow of the collimating system has to be minimized. It was set as a requirement for the endstation design that it should be possible to collect data up to $2\theta = 165°$ With the final collimating aperture at a few mm distance from the sample, this is unfeasible with a slit set, so conical holders for the collimating apertures were designed. These "bullets" allow placing the aperture very close to the sample with a low solid angle footprint. The variation of the beam size is achieved by changing the aperture to a different one using a carousel exchanger (Fig. 6). This setup easily allows optimizing the distance of the aperture from the crystal either using "bullets" of different lengths or using the xyz translation stage of the

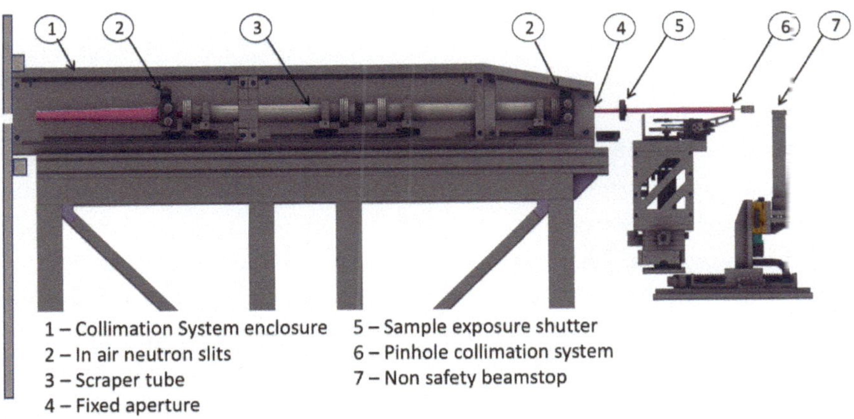

Fig. 6 The NMX collimation system and its components.

collimator exchanger. Using pre-made collimator apertures limits the flexibility in choosing precisely the size of the beam on the fly, but on the other hand the collimator "bullets" and the apertures therein are easy to fabricate, allowing for example custom shapes to match the shape of the crystal.

4.2 Sample exposure system

The sample will be mounted on a six-axis robotic goniometer. The principal driver for this choice is not automated sample changing—though that will also be possible—but rather the ability to change the sample orientation with respect to the beam as much as possible. The reason for this is that, due to the wavelength range used, there will be large blind regions of reciprocal space for any one crystal orientation, and to cover them the crystal orientation has to be changed significantly. The traditional way of dealing with this would be to use a four-circle diffractometer, but that is not very compatible with covering a large solid angle with detectors. A robotic arm offers the capability to reorient the crystal significantly with a small solid angle footprint, at least in a $\sim$50 cm radius around the sample. Even if the robot takes up space further out, this will not produce shadows on the detector.

4.3 Scattering characterization system

Another consequence of the long wavelengths used is that the detectors need to cover a large solid angle. This calls for surrounding the sample with detectors as much as possible. As discussed above this is not easily compatible with the idea of changing the crystal-to-detector distance to optimize the resolvable unit cell size, as the area needed to cover a given solid angle changes with the detector distance. Specifically this excludes detectors that are not flat, as the shape of the surface would have to change with detector distance, which is hardly realizable. This calls for flat detector units that can be moved in three dimensions.

Another consideration with surrounding the crystal with detectors is that it seriously complicates access to the sample position with both multi-axis goniometry—as discussed above—and sample environment equipment such as a cryostream. The approached adopted with NMX was to cover only the top hemisphere around the crystal with detectors, since for the vast majority of neutron data collections Friedel's law is true and hence the reciprocal space asymmetric unit is only a hemisphere. Therefore the floor space below the sample can be used to mount the goniometer as well as cryogenic and humidity control streams while leaving space to install various types of custom sample environment equipment like pump-probe setups.

While detector units could be moved in three dimensions on a mechanical rail system, it would be rather complicated and the footprint of such a system would be significant. Therefore it was also decided to also use six-axis robotic arms to move the detectors. In a geometric study it was realized that to fulfill the requirements (Table 3) it was not possible to find a mounting point that would allow the robots to provide overlapping detector coverage. Therefore it was necessary to mount the robots on high-precision rails and add an additional degree of freedom. The motion of the robotic arm on the rail is controlled by the robot controller, effectively leading to a seven-axis robot (Fig. 7).

For the detectors themselves the challenge was achieving submillimeter spatial resolution—in the requirements (Table 3) the goal was 0.2 mm, while 1 mm would be acceptable. An important boundary condition was that since the detector units needed to be flat, the angle of incidence of the incoming neutrons varies across the detector. This rules out several approaches for improving the spatial resolution. The ESS detector group set out to develop the technology to address this challenge focusing on Gd as an obvious candidate for the neutron absorber. Both ^{157}Gd and ^{155}Gd have very high neutron absorption cross-sections (Sears, 1992), but they have not been commonly used in neutron detectors. The reason is that upon absorbing a neutron these nuclei emit electrons, which are much more difficult to distinguish from the gamma-ray background that for example the alpha-particles emitted by ^{3}He. For the NMX application the selectivity for neutrons is less critical than a high detection efficiency, so Gd was chosen as the converting material. For the detection of the electrons the technology chosen was the gas-electron multiplier (GEM)—widely used in particle physics applications (Sauli, 2014). The GEM is based on copper-coated kapton foils with regularly spaced holes in them. When placed under a high voltage the electric field in the foil guides the charges generated in a counting gas—in this case Ar/CO_2—through the holes and effectively discretizes the position of the charge. In order to detect the conversion electrons from Gd, however, the counting gas thickness needed is $\sim$10 mm, which is a further complication in reaching the required spatial resolution. The problem is that, while each charge generated in the counting gas will drift linearly, the electrons generated in neutron absorption are not necessarily parallel to the neutron beam. This creates for each neutron absorption event a

Table 3 System requirements for the NMX scattering characterization system (SCS).

Id	Text	Rationale	Verification
13.6.4.3(1) Detection efficiency	The SCS should detect the scattered neutrons with an efficiency of 100% across the wavelength band	The higher the detection efficiency the higher the signal (see 13.6.4 (5))	Measurement of detection efficiency
13.6.4.3(2) Spatial resolution	The SCS shall detect the scattered neutrons with a *spatial resolution* <1 mm	The spatial resolution allows resolving the reflections from a crystal with large unit cells. It should correspond at least to the expected diffraction spot size of 1 mm (see 13.6.4. (1-4,6) and ConOps)	Imaging a reference grid
13.6.4.3(3) TOF resolution	The SCS shall detect the scattered neutrons with a *time-of-flight resolution* <0.1 ms	The time-of-flight resolution should be an order of magnitude smaller than the time width of a reflection, given by the pulse length (see 13.6.4.1 (1) and ConOps)	Measurement (TBD)
13.6.4.3(4) Neutron selectivity	The SCS should detect radiation incident to the detector system other than thermal neutrons with a 0% efficiency	Non-neutron originated detector signals (e.g. gamma-rays) contribute to background and should be minimized (see 13.6.4 (5))	Measurement of detection efficiency for gamma-rays
13.6.4.3(5) Accessible solid angle	The SCS shall be able to detect neutrons scattered by the sample within an angular range of ±165° from the neutron beam axis in the top hemisphere	To make use of the longer wavelength neutrons that are scattered to higher angles at a given d it has to be possible to detect these neutrons (see 13.6.4 (2) and ConOps)	Test
13.6.4.3(6) Sample-to-detector distance	The SCS shall have a sample-to-detector distance adjustable between >200 + 0.2 mm and <1000 + 0.2 mm	Changing the detector distance allows the maximum resolvable unit cell size to be traded for data collection time (see 13.6.4. (1) and ConOps)	Measurement of detector distance

13.6.4.3(7) Sample-to-detector distance step	The SCS shall have a sample-to-detector distance selectable in 1 mm steps	The step in detector distance has to allow choosing the optimal distance (see ConOps)	Measurement of detector distance
13.6.4.3(8) Detector stability	The SCS shall position the detectors with a *stability* of 0.1 mm	Vibrations or drift of the detectors should be small compared to the diffraction spot size (see 13.6.4.3(2) and ConOps)	Survey and measurement of acceleration
13.6.4.3(9) Detection solid angle	The SCS should simultaneously detect the scattered neutrons within a solid angle of 4π sr	The higher the solid angle coverage the shorter the data collection time (see ConOps)	Test
13.6.4.3(10) Detector noise	The SCS should record <0.001 counts/s in the absence of incoming radiation	The lower the background the higher the achievable S/B ratio, but the detector noise level must be lower than the background, otherwise the detector noise becomes limiting (see 13.6.4. (5))	Measurement with shutter closed
13.6.4.3(11) Spatial resolution goal	The SCS should detect the scattered neutrons with a *spatial resolution* <0.2 mm	The spatial resolution allows resolving the reflections from a crystal with large unit cells. A spatial resolution finer than the expected diffraction spot size of 1 mm allows integration by profile fitting (see 13.6.4. (1-4,6) and ConOps)	Imaging a reference grid

Fig. 7 A view of the NMX sample area with the goniometer robot and the three detector positioning robots on rails. Robotic system concept developed at Natx-Ray SAS, Saint Martin d'Hères, France—3D rendering image reprinted with permission from NatX-ray © 2016. All rights reserved.

trajectory of charges across several mm. For this issue the solution is the so-called micro time projection chamber (μTPC). This concept relies on the inherently very high time resolution of the read-out electronics, which allows to distinguish charges arriving to neighboring channels by their time of arrival in the nanosecond scale. This in turn allows the identification of the first charge—arriving to the GEM at the latest time—, which corresponds to the point of the original neutron absorption. This concept has been demonstrated to work and the achievable spatial resolution is given by the pitch of the readout strips, which in the current prototype is 400 μm (Pfeiffer et al., 2016). In practice an even more important parameter than the spatial resolution as such is the point spread function of the detector—i.e. how a point image is blurred by the detector.

5. Data collection strategies and optimizing the use of beam time

The multi-axis goniometry capability of the robotic goniometer together with the uniquely flexible detector geometry of NMX leads to rather complex data collection strategies. While it will be possible to collect good quality complete data with a naïve positioning of the detectors around the sample, it is not at all obvious what is the fastest way to collect an acceptable data set from a given crystal. In the Laue method with a 2–4% bandpass there is no need to oscillate the crystal during the exposure and hence there is also no need for consecutive rotation ranges. This means that optimization of the data collection strategy in terms of the crystal orientation and detector positions is highly redundant. Some development is needed in existing data collection strategy programs to account for the complex geometry and constraints of the robotic detector positioners. The robot control software will allow trajectories to be predicted and checked for physical obstacles, allowing a strategy software to find an optimal position for both the goniometer and detector positioning robots in a holistic way.

6. Supporting facilities—Deuteration and large crystal growth

As discussed in the chapter "Fundamentals of neutron crystallography in structural biology" by Helliwell (in this volume), the characteristics of the sample (crystal size, deuteration level) are decisive for the data collection time as well as data quality. Therefore the ESS identified early on the supporting facilities for sample preparation as a key for the successful scientific use of the neutron instruments. For NMX specifically the challenge is that large crystals will still be needed, while in X-ray crystallography the crystal sizes are less limiting. This means that new generations of crystallographers are no longer necessarily experts in growing as large crystals as possible. It is therefore critical to offer support for the potential users to grow their crystals as large as possible. For this purpose the ESS has set up the DEMAX platform, which can support the users of ESS in the techniques of crystal growth optimization described in the chapter "Large crystal growth for neutron protein crystallography" by Fisher and colleagues (in this volume).

7. Expected performance

It is far from straightforward to quantitate how a diffractometer performs, let alone before it has been built. From the user perspective the key questions are (a) how large a crystal is needed and (b) how long does the data collection take. In practice the answers to these questions depend very much on the properties of the crystal such as the space group, unit cell size, solvent content etc. so the contribution of the instrument itself has to be abstracted to some performance indicators.

For ESS NMX there are several performance indicators. The first one if the time-averaged spectral neutron brilliance, which defines the strength of a signal from a crystal of given size and diffracting power. The brilliance is essentially limited by the source (moderator) but the instrument design attempts to minimize any losses of the neutron brilliance. The spectral neutron brilliance is not trivial to measure, so usually the quantity that is used as a performance indicator is the neutron flux, which measures the number of neutrons per unit time and unit area over a given divergence and wavelength range. For NMX the simulated neutron flux at the sample is 2×10^9 neutrons cm^{-2} s^{-1} for a divergence of $\pm 0.2°$ and a wavelength range of 1.8–3.55 Å.

Based on experience at continuous source instruments like LADI-III at ILL, this neutron flux is estimated to allow a data set to be collected in less than 24 h from a crystal of ~0.1 mm^3 volume.

At a pulsed neutron source like the ESS the time-averaged neutron flux does not provide a full picture, because the signal for any one Bragg reflection will arise from one wavelength, while the background is to the first approximation uniform with wavelength. This means that the length of the neutron pulse defines the time during which the useful signal is accumulated, while the background is spread over the entire time between neutron pulse: the time frame. The ESS time frame is 71.4 ms, of which the 2.86 ms neutron pulse is ~1/25. Assuming this factor leads to a corresponding background reduction, the signal-to-background ratio would be 25 times better than at a continuous source, further reducing the required crystal size or data collection time.

The second performance aspect is the ability to resolve large unit cells, which is more complex to quantitate. The reason is that the ability to integrate two neighboring reflections depends on the size of a Bragg spot on the detector surface. This in turn depends on the size and divergence of the beam and the size and mosaicity of the crystal. The time-of-flight width of a reflection is equal to the length of the neutron pulse. Another consideration is the criterion for separating the Bragg spots, which depends also on the integration software used. For example the Daresbury Laue suite (Arzt, Campbell, Harding, Hao, & Helliwell, 1999) allows deconvolution of overlapped reflections, so even unit cells where the reflections cannot explicitly be separated can potentially be studied. As an example, the predicted diffraction pattern for bovine heart cytochrome *c* oxidase, a membrane protein with a unit cell of a = 182.59 Å, b = 205.40 Å, c = 178.25 Å in the space group $P2_12_12_1$ was analyzed for overlaps (Fig. 8). The beam size was assumed to be matched to the crystal size of 0.2 mm and the beam divergence to be $\pm 0.2°$.

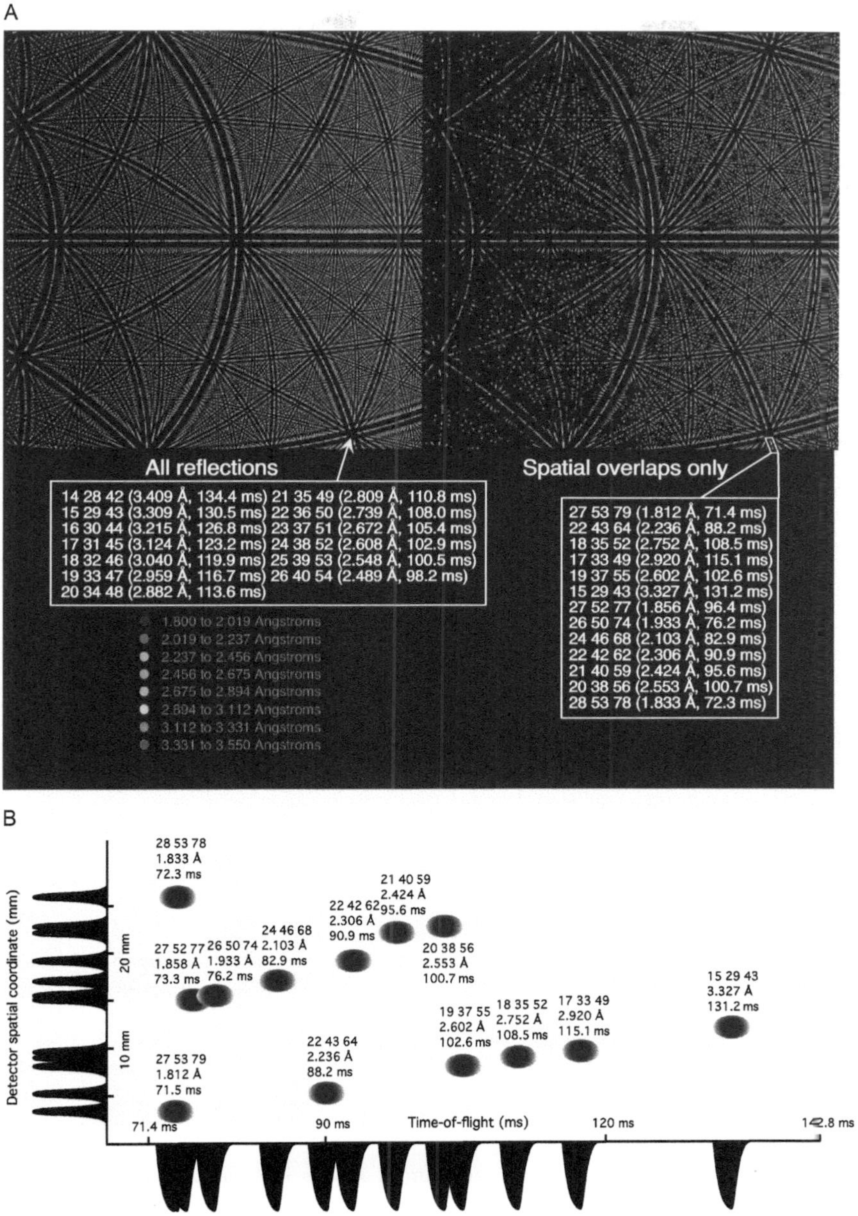

Fig. 8 (A) Predicted reflections from bovine heart cytochrome *c* oxidase at 1 m detector distance. Left: all reflections and right: the reflections classified as spatial overlaps by Lauegen. (B) The separation of some example high-resolution reflections in the spatial and time-of-flight dimensions.

The conclusion from this example was that, by using a combination of the detector spatial resolution and time-of-flight resolution, a unit cell of at least ~200 Å in all dimensions should be possible to resolve at ESS NMX without resorting to deconvolution.

8. Conclusion

The long-pulse neutron source and high brilliance moderators of the ESS are well suited for macromolecular crystallography. The NMX Macromolecular Diffractometer is designed with a rigorous systems engineering approach to make the best possible use of the source characteristics. The flexible geometry allows the data collection strategy to be optimized for each crystal. The instrument is expected to allow data collections from crystals of $<0.1\,mm^3$ volume in a day.

References

Aizawa, K., Watanabe, N., Suzuki, J., Niimura, N., Morii, Y., Katano, S., Osakabe, T., Teshigawara, M. & Mezei, F. (1997) A comparison between short pulse spallation source and long pulse spallation source. (JAERI-Conf--97-010). Tone, Tatsuzo (Ed.). Japan.

Andersen, K. H., Bertelsen, M., Zanini, L., Klinkby, E. B., Schonfeldt, T., Bentley, P. M., et al. (2018). Optimization of moderators and beam extraction at the ESS. *Journal of Applied Crystallography*, *51*, 264–281.

Arzt, S., Campbell, J. W., Harding, M. M., Hao, Q., & Helliwell, J. R. (1999). LSCALE—The new normalization, scaling and absorption correction program in the Daresbury Laue software suite. *Journal of Applied Crystallography*, *32*, 554–562.

Garoby, R., Vergara, A., Danared, H., Alonso, I., Bargallo, E., Cheymol, B., et al. (2017). The European spallation source design. *Physica Scripta*, *93*, 014001.

Jauch, W. (1997). Prospects of single-crystal diffraction at a long pulse spallation source. *Journal of Neutron Research*, *6*, 161–171.

Kai, T., Harada, M., Teshigawara, M., Watanabe, N., & Ikeda, Y. (2004). Coupled hydrogen moderator optimization with ortho/para hydrogen ratio. *Nuclear Instruments and Methods in Physics Research Section A: Accelerators, Spectrometers, Detectors and Associated Equipment*, *523*, 398–414.

Lechner, R. E. (1992). Optimization of the chopper system for the cold-neutron time-of-flight spectrometer NEAT at the HMI, Berlin. *Physica B: Condensed Matter*, *180-181*, 973–977.

Mezei, F. (1997a). The raison d'etre of long pulse spallation source. *Journal of Neutron Research*, *6*, 3–32.

Mezei, F. (1997b). Long pulse spallation sources. *Physica B: Condensed Matter*, *234-236*, 1227–1232.

Mezei, F., & Russina, M. (2002). Multiplexing chopper systems for pulsed neutron source instruments. *Proceedings of SPIE—The International Society for Optical Engineering*, *4785*, 24–33.

Niimura, N., Minezaki, Y., Nonaka, T., Castagna, J.-C., Cipriani, F., Høghøj, P., et al. (1997). Neutron Laue diffractometry with an imaging plate provides an effective data collection regime for neutron protein crystallography. *Nature Structural Biology*, *4*, 909–914.

Peggs, S. (Ed.), (2013). *ESS technical design report.* Lund: European Spallation Source.
Pfeiffer, D., Resnati, F., Birch, J., Etxegarai, M., Hall-Wilton, R., Höglund, C., et al. (2016). First measurements with new high-resolution gadolinium-GEM neutron detectors. *Journal of Instrumentation, 11,* P05011.
Pfeiffer, D., Resnati, F., Birch, J., Hall-Wilton, R., Höglund, C., Hultman, L., et al. (2015). The μTPC method: Improving the position resolution of neutron detectors based on MPGDs. *Journal of Instrumentation, 10,* P04004.
Sauli, F. (2014). Gas electron multiplier (GEM) detectors: Principles of operation and applications. *Comprehensive Biomedical Physics, 8,* 367.
Schoenborn, B. P., & Langan, P. (2004). Protein crystallography with spallation neutrons. *Journal of Synchrotron Radiation, 11,* 80–82.
Sears, V. F. (1992). Neutron scattering lengths and cross sections. *Neutron News, 3,* 26–37.
Sheldrick, G. M. (1990). Phase annealing in SHELX-90: Direct methods for larger structures. *Acta Crystallographica, A46,* 467–473.
Strobl, M., Bulat, M., & Habicht, K. (2013). The wavelength frame multiplication chopper system for the ESS test beamline at the BER II reactor—A concept study of a fundamental ESS instrument principle. *Nuclear Instruments and Methods in Physics Research Section A: Accelerators, Spectrometers, Detectors and Associated Equipment, 705,* 74–84.

CHAPTER EIGHT

Dynamic nuclear polarization enhanced neutron crystallography: Amplifying hydrogen in biological crystals

Joshua Pierce[a], Matthew J. Cuneo[a,†], Anna Jennings[a,c], Le Li[a,‡], Flora Meilleur[a,b], Jinkui Zhao[a,§], Dean A.A. Myles[a,*]

[a]Neutron Scattering Division, Oak Ridge National Laboratory, Oak Ridge, TN, United States
[b]Department of Molecular and Structural Biochemistry, North Carolina State University, Raleigh, NC, United States
[c]Oak Ridge Associated Universities, Oak Ridge, TN, United States
[*]Corresponding author: e-mail address: mylesda@ornl.gov

Contents

Abstract

Dynamic nuclear polarization (DNP) can provide a powerful means to amplify neutron diffraction from biological crystals by 10–100-fold, while simultaneously enhancing the visibility of hydrogen by an order of magnitude. Polarizing the neutron beam and aligning the proton spins in a polarized sample modulates the coherent and incoherent neutron scattering cross-sections of hydrogen, in ideal cases amplifying the coherent

† Present address: Department of Structural Biology, St. Jude Children's Research Hospital, Memphis, TN 38105, United States.
‡ Present address: Abilita Bio, Inc, 6888 Nancy Ridge Dr, San Diego, CA 92121, United States.
§ Present address: Beijing National Laboratory for Condensed Matter Physics, Institute of Physics, Chinese Academy of Sciences, Beijing 100190, China and Songshan Lake Materials Laboratory, Dongguan 523808, China.

Methods in Enzymology, Volume 634
ISSN 0076-6879
https://doi.org/10.1016/bs.mie.2019.11.018

scattering by almost an order of magnitude and suppressing the incoherent background to zero. This chapter describes current efforts to develop and apply DNP techniques for spin polarized neutron protein crystallography, highlighting concepts, experimental design, labeling strategies and recent results, as well as considering new strategies for data collection and analysis that these techniques could enable.

1. Introduction

Neutron macromolecular crystallography (NMC) continues to grow in importance in structural biology, and stands out in its ability to directly visualize hydrogen (H) atoms in biological structures (as this volume attests). NMC can provide a more complete and detailed picture of the atomic structures of bio-macromolecules, and can provide critical insights into biological structure, function and mechanism. Recent outstanding achievements include elucidation of protonation states at the active site of lytic polysaccharide monooxygenases (LPMOs) that activate molecular oxygen in breakdown of cellulose in plant biomass (O'Dell, Agarwal, & Meilleur, 2017); of the protonation state of ferryl heme in a peroxidase (Casadei et al., 2014); of the role of water in protein-ligand binding in trypsin (Schiebel et al., 2018); and of how a low-barrier hydrogen (LBHB) bond modulates enzyme activity and specificity in aminoglycoside acetyltransferases (Kumar et al., 2019), among many other examples. These achievements have been underpinned by advances in neutron instrumentation and sources that have delivered order of magnitude gains in capability, and parallel advances in deuterium labeling technologies that have enabled studies of smaller crystals than previously possible (for review, see Blakeley, 2009; O'Dell, Bodenheimer, & Meilleur, 2016 and references therein).

However, in comparison with other high-throughput structural techniques (NMR and X-ray crystallography), NMC remains flux limited and is challenged by the need for long measurement times (>days) and very large sample volumes (>0.1–$1.0\,mm^3$) that can be difficult to obtain. Moreover, NMC is intrinsically limited by a large hydrogen incoherent scattering ($\sigma_{inc}=79.8\,b$ ($1\,b=10^{-28}\,m^{-2}$)) background that can rapidly overwhelm the coherent ($\sigma_{coh}=1.8\,b$) elastic signal (i.e., diffraction is buried in the background from the sample itself) (Table 1). This problem is typically ameliorated by substituting hydrogen for deuterium ($\sigma_{inc}=2.0\,b$) in the crystal. For example, replacing H_2O with D_2O in solvent exchanged crystals can deliver ~two- to threefold improvements in signal to noise, while production of fully perdeuterated crystals can deliver up to ~10-fold

Table 1 Calculated background subtracted intensities for Bragg diffraction from a typical hydrogenated protein crystal measured under spin aligned and spin anti-aligned polarized conditions, and for unpolarized H-hydrogenated (native), D_2O-soaked and fully perdeuterated pD-crystals (following Schultz et al., 2005, and definitions therein).

		Deuterated sample		Polarized sample	
Sample	$H_{protein} + H_2O$	$H_{protein} + D_2O$	$D_{protein} + D_2O$	$H_{protein} + H_2O$	$H_{protein} + H_2O$
Polarization	–	–	–	+P	−P
H_{coh}	1.75	1.75		14.7	42.1
H_{inc}	80.26	80.26	–	0	106.5
D_{coh}	–	5.59	5.59	–	–
D_{inc}	–	2.05	2.05	–	–
Average reflection[a]					
Ave<I>	55	69	88	366	399
Background	3060	1374	79	39	4060
SigI[b]	56	38	13	14	67
<I>/SigI	1.0	1.8	6.8	11.6	6.0
Relative gain factor[c]					
[i] GF-H	1.0	3.2	46	134	36
[ii] GF-D_2O	–	1	14	41	11
[iii] 50% flux	–	–	–	(21)	(5.5)

Intensity gain factors (GF) are calculated relative to (i) GF-H—fully hydrogenated native samples (the default for DNP); (ii) GF-D_2O—D_2O soaked crystals, the most typical unpolarized sample; and (iii) assuming 50% reduction in flux for polarized neutron beams.

[a]Average reflection counts and background simulated for ideal detector placed 50 cm from sample, 90° from the incident beam.

[b]For weak reflections, SigI of an integrated peak is essentially the square root of background counts B.

[c]Gain factor in measurement time/sample size relative to:

(i) GF-H—$H_{protein} + H_20$; (ii) GF-$_{D2O}$ $H_{protein} + D_2O$; (iii) GF-$_{50\%}$—$H_{protein} + D_2O$ with 50% incident flux.

$$I_\tau = D_{expt} q I_L(\lambda)\varepsilon(\lambda) N_s \kappa \frac{\lambda^4}{2v_{uc}\sin^2\theta}|F_\tau|^2$$

$$B_{inc} = D_{expt} q I_L(\lambda)\Delta\lambda\varepsilon(\lambda) N_s \kappa \frac{\sum \sigma_{inc}}{4\pi}\frac{A_{det}}{1_2^2}$$

improvement (Blakeley et al., 2008; Shu, Ramakrishnan, & Schoenborn, 2000). However, while D_2O solvent exchange is generally routine, the production of large perdeuterated protein crystals is not, and is generally limited to recombinant proteins that can be readily expressed in *Escherichia coli* under deuterated conditions (Meilleur, Weiss, & Myles, 2009). For other proteins, purified from endogenous sources or overexpressed in yeast, insect or human cell systems, the large hydrogen incoherent scattering term remains a significant impediment and either new labeling strategies or new experimental techniques are required.

One way to overcome the intrinsic background noise, and to amplify the scattered signal from hydrogen, is to modulate and tune the neutron scattering lengths of hydrogen using dynamic nuclear polarization (DNP) (Stuhrmann, 2004). This approach uses the fact that neutron interactions with spin ½ nuclei, such as hydrogen, depends upon their relative spin orientations. In today's NMC experiments, the spins are disordered, which gives rise to the large incoherent scattering. In DNP-NMC, the neutron and nuclei spins are aligned. In a DNP experiment (Abragam, 1961), the polarization of high-gyromagnetic ratio (γ) unpaired electrons in a paramagnetic center is transferred to surrounding hydrogen nuclei (protons) by saturating the electron paramagnetic resonance (EPR) transitions using microwave (MW) irradiation. Under idealized conditions, the spin polarization for protons can be enhanced by ~660-fold. With protein crystals, suitable paramagnetic centers, such as paramagnetic metal ions (e.g., Mn^{2+}, Fe^{2+}, Co^{2+}, Ni^{2+}, Cu^{2+}) or stable free radicals (e.g., nitroxides), can be simply added to the sample. The DNP technique then uses a combination of low temperatures (0.2–1.0 K), high magnetic field (2.5–5 T), and microwave (70–140 GHz) irradiation of the EPR resonances to produce very high positive or negative nuclear polarizations that then propagate throughout the sample by proton-proton spin diffusion.

In the ideal case, polarizing the neutron beam and aligning the proton spins in DNP-polarized samples changes the neutron scattering cross-sections of hydrogen dramatically (see Fig. 1), amplifying the coherent scattering component by almost an order of magnitude ($\sigma_{coh} = 14.8$ b) while suppressing the incoherent background to zero. Moreover, when the spins are anti-aligned, the coherent cross section is amplified further, by ~24-fold ($\sigma_{coh} = 40$ b), while the incoherent cross section is increased just marginally, by ~1.3-fold ($\sigma_{inc} = 106$ b). Polarizing the hydrogen nuclei in a sample could therefore deliver 10–100-fold gains in performance for biological crystallography, shortening data collection times dramatically (from days

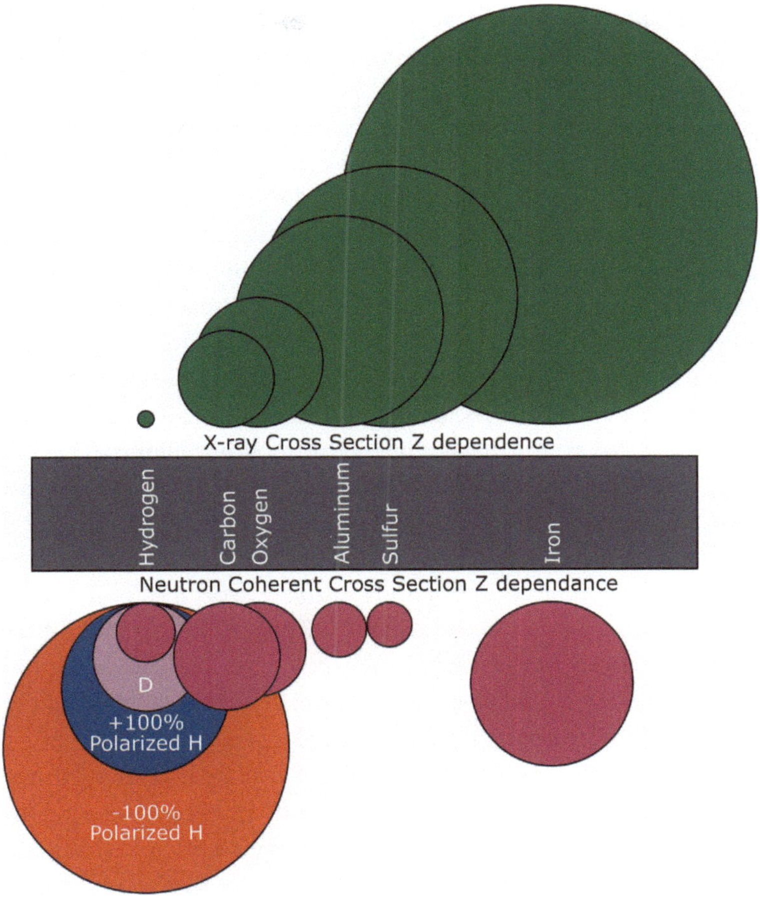

Fig. 1 The relative change in cross section for neutrons and X-rays with increasing Z, and in the case of neutrons, isotopic substitution and polarization. In the special case of hydrogen, substituting D for H in the sample enables solvent contrast techniques and enhances the visibility of D in labeled protein structures. Polarizing the neutron beam and the hydrogen nuclei in the sample changes these cross-sections dramatically, changing the sign and magnitude of coherent scattering component significantly, and enabling the scattering from hydrogen to be modulated and tuned between these extremes, in situ, within a single polarized sample.

to hours), and/or allowing analysis of radically smaller (0.01 mm^3) hydrogenated samples than has been possible to date. This creates the exciting prospect of visualizing hydrogen atoms using crystals similar in size ($200 \times 200 \times 200\,\mu m^3$) to those that can be grown routinely for home source X-ray diffraction experiments today, breaking through the sample size limitations that hamper NMC analysis.

Realizing this potential, however, remains challenging. The advantages of DNP-enhanced polarized neutron diffraction and scattering have long been realized, with first demonstration by Hayter, Jenkin, and White (1974), using polarized crystals of neodymium-doped lanthanum magnesium nitrate hydrate (LMN:Nd) (Hayter et al., 1974). Stuhrmann and colleagues then developed and demonstrated DNP applications using Small Angle Neutron Scattering (SANS) (Stuhrmann et al., 1986), most notably in the analysis of spin labeled subunits in the ribosome (Zhao et al., 1995) and recently in location of tyrosyl radicals in spin polarized proteins (Zimmer, Jouve, & Stuhrmann, 2016). However, while opportunities to exploit DNP for neutron protein crystallography have continued to be investigated at several facilities, including tests with small organic molecules (Piegsa et al., 2013) and development of ancillary methods for protein labeling, cooling and polarization (Tanaka et al., 2018; Tanaka, Kusaka, Chatake, & Niimura, 2013; Zhao, Robertson, Herwig, & Crabb, 2013), many technical hurdles and challenges remain. Here, we describe the development and application of a miniaturized "frozen spin" DNP sample environment, designed to provide a large $>2\pi$ opening angle for diffraction, and how this has enabled first wide angle polarized neutron diffraction data to be collected from paramagnetically labeled sub-mm^3 protein crystals, polarized at sub-K temperatures in a 5 T field.

Harnessing DNP for NMC could be transformative, opening the way to new classes of experiments on smaller crystals of more complex systems, and providing a wealth of information at the hydrogen atom level of detail. The necessary DNP technologies (high power gyrotron MW sources and optimized radical species) have developed rapidly in the last decade. In solid state and solution NMR and imaging, such microwave-driven DNP experiments have improved sensitivity by 10^2 and reduced measurement times by 10^4, and DNP-NMR systems are now commercially available. Here we consider progress toward the development and use of analogous DNP techniques to enhance NMC, highlighting concepts, experimental design and labeling strategies, detailing initial proof of concept results on spin-polarized protein crystals, and then considering new strategies for data collection and analysis that these DNP-NMC techniques may enable.

2. Spin polarized neutron diffraction

Neutrons are fundamental particles that are scattered by atomic nuclei (leaving aside magnetic scattering from unpaired electrons), and neutron scattering lengths (*b*) show little uniformity across the periodic table (Fig. 1). The likelihood of interaction between an incident neutron and a target nucleus is expressed as the neutron scattering cross section σ, measured in units of barns ($1\,\mathrm{b} = 10^{-24}\,\mathrm{cm}^2$). The total neutron scattering cross section is the sum of two terms, coherent and incoherent scattering. Coherent scattering arises from correlations between different nuclei in a material, giving rise to interference and diffraction, and providing information on molecular structure. Incoherent scattering arises from the correlations of individual nuclei through time, is spatially isotropic, and gives information on molecular dynamics of materials. Most importantly for studies of biological systems, neutrons are extremely sensitive to hydrogen, which accounts for more than half the number of atoms in a typical biological macromolecule, and hence dominates neutron scattering in biological materials. In NMC, the large incoherent scattering from hydrogen gives rise to a large experimental background that can rapidly swamp Bragg diffraction. This has required that in almost all crystals studied with NMC, the hydrogen atoms in the solvent, and when feasible the protein, be exchanged with deuterium in order to enhance the signal to noise ratio of the data.

The origin of the large hydrogen incoherent scattering cross-section lies at the unordered spin states of the neutrons and protons (Abragam & Goldman, 1982). Polarizing the neutron beam and the hydrogen nuclei drastically changes these cross-sections. This can be seen from rewriting the scattering amplitude *a*, in terms of the spin independent component b_0, and spin dependent component *b*, as shown in Eq. (1), where $\boldsymbol{I}$ is the spin of the nucleus and $\boldsymbol{s}$ is the spin of the neutron.

$$a = b_0 + b\boldsymbol{I} \cdot \boldsymbol{s} \tag{1}$$

These components can be related to the conventional amplitudes b_- and b_-, by the following relations.

$$b_+ = b_0 + (\boldsymbol{I} \cdot \boldsymbol{s})_+ = b_0 + b\frac{I}{2};\, b_- = b_0 + (\boldsymbol{I} \cdot \boldsymbol{s})_- = b_0 - b\left(\frac{I+1}{2}\right); \tag{2}$$

$$\text{With } (\boldsymbol{I} \cdot \boldsymbol{s})_\pm = \frac{1}{2}\left\{\left(I \pm \frac{1}{2}\right)\left(I \pm \frac{1}{2} + 1\right) - I(I+1) - \frac{3}{4}\right\}; \tag{3}$$

In the case of a sample composed of non-correlated spins of a single isotope, the cross section can be written as below, where $\boldsymbol{\kappa}$ is the momentum transfer and $\boldsymbol{r}_i$ is the position of an atom. For spin polarized samples with a polarization *P*, and a neutron beam with polarization *p* along the same axis as the sample polarization, the following relations can be written.

$$\frac{d\sigma}{d\Omega} = \sum_i \left\{ \langle |a_i|^2 \rangle - \langle a_i a_j^* \rangle_{i \neq j} \right\} + \langle a_i a_j^* \rangle_{i \neq j} \sum_{ij} e^{(-i\boldsymbol{\kappa} \cdot (\boldsymbol{r}_i - \boldsymbol{r}_j))}; \tag{4}$$

$$\langle \boldsymbol{I}_i \cdot \boldsymbol{I}_i \rangle = \langle \boldsymbol{I}_i \cdot \boldsymbol{I}_j \rangle = P^2 I^2; \langle \boldsymbol{I}_i \times \boldsymbol{I}_i \rangle = 0; \langle \boldsymbol{s} \cdot \boldsymbol{I}_i \rangle = \frac{pPI}{2}; \tag{5}$$

$$\begin{aligned} \langle a_i a_j^* \rangle_{i \neq j} &= b_0^2 + bb_0 IpP + \frac{b^2 I^2 P^2}{4}; \langle |a_i|^2 \rangle \\ &= b_0^2 + bb_0 IpP + \frac{b^2}{4}\{I(I+1) - pPI\}; \end{aligned} \tag{6}$$

$$\frac{d\sigma}{d\Omega} = N\left(\frac{d\sigma}{d\Omega}\right)_{inc} + N\left(\frac{d\sigma}{d\Omega}\right)_{coh} \sum_i e^{(-i\boldsymbol{\kappa} \cdot \boldsymbol{r}_i)}; \tag{7}$$

When combined with the relation for the cross section for Bragg scattering, this allows the polarization dependence of the coherent and incoherent cross sections to be written as shown below.

$$\begin{aligned} \left(\frac{d\sigma}{d\Omega}\right)_{inc} &= \frac{b^2}{4}\{I(I+1) - pPI - P^2 I^2\}; \left(\frac{d\sigma}{d\Omega}\right)_{coh} \\ &= \left\{ b_0^2 + bb_0 IpP + \frac{I^2 P^2 b^2}{4} \right\}; \end{aligned} \tag{8}$$

An immediate result of this spin dependence is that for the case of $P=1$ and $p=1$, the incoherent cross section reduces to zero, effectively eliminating this major source of experimental background. The effect is largest for hydrogen, and in the limiting case where proton spins are fully aligned in the same direction as that of the neutron beam, the coherent and incoherent hydrogen neutron scattering cross-sections become 14.7 and 0 b, respectively (Fig. 2).

Moreover, the coherent and incoherent cross sections vary continuously as a function of spin polarization in the sample (Fig. 2). Flipping the spin to negative changes the cross sections dramatically, and in the limiting case the coherent cross section increases ~24-fold, while the incoherent cross section increases marginally by just ~1.3-fold. For a typical protein sample, measured on a generic NMC diffractometer, this translates into calculated gain factors of ~40 and ~100-fold in average background subtracted intensities under spin aligned and anti-aligned conditions, respectively, relative to fully hydrogenated native samples (the default for DNP), and of ~10- and ~40-fold, respectively, relative to the more typical case of D_2O exchanged crystals (see Table 1 and definitions therein, following Schultz et al., 2005).

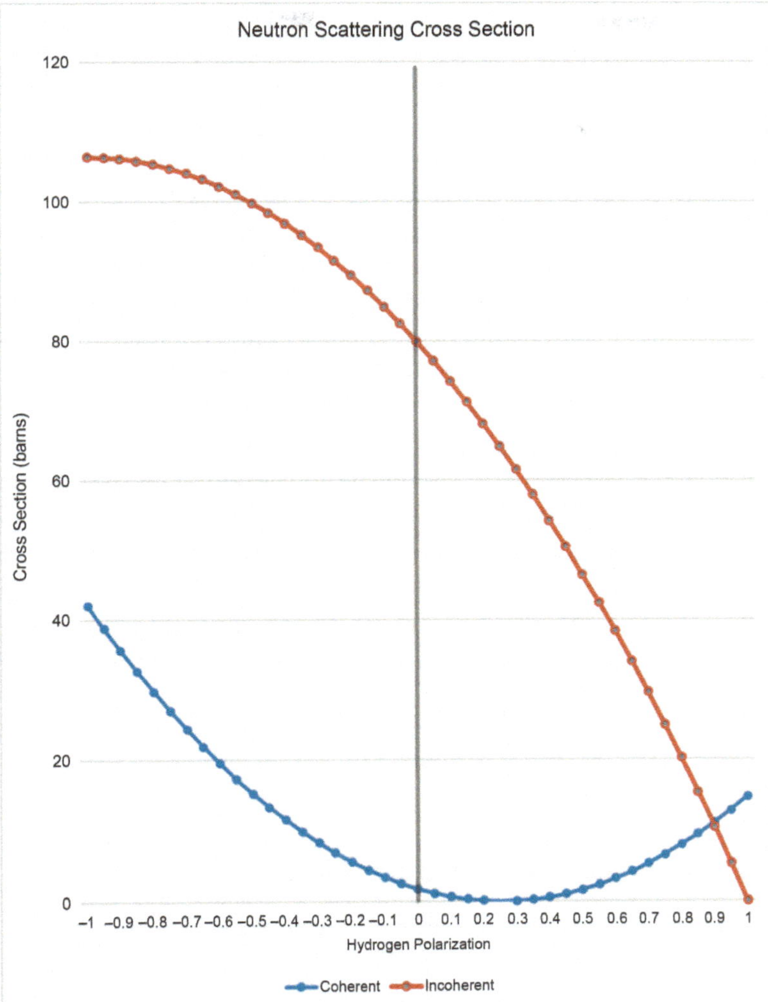

Fig. 2 Proton scattering cross-section from polarized neutrons as a function of proton polarization. In the general case, protons are un-polarized (0) and the incoherent cross-section is 80.3 b, while the coherent component is 1.8. In DNP, when protons are fully polarized parallel (+1) to the polarization of neutrons, the incoherent cross-section reduces to 0 b while the coherent cross-section increases to 14.7 b. When protons are fully polarized anti-parallel (−1) to the neutron polarization, the incoherent cross-section increases to 106.5 b scattering, while the coherent component is maximized to 42.1 b.

Polarizing the hydrogen nuclei in a sample can therefore deliver order of magnitude increases in signal to noise ratio of the data, which should shorten data collection times significantly, and allow analysis of radically smaller crystals of hydrogenated materials than is possible today.

A further, and significant, benefit is that polarized hydrogen atoms are then much more visible in the resulting Fourier maps (since the polarization dependence arises from changes in the neutron scattering length b (expressed in units of a Fermi (1 fm = 10^{-15} m))), amplifying the contribution and visibility of hydrogen atoms in the structure—and quite literally making hydrogen atoms stand out (Fig. 3). The results could be dramatic. For example, for fully polarized protons in a polarized neutron beam, the neutron scattering length of every hydrogen atom could be tuned between maximal values of −18.3 fm (for the spins anti-aligned) to 10.82 fm (for the spins aligned). By comparison, in the normal unpolarized case, the coherent scattering length of hydrogen is just −3.74 fm while that of an isotopically substituted deuterium is 6.6 fm. Moreover, in a DNP-NMC experiment, the location and identity of any hydrogen atom of interest could then be verified and confirmed simply by changing the relative spin polarization in the experiment, which will tune its neutron scattering length accordingly. Further, the ability to vary both the sign and amplitude of polarized scattering in situ, within the same single crystal, could/would open the way to new classes of spin-flipped difference experiments that can exploit this potential 29.12 fm difference between spin aligned and spin anti-aligned conditions (Stuhrmann, 2004). Such experiments can be conceived and implemented at pulsed spallation neutron sources, providing tunable control of hydrogen atom scattering in situ within the same, perfectly isomorphous sample (Fig. 3D).

3. Dynamic nuclear polarization

Polarizing the neutron beam is straightforward (Coulter, McDonald, Happer, Chupp, & Wagshul, 1988; Mezei, 1976), though at the cost of 50% reduction in flux. Polarizing the sample is more challenging. At any given field and temperature, the equilibrium polarization (for spin ½ nuclei like hydrogen) can be simply derived from Eq. (9), where μ is the magnetic moment of the nucleus, B is the magnetic field, T is temperature, and k is Boltzmann's constant.

$$P = \tanh\left(\frac{\mu B}{kT}\right); \tag{9}$$

For a 5 T field at 1 K, the thermodynamic equilibrium polarization of hydrogen nuclei is just ~0.5%, which makes direct polarization using high field, brute force techniques difficult. By contrast, under these same conditions, the spin polarization of unpaired electrons is ~99%. Moreover, the electron spin is strongly coupled to the surrounding nuclei through

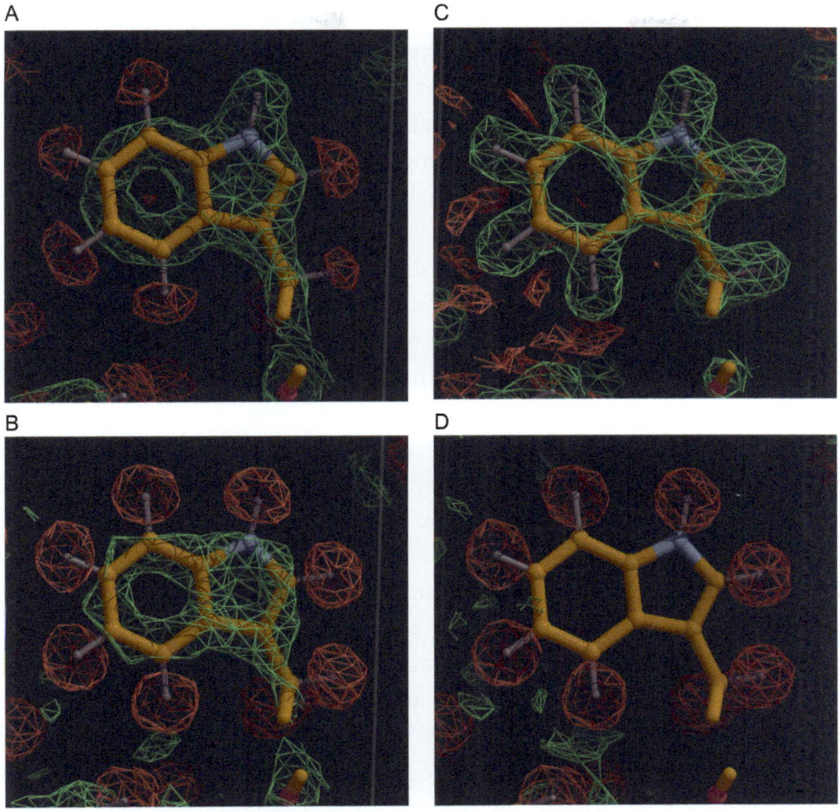

Fig. 3 Hydrogen atoms can be located directly in neutron Fc maps as illustrated for residue Trp-158 in wtT4L, using simulated polarized diffraction data calculated in Phenix (Adams et al., 2011) with polarized neutron scattering length b for spin aligned (10.82 fm) and anti-aligned (−18.3 fm) H atoms. (A) The Fc neutron map of unpolarized T4L, with hydrogenated protein in D_2O solvent, using 10.0- to 1.5-Å data contoured at +1.5 σ (1.12 fermi/Å3). (B) The Fc neutron map of fully polarized, spin anti-aligned T4L using 10.0- to 1.5-Å data, contoured at +1.5 σ (2.41 fermi/Å3). (C) The Fc neutron map of fully polarized, spin aligned T4L using 10.0- to 1.5-Å data, contoured at +1.5 σ (1.71 fermi/Å3). (D) The Fc-FC isomorphous difference map of fully polarized, spin aligned and anti-aligned T4L using 10.0- to 1.5-Å data, contoured at +1.5 σ (3.63 fermi/Å3), showing potential result of a spin-flipped perfect isomorphous difference experiment.

hyperfine coupling. In a DNP experiment, spin polarization is transferred from electrons to surrounding nuclei by irradiating the electron spin transitions with microwave irradiation near the Larmor frequency. Under ideal conditions, spin transfer from high-gyromagnetic ratio (γe) electrons to the surrounding nuclei (γn) can reach a theoretical maximum of $\gamma e/\gamma n > 657$. Polarization then propagates out across other protons in the solvent/sample matrix through proton-proton spin diffusion

(Bloembergen, 1949), spreading polarization out from a paramagnetic center. With well optimized samples, hydrogen polarizations of >90% have been achieved at 5 T and 1 K (Abe et al., 1997).

In practice, spin polarization is transferred and propagated through several mechanisms. The concept, proposed by Overhauser (1953), was demonstrated first in metals (Carver & Slichter, 1953), then in liquids (Carver & Slichter, 1956; Hausser & Stehlik, 1968) and is now rapidly evolving as a transformative technology for solid state and solution NMR and imaging (see Thankamony, Wittmann, Kaushik, & Corzilius, 2017, and references therein). The dominant spin transfer processes (thermal mixing (Borghini, De Boer, & Morimoto, 1974) and the cross-effect (Hwang & Hill, 1967; Wollan, 1976)) are most efficient near the EPR line, which can be shifted and optimized by varying the concentration, gyromagnetic ratio, relaxation properties and type of paramagnetic centers in the sample (because the electron spin-spin interaction changes the EPR spectrum). A variety of radicals with differing spectroscopic properties have been developed and optimized for DNP in the last several decades, most notably for EPR and NMR applications. Under ideal conditions and continuous MW irradiation, all spin ½ nuclei in the sample are polarized, depending upon type and abundance, with hydrogen nuclei usually polarizing the most rapidly. In practice, DNP enhancements are a function of temperature and magnetic field. While enhancements of 10–25 are possible in some materials at room temperature (Afeworki, Mckay, & Schaefer, 1992; Becerra, Gerfen, Temkin, Singel, & Griffin, 1992), DNP experiments are more typically performed at cryogenic temperatures (≤90 K), which extends the nuclear spin-lattice proton relaxation time (T1) and increases the efficiency of spin transfer and propagation through the sample by spin diffusion.

DNP requires a suitable paramagnetic center as a source of unpaired electrons, which can be either intrinsic to the sample, or added to it. In metalloproteins, the presence (or substitution) of an intrinsic metal center that is paramagnetic (e.g., Mn^{2+}, Fe^{2+}, Co^{2+}, Ni^{2+}, Cu^{2+} or most of the lanthanides) can allow direct DNP polarization of the sample. However, in most systems of interest, DNP will typically require addition of an exogenous electron spin ½ label (e.g., stable organic radicals) as a polarization source. Paramagnetic centers, such as the commercially available nitroxide dopant TEMPO and its derivatives, can be co-crystallized or soaked in protein crystals (Tanaka et al., 2018; Zhao et al., 2013), which should be sufficient to distribute centers uniformly through the sample. In practice, most DNP-NMR experiments use partially deuterated solvents to optimize and balance proton relaxation and proton-proton spin diffusion rates across the sample, most typically using a D_2O/H_2O/D-glycerol glassy matrix to propagate polarization (Hu, Song, Yu, Swager, & Griffin, 2008), while also serving as a cryoprotectant. Such protocols are fortuitously well matched to experimental analysis of biological crystals, which typically contain large solvent fractions (35–95%, v/v) and are routinely flash-cooled to liquid nitrogen temperature for crystallographic analysis using similar cryoprotectant mixtures.

4. DNP apparatus and operation

We have completed a prototype DNP system (Pierce et al., 2019) designed for use on the IMAGINE beamline (Meilleur et al., 2013) at the High Flux Isotope Reactor in Oak Ridge National Laboratory. The key components of DNP system are (i) the 5 T magnet; (ii) dilution refrigerator; (iii) 140 GHz microwave system; (iv) 213 MHz NMR system, to measure the sample polarization; and (v) a custom sample cold space. The 5 T magnet field is designed to provide field homogeneity of delta $B/B < 10^{-4}$ T cm^{-1}. The microwave frequency required is a function of the magnetic field, e.g., for a 5 T magnet 2 mm microwaves are needed. The dilution refrigerator system is designed to provide high cooling power at the sample when polarizing at 1 K (to remove the heat from the microwaves) and to provide a base temperature as low as achievable (<100 mK) during data collection, thus ensuring that nuclear spins will be "frozen" in place for an extended period of time. The nuclear polarization is measured by the NMR system. Fig. 4A shows a schematic of the system and mode of operation. Detailed technical design specifications and descriptions of the sample space configuration, the sample loading scheme and polarization protocols are described elsewhere (Pierce et al., 2019). Fig. 4B shows the apparatus assembled and installed on the IMAGINE beamline at HFIR.

Neutron diffractometers for NMC typically operate at long wavelength (2–5 Å) and are equipped with detectors that have a large (>2π) solid angle acceptance, hence the use of large, high field magnets for DNP is problematic, because they occlude large fractions of the detector. In order to match the maximum horizontal (±180°) and vertical (> ±47°) opening angles for diffraction on IMAGINE, the DNP apparatus is designed to operate as a two-step "Frozen Spin Target" system (Bradtke et al., 1999; Niinikoski & Udo, 1976). In step one, a

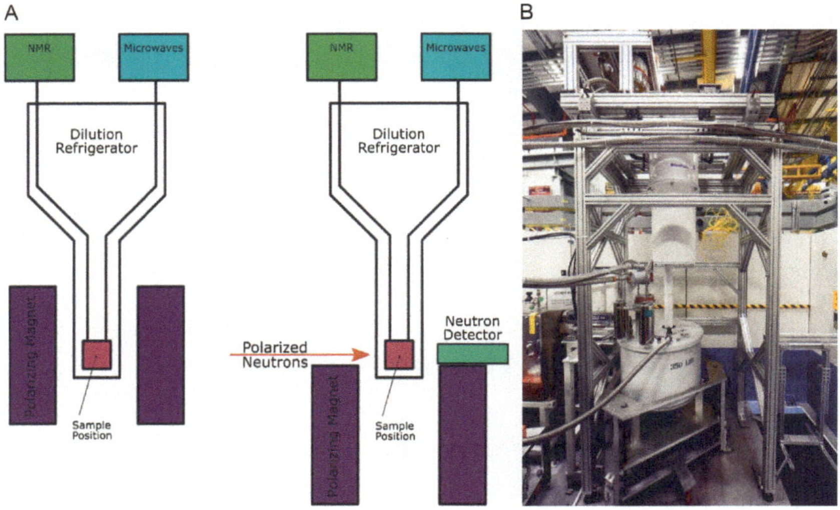

Fig. 4 Schematic of a frozen spin DNP system for NMC. (A) Left: the system in polarizing mode, installed in a 5 Tesla magnet; Right: the system in Frozen spin mode, with sample aligned in the neutron beam. (B) The DNP system installed on the IMAGINE beamline for data taking. NMR is used to determine the degree of polarization.

paramagnetically labeled sample is cooled to 1 K in a 5 T magnetic field and the hydrogen is polarized via DNP. In step two, the sample temperature is reduced as low as possible, which extends the T_1 polarization decay time for data collection. The 5 T magnet is then lowered, allowing the polarized neutron beam to pass through the sample, and the 5 T magnet is positioned to maintain the sample in the 0.5 T fringe field for polarized neutron diffraction.

5. DNP-NMC: Polarized neutron diffraction

Here, we summarize the experimental protocols that have been developed, discuss some of the technical challenges that remain, and highlight the key questions, developments and opportunities ahead. We have performed first DNP enhanced neutron diffraction experiments on a spin polarized protein sample, namely the cysteine-free pseudo wild-type variant of T4 lysozyme—wt*T4L (Matsumura & Matthews, 1989). Expression, purification and crystallization of wt*T4L followed published procedures (Fleissner, Cascio, & Hubbell, 2009). Paramagnetic centers were incorporated into the protein crystals using three approaches: (i) soaking crystals in solutions containing 10–200 mM TEMPOL for between 10 and 150 min; (ii) incubating and co-crystallizing proteins with 10-200 mM TEMPOL in crystallization drops; and (iii) using amino acid mutagenesis to incorporate proxyl-methanethiosulfonate (MTSL) spin-labeled targets at site-specific positions at the surface of the protein, which were selected to avoid perturbation of the structure or of the crystal lattice contacts. The TEMPOL soaked and co-crystallized crystals showed no deleterious effects on the X-ray diffraction quality of the crystals (data not shown). Similarly, the two site-specific MTSL-labeled mutants, covalently labeled at cysteine mutants at positions Lys65 and Val75, diffracted X-rays to similar resolution as wt*T4L crystals (data not shown).

For DNP measurements at ultra-low temperatures of 1 K and less, protein crystals (which contain ~50% solvent, v/v) must be rapidly cooled and maintained below the glass transition temperature (~146 K) to avoid ice formation and disruption of the crystal lattice, which can be problematic for large, neutron size crystals. To achieve this, we used cooling and transfer protocols similar to those previously used for NMC with closed cycle refrigerators (Blakeley, Kalb (Gilboa), Helliwell, & Myles, 2004; Myles, Dauvergne, Blakeley, & Meilleur, 2012), and optimized this approach for the T4L crystals, which enabled a 1.7 Å neutron structure to be determined at <85 K on IMAGINE (Li et al., 2017). For DNP, selected crystals were first flash cooled to 77 K simply by plunging in liquid nitrogen, then transferred under liquid nitrogen into a custom DNP sample holder (machined from neutron-transparent PCTFE (polychlorotrifluoroethylene)), and then rapidly inserted into a pre-cooled (at <100 K) DNP apparatus and aligned in the neutron beam. Once the fridge cooled to base temperature (below 1 K), the magnet was raised to center the sample in the 5 T field. Samples were polarized in both spin aligned and anti-aligned configurations, by changing the frequency of the microwaves, and polarized using MW powers in the range 10–20 mW, typically for 6–8 h. The maximum sample polarization was estimated to be ~50%, with T_1 ranging between 1 and 3 h. Since the relaxation time was rather short, the average in-beam polarization was about half the maximum polarization, and data were collected for 1–2 h at a time. The sample was then repolarized, and the cycle repeated

Polarized neutron diffraction images were collected using the broad bandpass (2.8–10 Å) Laue configuration on IMAGINE. Polarized neutrons were produced using a "V-cavity" supermirror polarizer, installed just after the focusing mirror optics. The measured flipping ratio for the V-cavity was ~56, which corresponds to an average neutron polarization

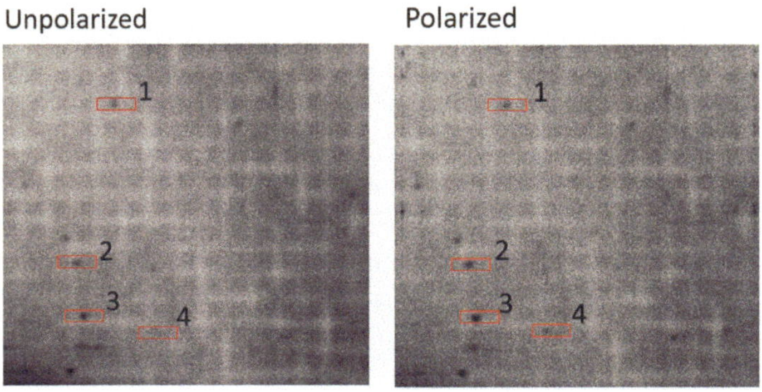

Fig. 5 Polarized Laue neutron diffraction (dλ/λ=2.8–10 Å) collected from a 100 mM TEMPOL soaked ~0.6 mm^3 wtT4L crystal under two conditions; (Left) unpolarized, and (Right) polarized with protons spins aligned antiparallel to the neutron polarization. The number of discernable peaks is doubled on the polarized image, with average enhancements in diffracted intensity of threefold across all reflections.

of >96.5% (or higher) over the entire wavelength band. Diffraction data were recorded on a prototype Anger camera system, mounted at angles between 30° and 60° on a theta rotation stage and set 200 mm from the crystal. Laue diffraction images for a typical polarized crystal experiment are shown in Fig. 5. The ~0.6 mm^3 crystal was soaked in 100 mM TEMPOL for 30 min and flash-cooled in liquid N_2. Data were collected under three limiting conditions; (i) unpolarized; and polarized with protons spins (ii) parallel; and (iii) antiparallel to the neutron polarization. Owing to the small acceptance of the single detector, and limited neutron beam time that was available, we could not collect a complete dataset, but rather focused on optimizing polarization conditions that provided maximal polarization and amplification and modulation of intensity over a subset of reflections at a single, fixed detector position. The measured images show clear, reproducible changes and enhancements in diffraction under polarizing conditions, both in the number and relative intensities of observable reflections (measured using ImageJ software (Schneider, Rasband, & Eliceiri, 2012)). We note the following principle results and observations from these initial, proof of concept experiments:

- The background subtracted reflection intensities show an overall ~threefold average spin-dependent amplification between polarized and unpolarized states, which is consistent with a ~50% initial hydrogen polarization.
- The number of reflections observed in the anti-aligned spin state is almost double that observed in the unpolarized state.

- Reversing the proton spin-polarization in the crystal changes the measured intensities of individual reflections by up to a factor 4, demonstrating in situ tuning and control of the coherent scattering length of hydrogen.

The results are highly encouraging, demonstrating that spin-controlled enhancement of neutron diffraction in paramagnetically labeled protein crystals is both practical and achievable, and that the long-understood potential for DNP techniques to transform NMC may now be within reach

6. Future development and perspectives

The challenges ahead are to improve the initial extent of polarization (toward ~90%) and to maintain high levels of polarization through the time frame of data collection (>hours or days). DNP polarization is primarily affected by the temperature, magnetic field strength and microwave power of the system, and by the radical type and concentration, polarization time and the nature of the sample. Results so far indicate that operating temperatures <100 mK and extended T1 lifetimes >10 h are required, and are within reach using the prototype device. Several modifications in the apparatus will enhance performance, including design of a new sample loading and manipulation system compatible with the low temperature "frozen spin" state of the system, which is expected to extend the spin relaxation time significantly. An alternative concept that should be fully explored, is to design a continuously pumped DNP apparatus that will operate at 1 K and that will maximize and maintain high, steady state levels of polarization throughout the duration of the experiment.

Maximizing sample polarization also requires optimization of the type, nature, spectral properties, concentration and distribution of the paramagnetic centers in the sample. Because protein crystals are porous (and contain large (10–40 Å) solvent-filled channels), simply soaking crystals in mixtures of 1–100 mM paramagnetic spin labels (such as TEMPOL) may serve to distribute sufficient quantities of the spin-label throughout the volume of the crystal. In a (limited) survey of soaking conditions, we find that addition of paramagnetic labels at such concentrations has minimal effect on the diffraction quality of the protein crystal systems screened to date (to be reported). For more precise control, and in special cases, paramagnetic centers such as MSTL can be covalently tagged at site specific positions in target proteins. Several MSTL-labeled samples have been produced, which we intend to use to quantify how polarization at an individual center propagates out and across the neighboring local nuclear structure. Other potentially useful labeling strategies, including use of paramagnetically labeled reagents for DNP-MNC analysis of membrane proteins, and substitution of suitable paramagnetic metal centers in metalloprotein systems of interest, should all be explored. Optimizing these DNP-NMC protocols and reagents are important areas that are ripe for development.

Other practical considerations and some potential complications arise. As in any MNC experiment, success depends upon the intrinsic properties and diffraction quality of the crystal at hand, which may be sensitive to addition of exogenous spin label reagents (akin to heavy atom soaks for MIR phasing). Further, in a DNP-NMC experiment, the paramagnetic center used for polarization may also provide pathways for depolarization and relaxation, such that the nature, type and spectral properties of the radical center itself may need to be matched and optimized for the particular system at hand, while the concentration and spatial distribution of radicals within the protein/solvent matrix all become important experimental variables. The H/D composition and proton concentration within the protein/solvent matrix itself is also important, with specific ratios (typically 10/30/60) of H_2O/D_2O/D-Glycerol reported optimal in many NMR applications.

Moreover, it may also be necessary to enhance and control the through space propagation of polarization within the crystal, and to evaluate how polarization is affected or perturbed by local structure or by alternate relaxation pathways (through other co-factors or metal centers). In such cases, the ability to paramagnetically tag or label individual residues near to a site of specific functional interest may help maximize polarization locally and highlight hydrogen atoms of particular interest. Similarly, if build-up of polarization, or subsequent relaxation, proves spatially or temporally anisotropic and/or dependent on local structure or environment, then proper account of the variation of the neutron scattering factor of individual polarized H atoms may be required in the software packages used for structure refinement and analysis. Otherwise, we envisage that the reduction and analysis of DNP-NMC data will be routinely achieved using currently available software, and that only minor adaption of current industry standard refinement programs (such as Phenix (Afonine et al., 2012); or SHELXL (Gruene, Hahn, Luebben, Meilleur, & Sheldrick, 2013)) will be required to allow for tunable refinement of hydrogen coherent scattering factors in polarized structures, which will likely be similar to current protocols that allow for variable occupancy refinement of H and D in solvent exchanged structures. This latter scenario is most likely to be true for the continuously pumped configuration mentioned above, which will maintain hydrogen polarization in "steady state" throughout the experiment.

Finally, we make the following observations on the potential advantages and future use of DNP-NMC to enhance neutron diffraction analysis of hydrogenated protein crystals if, or when, the remaining technical challenges have been resolved.

(1) DNP offers large 10–100-fold increases in S/N of the data, through either: (i) amplifying the diffracted intensity and minimizing incoherent scattering background when the spins are aligned or and (ii) maximizing the diffracted intensity when the spins are anti-aligned, although accompanied by a ~30% increase in incoherent background noise. (We note that the gains achieved in the spin-aligned condition will be limited by the point at which sample background falls below the intrinsic instrument noise (as is observed for analogous experiments with perdeuterated crystals).)

(2) Gaining tunable control of the scattering length of hydrogen enables the visibility of hydrogen atoms in Fourier maps to be amplified and enhanced (Fig. 3), making hydrogens stand out. Moreover, these hydrogens are then verifiable, by using spin flipping to switch hydrogen between maximal positive and negative scattering contributions within the same protein crystal.

(3) The ability to tune and control the spin-dependent coherent scattering of hydrogen in situ creates new opportunities for spin-dependent measurement and analysis. For example, spin polarized neutron diffraction presents a path to a new class of perfectly isomorphous difference experiments in which the scattering length of hydrogen can be varied across the range from −18.3 to +10.8 fm, a net eightfold amplification in the scattering signal of hydrogen (see Fig. 3D). This suggests new ways to collect and analyze neutron diffraction data from hydrogenous crystals, especially at pulsed neutron sources, and could open up new fields of neutron research and discovery in the decade ahead.

7. Conclusion

The proof of concept DNP-NMC experiments and approaches described here are foundational and are a first step toward order-of-magnitude enhancements in the neutron analysis of biological materials. Harnessing this new technology in neutron crystallography would remove the intrinsic limitations of sample background limited diffraction while simultaneously enabling the scattering from—and visibility of—hydrogen atoms to be tuned and maximized in biological materials. This would breakthrough a critical threshold in application of neutrons in biology, by enabling DNP-NMC analysis of radically smaller crystals ($0.01\,mm^3$) that are similar to those routinely produced for X-ray diffraction experiments today. Moreover, the ability to tune and control the scattering from hydrogen, in situ, within a single, perfectly isomorphous sample suggests new experimental approaches for data collection, analysis and phasing at next generation facilities.

Acknowledgments

The authors would like to acknowledge the help of the IMAGINE instrument team and support provided by the HFIR in the installation of the DNP system on the IMAGINE beam-line. The Neutron Sciences Directorate Detector group provided the prototype detector and assisted with the data acquisition. The Jefferson Lab Target Group offered insight into the design of the sample space. Research was sponsored by the Laboratory Directed Research and Development Program of Oak Ridge National Laboratory, managed by UT-Battelle, LLC, for the U.S. Department of Energy under Contract No. DE-AC05-00OR22725. Research conducted at ORNL's High-Flux Isotope Reactor was sponsored by the US Department of Energy's Office of Basic Energy Sciences, Scientific User Facilities Division. The construction and installation of the IMAGINE beam line was partly supported by NSF grant CHE-0922719.

References

Abe, K., Akagi, T., Anderson, B. D., Anthony, P. L., Arnold, R. G., Averett, T., et al. (1997). Precision determination of the neutron spin structure function $g1^n$. *Physical Review Letters*, *79*, 26.

Abragam, A. (1961). *The principles of nuclear magnetism*. Oxford University Press.

Abragam, A., & Goldman, M. (1982). *Nuclear magnetism*. Clarendon Press.

Adams, P. D., Afonine, P. V., Bunkóczi, G., Chen, V. B., Echols, N., Headd, J. J., et al. (2011). The Phenix software for automated determination of macromolecular structures. *Methods*, *55*, 94–106.

Afeworki, M., Mckay, R. A., & Schaefer, J. (1992). Selective observation of the interface of heterogeneous polycarbonate/polystyrene blends by dynamic nuclear polarization carbon-13 NMR spectroscopy. *Macromolecules*, *25*, 4084–4091.

Afonine, P. V., Grosse-Kunstleve, R. W., Echols, N., Headd, J. J., Moriarty, N. W., Mustyakimov, M., et al. (2012). Towards automated crystallographic structure refinement with phenix.refine. *Acta Crystallographica. Section D, Biological Crystallography*, ***D68***, 352–367.

Becerra, L. R., Gerfen, G. J., Temkin, R. J., Singel, D. J., & Griffin, R. G. (1992). Dynamic nuclear polarization with a cyclotron resonance maser at 5 T. *Physical Review Letters*, *71*, 3561–3564.

Blakeley, M. P. (2009). Neutron macromolecular crystallography. *Crystallography Reviews*, *15*(3), 157–218.

Blakeley, M. P., Kalb (Gilboa), A. J., Helliwell, J. R., & Myles, D. A. A. (2004). The 15-K neutron structure of saccharide-free concanavalin A. *Proceedings of the National Academy of Sciences*, *101*, 16405–16410.

Blakeley, M. P., Ruiz, F., Cachau, R., Hazemann, L., Meilleur, F., Mitschler, A., et al. (2008). Quantum model of catalysis based on a mobile proton revealed by subatomic x-ray and neutron diffraction studies of h-aldose reductase. *Proceedings of the National Academy of Sciences of the United States of America*, *105*, 1844–1848.

Bloembergen, N. (1949). On the interaction of nuclear spins in a crystalline lattice. *Physica (Utrecht)*, *15*, 386–426.

Borghini, M., De Boer, W., & Morimoto, K. (1974). Nuclear dynamic polarization by resolved solid-state effect and thermal mixing with an electron spin-spin interaction reservoir. *Physics Letters A*, *48*, 244–246.

Bradtke, C., Dutz, H., Peschel, H., Goertz, S., Harmsen, J., Hasegawa, S., et al. (1999). A new frozen-spin target for 4π particle detection. *Nuclear Instruments and Methods in Physics Research Section A: Accelerators, Spectrometers, Detectors and Associated Equipment*, *436*, 430–442.

Carver, T. R., & Slichter, C. P. (1953). Polarization of nuclear spins in metals. *Physics Review*, *92*, 212–213.

Carver, T. R., & Slichter, C. P. (1956). Experimental verification of the overhauser nuclear polarization effect. *Physics Review*, *102*, 975–980.

Casadei, C. M., Gumiero, A., Metcalfe, C. L., Murphy, E. J., Basran, J., Concilio, M. G., et al. (2014). Neutron cryo-crystallography captures the protonation state of ferryl heme in a peroxidase. *Science*, *345*, 193–197.

Coulter, K. P., McDonald, A. B., Happer, W., Chupp, T. E., & Wagshul, M. E. (1988). Neutron polarization with polarized ^{3}He. *Nuclear Instruments and Methods in Physics Research Section A: Accelerators, Spectrometers, Detectors and Associated Equipment*, *270*, 90–94.

Fleissner, M. R., Cascio, D., & Hubbell, W. L. (2009). Structural origin of weakly ordered nitroxide motion in spin-labeled proteins. *Protein Science*, *18*, 893–908.

Gruene, T., Hahn, H. W., Luebben, A. V., Meilleur, F., & Sheldrick, G. M. (2013). Refinement of macromolecular structures against neutron data with SHELXL2013. *Journal of Applied Crystallography*, *47*, 462–466.

Hausser, K. H., & Stehlik, D. (1968). In J. S. Waugh (Ed.), *3. Advances in magnetic resonance* (pp. 79–139): Academic Press.

Hayter, J. B., Jenkin, G. T., & White, J. W. (1974). Polarized-neutron diffraction from spin-polarized protons: A tool in structure determination? *Physical Review Letters*, *33*, 696–699.

Hu, K.-N., Song, C., Yu, H.-h., Swager, T. M., & Griffin, R. G. (2008). High-frequency dynamic nuclear polarization using biradicals: A multifrequency EPR lineshape analysis. *The Journal of Chemical Physics*, *128*, 052321.

Hwang, C. F., & Hill, D. (1967). New effect in dynamic polarization. *Physical Review Letters*, *18*, 110.

Kumar, P., Agarwal, P. K., Waddell, M. B., Mittag, T., Serpersu, E. H., & Cuneo, M. J. (2019). Low-barrier and canonical hydrogen bonds modulate activity and specificity of a catalytic triad. *Angewante Chemie International Edition*, *131*, 16406–16412.

Li, L., Shukla, S., Meilleur, F., Standaert, R. F., Pierce, J., Myles, D. A. A., et al. (2017). Neutron crystallographic studies of T4 lysozyme at cryogenic temperature. *Protein Science*, *26*, 2098–2104.

Matsumura, M., & Matthews, B. W. (1989). Control of enzyme activity by an engineered disulfide bond. *Science*, *243*, 792–794.

Meilleur, F., Munshi, P., Kovalevsky, A., Koritsanszky, T., Chakoumakos, B., Blessing, R., et al. (2013). The IMAGINE instrument: First neutron protein structure and new capabilities for neutron macromolecular crystallography. *Acta Crystallographica Section D*, *69*, 2157–2160.

Meilleur, F., Weiss, K. L., & Myles, D. A. A. (2009). Methods in molecular biology. In J. W. Lee & R. S. Foote (Eds.), *Vol. 544. Micro and nanotechnologies in separation, detection and analysis of biomolecules* (pp. 281–292). New York, NY: Springer.

Mezei, F. (1976). Novel polarized neutron devices: supermirror and spin component amplifier. *Communications on Physics (London)*, *1*, 81–85.

Myles, D. A. A., Dauvergne, F., Blakeley, M. P., & Meilleur, F. (2012). Neutron protein crystallography at ultra-low (< 15 K) temperatures. *Journal of Applied Crystallography*, *45*, 686–692.

Niinikoski, T., & Udo, F. (1976). "Frozen spin" polarized target. *Nuclear Instruments and Methods*, *134*, 219–233.

O'Dell, W. B., Agarwal, P. K., & Meilleur, F. (2017). Oxygen activation at the active site of a fungal lytic polysaccharide monooxygenase. *Angewandte Chemie (International Ed. in English)*, *56*(3), 767–770.

O'Dell, W. B., Bodenheimer, A. M., & Meilleur, F. (2016). Neutron protein crystallography: A complementary tool for locating hydrogens in proteins. *Archives of Biochemistry and Biophysics*, *602*, 48–60.

Overhauser, A. W. (1953). Polarization of nuclei in metals. *Physics Review*, *92*, 411–415.

Piegsa, F. M., Karlsson, M., Brandt, B., Carlile, C. J., Forgan, E. M., Hautle, P., et al. (2013). Polarized neutron Laue diffraction on a crystal containing dynamically polarized proton spins. *Journal of Applied Crystallography*, *46*, 30–34.

Pierce, J., Crow, L., Cuneo, M. J., Edwards, M., Herwig, K. W., Jennings, A. R., et al. (2019). A prototype system for dynamically polarized neutron protein crystallography. *Nuclear Instruments and Methods in Physics Research Section A*, *940*, 430–434.

Schiebel, J., Gaspari, R., Wulsdorf, T., Ngo, K., Sohn, C., Schrader, T. E., et al. (2018). Intriguing role of water in protein-ligand binding studied by neutron crystallography on trypsin complexes. *Nature Communications*, *9*, 3559.

Schneider, C. A., Rasband, W. S., & Eliceiri, K. W. (2012). NIH Image to ImageJ: 25 years of image analysis. *Nature Methods*, *9*(7), 671–675.

Schultz, A. J., Thiyagarajan, P., Hodges, J. P., Rehm, C., Myles, D. A. A., Langan, P., et al. (2005). Conceptual design of a macromolecular neutron diffractometer (MaNDi) for the SNS. *Journal of Applied Crystallography*, *38*, 964–974.

Shu, F., Ramakrishnan, V., & Schoenborn, B. P. (2000). Enhanced visibility of hydrogen atoms by neutron crystallography on fully deuterated myoglobin. *Proceedings of the National Academy of Sciences of the United States of America*, *97*, 3872–3877.

Stuhrmann, H. B. (2004). Unique aspects of neutron scattering for the study of biological systems. *Reports on Progress in Physics*, *67*, 1073–1115.

Stuhrmann, H. B., Scharpf, O., Krumpolc, M., Niinikoski, T. O., Rieubland, M., & Rijlart, A. (1986). Dynamic nuclear polarization of biological matter. *European Biophysics Journal*, *14*, 1–6.

Tanaka, I., Komatsuzaki, N., Yue, W. X., Chatake, T., Kusaka, K., Niimura, N., et al. (2018). Cryoprotectant-free high-pressure cooling and dynamic nuclear polarization for more sensitive detection of hydrogen in neutron protein crystallography. *Acta Crystallographica Section D*, *74*, 787–791.

Tanaka, I., Kusaka, K., Chatake, T., & Niimura, N. (2013). Fundamental studies for the proton polarization technique in neutron protein crystallography. *Journal of Synchrotron Radiation*, *20*, 958–961.

Thankamony, A. S. L., Wittmann, J. J., Kaushik, M., & Corzilius, B. (2017). Dynamic nuclear polarization for sensitivity enhancement in modern solid-state NMR. *Progress in Nuclear Magnetic Resonance Spectroscopy*, *102–103*, 120–195.

Wollan, D. S. (1976). Dynamic nuclear polarization with an inhomogeneously broadened ESR line. I. Theory. *Physical Review B, 13*, 3671.
Zhao, J., Meerwinck, W., Niinikoski, T., Rijllart, A., Schmitt, M., Willumeit, R., et al. (1995). The polarized target station at GKSS. *Nuclear Instruments and Methods in Physics Research Section A: Accelerators, Spectrometers, Detectors and Associated Equipment, 356*, 133–137.
Zhao, J. K., Robertson, L., Herwig, K., & Crabb, D. (2013). Polarized neutron in structural biology–present and future outlook. *Physics Procedia, 42*, 39–45.
Zimmer, O., Jouve, H. M., & Stuhrmann, H. B. (2016). Polarized proton spin density images the tyrosyl radical locations in bovine liver catalase. *IUCrJ, 3*, 326–340.

Further reading

Berman, H. M., Westbrook, J., Feng, Z., Gilliland, G., Bhat, T. N., Weissig, H., et al. (2000). The protein data bank. *Nucleic Acids Research, 28*, 235.
Matthews, B. W., Dahlquist, F. W., & Maynard, A. Y. (1973). Crystallographic data for lysozyme from bacteriophage T4. *Journal of Molecular Biology, 78*, 575–576.
Schröder, G. C., O'Dell, W. B., Myles, D. A. A., Kovalevsky, A., & Meilleur, F. (2018). IMAGINE: Neutrons reveal enzyme chemistry. *Acta Crystallographica Section D, Biological Crystallography, D74*, 778–786.
Stuhrmann, H. B., Burkhardt, N., Dietrich, G., Jünemann, R., Meerwinck, W., Schmitt, M., et al. (1995). Proton-and deuteron spin targets in biological structure research. *Nuclear Instruments and Methods in Physics Research Section A: Accelerators, Spectrometers, Detectors and Associated Equipment, 356*, 124–132.
Zhao, J. K. (2005). Dynamically polarized hydrogenous materials for neutron scattering. *Physica B: Condensed Matter, 356*, 168–173.
Zhao, J., & Stuhrmann, H. B. (1993). The in situ structure of the L3 and L4 proteins of the large subunit of E. coli ribosomes as determined by nuclear spin contrast variation. *Journal de Physique IV, 3*, 233–236.

CHAPTER NINE

Implementation of the riding hydrogen model in *CCTBX* to support the next generation of X-ray and neutron joint refinement in *Phenix*

Dorothee Liebschner[a,*], Pavel V. Afonine[a], Alexandre G. Urzhumtsev[b,c], Paul D. Adams[a,d]

[a]Molecular Biophysics and Integrated Bioimaging Division, Lawrence Berkeley National Laboratory (LBNL), Berkeley, CA, United States
[b]Centre for Integrative Biology, Institut de Génétique et de Biologie Moléculaire et Cellulaire, Illkirch, France
[c]Faculté des Sciences et Technologies, Université de Lorraine, Nancy, France
[d]Department of Bioengineering, University of California Berkeley, Berkeley, CA, United States
[*]Corresponding author: e-mail address: dcliebschner@lbl.gov

Contents

Methods in Enzymology, Volume 634
ISSN 0076-6879
https://doi.org/10.1016/bs.mie.2020.01.007

Abstract

A fundamental prerequisite for implementing new procedures of atomic model refinement against neutron diffraction data is the efficient handling of hydrogen atoms. The riding hydrogen model, which constrains hydrogen atom parameters to those of the non-hydrogen atoms, is a plausible parameterization for refinements. This work describes the implementation of the riding hydrogen model in the *Computational Crystallography Toolbox* and in *Phenix*. Riding hydrogen atoms can be found in several different configurations that are characterized by specific geometries. For each configuration, the hydrogen atom parameterization and the expressions for the gradients of refinement target function with respect to non-hydrogen parameters are described.

1. Introduction

X-ray crystallography is the leading method for obtaining three-dimensional structures of macromolecules. Oftentimes, hydrogen (H) atoms play an important biological role and may be the subject of a structural study, in particular when trying to understand enzyme mechanism. Accurate determination of the location of key H atoms therefore requires experimental information, but as hydrogen has a weak X-ray scattering cross section, its electron density is not observed unless very high quality, and very high-resolution data are available. Even then, it is not guaranteed that all H atoms will be observed: Petrova and Podjarny (2004) showed that only about half of the hydrogen atom positions can be determined experimentally with high resolution electron density maps of macromolecules. By contrast, neutron crystallography allows the observation of hydrogen atoms (or deuterium substitutes) at resolutions better than 3 Å (Ostermann, Tanaka, Engler, Niimura, & Parak, 2002; Shu, Ramakrishnan, & Schoenborn, 2000).

The determination of macromolecular structures with neutron diffraction has two major complications. First, since H atoms have a similar scattering contribution as other atoms, they need to be refined individually in the atomic model. Also, as a result of partial deuteration, some H atom positions are shared with deuterium atoms (D) meaning that their relative occupancies need to be refined. Given that H atoms constitute about 50% of the atoms in a typical macromolecular structure, refining them individually adds a substantial number of refined parameters. Second, the experimental data for neutron diffraction data are often poorer than those of corresponding X-ray data: the low flux of available neutron beams and incoherent scattering of hydrogen (if present) lead to low signal-to-noise ratios, and the limited data-collection time on oversubscribed instruments result in low completeness, averaging about 80% (Liebschner, Afonine, Moriarty, Langan, & Adams, 2018). In contrast, the completeness of a typical X-ray dataset is expected to be greater than 95% (Dauter & Dauter, 2017).

With rare exceptions, X-ray data and the corresponding atomic models are almost always available prior to structure determination with neutrons. The availability of two sources of information, namely X-ray and neutron data, may help alleviate some of the aforementioned challenges of refinement against neutron data: the joint X-ray and neutron (joint XN) refinement method enables building more complete atomic models that include experimentally observed hydrogen and non-hydrogen atoms (Adams, Mustyakimov, Afonine, & Langan, 2009; Afonine et al., 2010; Coppens, 1967; Wlodawer, 1980; Wlodawer & Hendrickson, 1982; Wlodawer, Miller, & Sjölin, 1983). The currently used joint XN refinement method optimizes a single atomic model simultaneously against two datasets using a combined refinement target function:

$$T = w_x \bullet T_{X-\text{ray}} + w_n \bullet T_{\text{neutron}} + T_{\text{restraints}} \tag{1}$$

where $T_{\text{X-ray}}$ and T_{neutron} are target functions relating the model and respective data, $T_{\text{restraints}}$ is a restraint term that adds a priori information, and w_{x} and w_{n} are empirical scale factors.

Using the refinement target (1) with a single atomic model relies on the assumption that the X-ray and neutron crystals and respective structures are isomorphous. This hypothesis is only approximate, because:

(a) If different crystals are used for the neutron and X-ray experiments, the unit cell parameters of the crystals may be different.

(b) Ordered and semi-ordered parts of the structures may be different.

(c) X-rays are scattered by electrons while neutrons are scattered by atomic nuclei. The datasets therefore convey different information about covalent X-H bond lengths: the electron distribution of the H atom, which has only one valence electron and no core electron, is shifted along the X-H bond away from the H atom nucleus toward atom X. Accordingly, the locations of H atom electron density peaks and nuclear scattering length density peaks are different and the X-H bond lengths appear about 10–20% shorter for X-rays (Allen, 1986; Allen & Bruno, 2010).

(d) The X-ray and neutron datasets may be collected under different conditions. For example, X-ray data are typically collected at cryogenic temperatures to mitigate radiation damage while neutron diffraction experiments are carried out at room temperature because neutrons do not damage the sample. Atomic models of structures derived from room-temperature experiments are likely to exhibit more variability due to thermal motion (Fraser et al., 2011).

(e) Data collection times are very different: X-ray data can be collected in minutes at synchrotron beamlines (Turkenburg & McAuley, 2013) while neutron data collections typically last days or weeks (Chen & Unkefer, 2017; Coates et al., 2015). Atomic models derived from diffraction data are averaged over space (unit cells of the crystal) and time (duration of the data collection). It may be that data collected over substantially different periods of time result in different models.

Analysis of atomic models derived from joint XN refinement (see Chapter "What are the current limits on determination of protonation state using neutron macromolecular crystallography?" by Liebschner et al.) shows that in practice there often are differences between the neutron and X-ray datasets (for the reasons just described). Therefore, it is desirable for a joint XN refinement algorithm to yield two models that are related to each other and contain H atoms; each model should fit their respective dataset and reflect its properties while at the same time it should benefit from the complementarity of the X-ray and neutron data. Implementation details of such an approach will be described elsewhere. In this chapter we describe a fundamental prerequisite for implementing such a procedure in the software package *Phenix* (Liebschner et al., 2019), namely efficient handling of H atoms.

The contribution of H atoms can be modeled in several ways. One possibility is to include the H atoms explicitly in the input model and to refine them independently, similar to the other non-H atoms. Alternatively, while also being modeled explicitly, their parameters can be constrained by those of the non-H atoms. This approach does not increase the number of parameters even when the model contains H atoms. This is because only the positions of the non-H atoms are refined explicitly while the positions of H atoms are recalculated based on geometric constraints. This riding hydrogen model (Busing & Levy, 1964; Sheldrick & Schneider, 1997) is almost universally used in X-ray refinements. It may also be a plausible parameterization for refinements against neutron data (Gruene, Hahn, Luebben, Meilleur, & Sheldrick, 2014), especially for low-resolution or poorly complete datasets. It is also conceivable to mix different hydrogen modeling approaches, such as refining H atoms with geometric degrees of freedom individually while treating the remaining H atoms as riding.

Below we describe algorithmic details of the implementation of the riding hydrogen model in the Computational Crystallography Toolbox (*CCTBX*; Grosse-Kunstleve, Sauter, Moriarty, & Adams, 2002) and in *Phenix*.

2. Parameterizing the riding hydrogen atom model for typical geometrical configurations

2.1 Riding H definition

The concept of riding hydrogen model ('riding H' in what follows) relies on the fact that the coordinates of most (but not all) H atoms in proteins can be unambiguously expressed through the coordinates of their covalently connected non-H atoms. To define the riding H atom position, at least three neighboring non-H atoms are necessary. These are typically the 'parent' atom (A_0), to which the H atom is covalently bound, and two other atoms, that we note by A_1 and A_2, bound to the parent atom (neighbors). An example configuration of a hydrogen atom is shown in Fig. 1 (left). In this example, the H atom is in a plane with A_0, A_1 and A_2. This geometry can be found in the peptide unit or in aromatic rings.

Riding H atoms can be found in several different configurations; for example, three non-H atoms may define simultaneously several H atoms; an example is the Hγ11-Hγ12-Hγ13 group in valine residue; all of its H atoms are bound to the Cγ atom. Other typical examples are shown in Fig. 2. In each configuration, the number of neighbors and their ideal bond and angle values may vary. The algorithmic challenge is to define the coordinates of the H atom from the coordinates of a group of its non-H neighbors for each configuration. It is desirable that the algorithm is independent of eventual imperfections in the geometry of non-H atoms as much as possible.

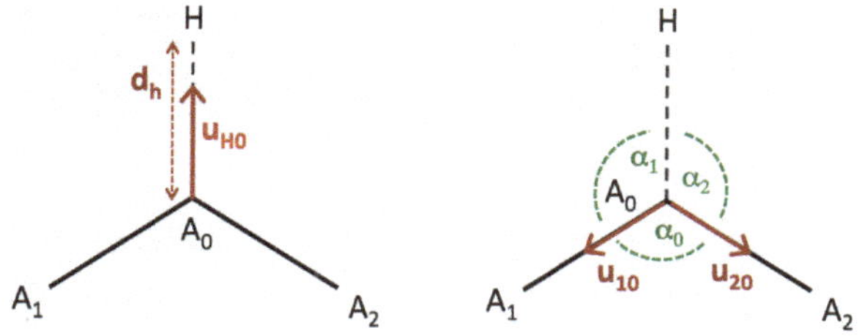

Fig. 1 Example of a hydrogen atom H in a coplanar configuration. To improve visibility the parameters are shown in two figures. H is covalently bound to its parent atom A_0, the neighbors are A_1 and A_2.

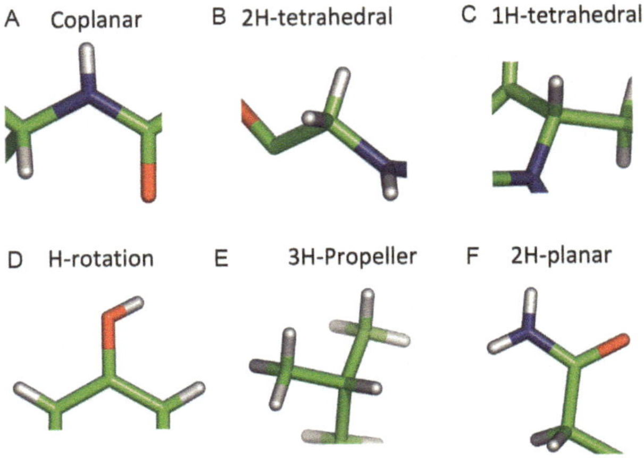

Fig. 2 Examples of hydrogen atom configurations. While the examples shown in the figure refer to amino acid residues, the implemented riding-H algorithm is general and not specific to proteins only.

For each configuration the H atom position is parameterized using coordinates of (at least) three non-H neighbors and by using $\vec{r}_H = \vec{r}_0 + \vec{r}_{H0} = \vec{r}_0 + d_H \bullet \vec{u}_{H0}$, where $\vec{u}_{H0}$ is the unit vector along the A_0-H direction $\vec{r}_{H0}$ and d_H is the ideal H-A_0 bond length (Fig. 1). Since d_H is known for each kind of models (and is different for the X-ray and neutron data), the main task is to define $\vec{u}_{H0}$ from the set of non-H neighbors and ideal bond lengths and angles. A chain of transformations is developed for all possible configurations. Each step passes from a set of input parameters to a set of output parameters. This way, the first step in the chain of transformations uses the coordinates of atoms parameterizing the H atom and respective ideal bond lengths and angles. The last step of the chain of transformations provides $\vec{u}_{H0}$. This way, it is easier to apply the chain rule for gradient calculations (see Section 3 and Appendix) and the procedure can be carried out in the most efficient way (Lunin & Urzhumtsev, 1985).

We note that configurations described below are general and not specific to protein residues only.

2.2 Coplanar configuration

The first configuration describes the H atom linked to the N atom in the peptide unit (Fig. 2A) or the H atoms in aromatic rings. The theoretical configuration is represented in Fig. 1. In the ideal conformation, the unit vectors in the directions A_0-A_1 ($\vec{u}_{10}$) and A_0-A_2 ($\vec{u}_{20}$) form the angle α_0. Being coplanar, the unit vector $\vec{u}_{H0}$ can be expressed through these two vectors as $\vec{u}_{H0} = a \bullet \vec{u}_{10} + b \bullet \vec{u}_{20}$ forms the ideal angles α_1 and α_2 with $\vec{u}_{10}$ and $\vec{u}_{20}$, respectively. We note $c_0 = \cos\alpha_0$, $c_1 = \cos\alpha_1$ and $c_2 = \cos\alpha_2$.

The coefficients a and b can be found from the conditions $a + b \bullet c_0 = c_1$, $a \bullet c_0 + b = c_2$ that give an analytic solution $a = (1 - c_0^2)^{-1}(c_1 - c_0 \bullet c_2)$, $b = (1 - c_0^2)^{-1}(c_2 - c_0 \bullet c_1)$. Thus, the values of parameters a and b for the ideal coplanar conformation can be calculated in advance.

The chain of transformations to obtain the H-atom position is the following:

(a) $\vec{r}_{10} = \vec{r}_1 - \vec{r}_0$, $\vec{r}_{20} = \vec{r}_2 - \vec{r}_0$
(b) $\vec{u}_{10}$ *is* normalized $\vec{r}_{10}$, $\vec{u}_{20}$ *is* normalized $\vec{r}_{20}$
(c) $\vec{r}_{H0} = a \bullet \vec{u}_{10} + b \bullet \vec{u}_{20}$
(d) $\vec{u}_{H0} = \|\vec{r}_{H0}\|^{-1}\vec{r}_{H0}$ with $\|\vec{r}_{H0}\| = (\vec{r}_{H0} \bullet \vec{r}_{H0})^{1/2}$

This procedure positions the H atom in the plane and at a correct distance to its parent atom. We note that the input model does not necessarily possess an ideal geometry, which means that the angle between A_0-A_1 and A_0-A_2 is not necessarily equal to the ideal value α_0, and similarly for α_1 and α_2. However, the riding procedure uses the ideal angle values to calculate the coefficients a and b. This ensures that the H atom geometry is regularized to match the ideal values.

In summary, this configuration requires the following parameters: A_0, A_1, A_2, d_H, a, b.

2.3 2H-tetrahedral configuration

Examples of this configuration are the Hα1 and Hα2 atoms bound to the glycine Cα atom (Fig. 2B). The configuration is represented schematically in Fig. 3. The atoms H_1 and H_2 are at the given distance d_H from atom A_0.

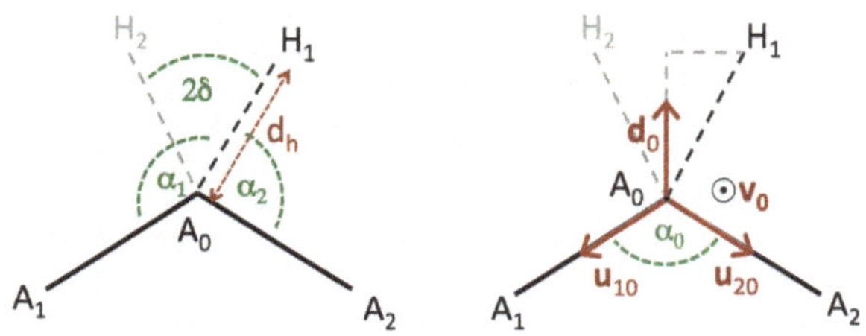

Fig. 3 H-atoms in 2H-tetrahedral configuration. To improve visibility the parameters are shown on two figures. In the ideal configurations, the positions of H_1 and H_2 are symmetric with respect to the plane A_1-A_0-A_2. The vector d_0 is bisector of the angle H_1-A_0-H_2 and is in the plane A_1-A_0-A_2.

In the ideal configuration, they are symmetric with respect to the plane A_1-A_0-A_2 and form an angle H_1-A_0-H_2 equal to 2δ. The unit vector $\vec{u}_{H1}$ in the direction A_0-H_1 forms angles α_1 and α_2 with A_0-A_1 ($\vec{u}_{10}$) and A_0-A_2 ($\vec{u}_{20}$), respectively. If the coefficients a and b are defined as in Section 2.2, the vector $\vec{d} = a \bullet \vec{u}_{10} + b \bullet \vec{u}_{20}$ is collinear with the orthogonal projection of A_0-H_1 and A_0-H_2 onto the plane A_1-A_0-A_2.

Steps (a) and (b) are defined as in Section 2.2. The chain of additional transformations is as follows:

(c) $\vec{v} = \vec{u}_{10} \times \vec{u}_{20}$

(d) $\vec{v}_0$ *is* normalized $\vec{v}$

(e) $\vec{d} = a \bullet \vec{u}_{10} + b \bullet \vec{u}_{20}$

(f) $\vec{d}_0$ *is* normalized $\vec{d}$

(g) $\vec{u}_{H1} = \cos\delta \bullet \vec{d}_0 + \sin\delta \bullet \vec{v}_0$, $\vec{u}_{H2} = \cos\delta \bullet \vec{d}_0 - \sin\delta \bullet \vec{v}_0$

In summary, this configuration requires the following parameters: A_0, A_1, A_2, d_H, a, b, δ.

2.4 1H-tetrahedral configuration

An example is the H_α atom bound to the Cα atom of non-glycine residues (Fig. 2C). The configuration is represented schematically in Fig. 4. The H atom forms a tetrahedron with A_1, A_2 and A_3. The unit vectors $\vec{u}_{10}$, $\vec{u}_{20}$ and $\vec{u}_{30}$ are defined along the directions A_0-A_1, A_0-A_2 and A_0-A_3. The vector $\vec{r}_{H0}$ forms angles α_1, α_2, α_3 with the vectors $\vec{u}_{10}$, $\vec{u}_{20}$ and $\vec{u}_{30}$, respectively. $\vec{r}_{H0}$ can be expressed as a linear combination of these vectors, $\vec{r}_{H0} = a \bullet \vec{u}_{10} + b \bullet \vec{u}_{20} + c \bullet \vec{u}_{30}$. The coefficients can be found from the conditions $\vec{r}_{H0} \bullet \vec{u}_{10} = c_1$, $\vec{r}_{H0} \bullet \vec{u}_{20} = c_2$, $\vec{r}_{H0} \bullet \vec{u}_{30} = c_3$ where $c_1 = \cos\alpha_1$,

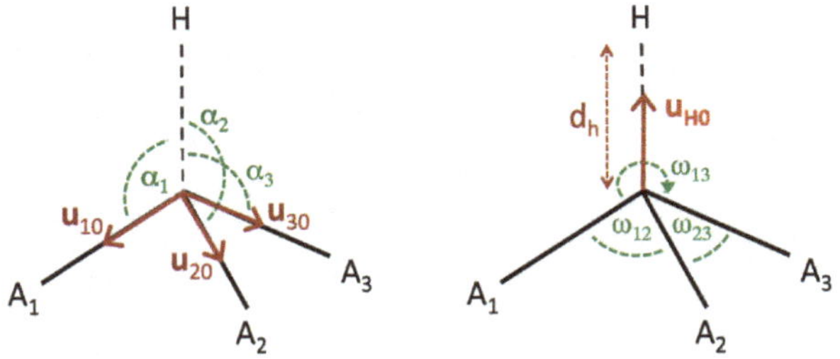

Fig. 4 H atom in a 1H-tetrahedral position. The vector A_0-H makes angles α_1, α_2, α_3 with vectors A_0-A_1, A_0-A_2, A_0-A_3, which form angles ω_{12}, ω_{13}, ω_{23} between themselves.

$c_2 = \cos\alpha_2$, $c_3 = \cos\alpha_3$. If ω_{12}, ω_{23} and ω_{13} are angles between respective pairs of $\vec{u}_{10}$, $\vec{u}_{20}$ and $\vec{u}_{30}$, then these conditions become a system of linear equations,

$$\begin{cases} a + \cos\omega_{12}\bullet b + \cos\omega_{13}\bullet c = c_1 \\ \cos\omega_{12}\bullet a + b + \cos\omega_{23}\bullet c = c_2 \\ \cos\omega_{13}\bullet a + \cos\omega_{23}\bullet b + c = c_3 \end{cases}$$

The system can be solved analytically (Cramer, 1750; Kosinski, 2001), yielding the parameters a, b and c that can be calculated using ideal angles α_i and ω_{ij}.

This geometry is an example where four non-H atoms are used to parameterize the H atom. In principle, A_0 and any two atoms of A_1, A_2, A_3 are sufficient to determine the H atom position. However, if the conformation of the tetrahedron is distorted, this can lead to suboptimal H atom coordinates. The use of all three neighbors is therefore preferable.

The chain of transformations is the following:

(a) $\vec{r}_{10} = \vec{r}_1 - \vec{r}_0$, $\vec{r}_{20} = \vec{r}_2 - \vec{r}_0$, $\vec{r}_{30} = \vec{r}_3 - \vec{r}_0$
(b) $\vec{u}_{10}, \vec{u}_{20}, \vec{u}_{30}$ are normalized vectors $\vec{r}_{10}$, $\vec{r}_{20}$, $\vec{r}_{30}$, respectively
(c) $\vec{r}_{H0} = a\bullet\vec{u}_{10} + b\bullet\vec{u}_{20} + c\bullet\vec{u}_{30}$
(d) $\vec{u}_{H0} = \|\vec{r}_{H0}\|^{-1}\vec{r}_{H0}$ with $\|\vec{r}_{H0}\| = (\vec{r}_{H0}\bullet\vec{r}_{H0})^{1/2}$

In summary, this configuration requires the following parameters: A_0, A_1, A_2, A_3, d_H, a, b, c

2.5 Single H with rotational degree of freedom

An example is the Hη atom bound to the Oη atom of tyrosine (Fig. 2D), represented schematically in Fig. 5. The ideal angle between A_0-H and

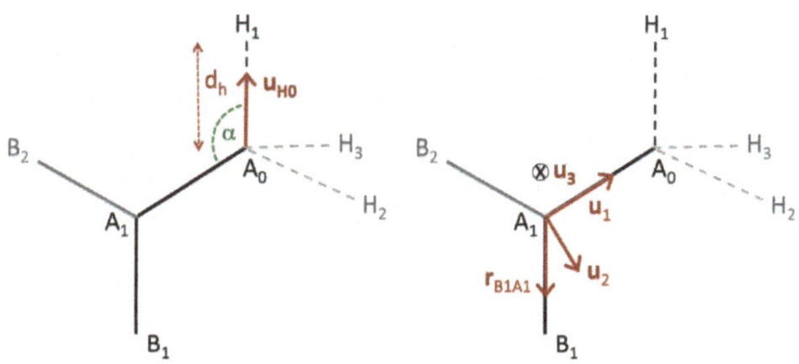

Fig. 5 H-atom in the free-rotation configuration. Alternative positions are indicated with H_2 and H_3. Only one atom among B_1 and B_2 defines a dihedral angle with A_1-A_0-H_1 (in this example: B_1) and is therefore used to parameterize the H atom.

A_0-A_1 is α, and there is a dihedral angle φ between the planes H-A_0-A_1 and A_0-A_1-B_1. We note $a=\cos(\pi-\alpha)=-\cos\alpha$, $b=\sin\alpha\cos\varphi$ and $c=\sin\alpha\sin\varphi$.

This configuration has one degree of freedom, the dihedral angle φ. Its value can be arbitrarily set to some a priori known value or according to the angle found in the input model, otherwise it can be optimized to minimize clashes or maximize hydrogen bonding interactions.

The chain of transformations is the following:

(a) $\vec{r}_{01}=\vec{r}_0-\vec{r}_1$, $\vec{r}_{B1A1}=\vec{r}_{B1}-\vec{r}_1$—for this configuration we calculate vectors with respect to A_1

(b) $\vec{u}_1$ *is* normalized $\vec{r}_{01}$

(c) $\vec{v}=\vec{r}_{B1A1}-(\vec{r}_{B1A1}\bullet\vec{u}_1)\bullet\vec{u}_1$—vector in the plane A_0-A_1-B_1 and normal to $\vec{u}_1$

(d) $\vec{u}_2$ *is* normalized $\vec{v}$

(e) $\vec{u}_3=\vec{u}_1\times\vec{u}_2$

(f) $\vec{u}_{H0}=a\bullet\vec{u}_1+b\bullet\vec{u}_2+c\bullet\vec{u}_3$

In summary, this configuration requires the following parameters: A_0, A_1, B_1, d_H, a, b, c.

2.6 3H-propeller configuration

An example is the group of Hγ11, Hγ12 and Hγ13 atoms bound to Cγ1 of valine (Fig. 2E). In the ideal configuration, we note the hydrogen atoms H_1, H_2, H_3. Each atom forms an ideal angle α between A_0-H and A_0-A_1. A dihedral angle φ is defined between the planes H_1-A_0-A_1 and A_0-A_1-B_1. For H_2 and H_3 the dihedral angle is $\varphi+\frac{2\pi}{3}$ and $\varphi+\frac{4\pi}{3}$, respectively.

The chain of transformations and the schematic representation is similar to that for the free-rotation configuration, Section 2.5 and Fig. 5. The parameters are determined for the H-atom, H_1, that possesses the ideal dihedral angle. The parameters are identical for the other two H-atoms, with exception of the angle φ, which is $\varphi+n\bullet\frac{2\pi}{3}$ $(n=1,2)$ for the second and third H-atom. Steps (a)–(e) are defined as in Section 2.5, step (f) is given by:

(f) $\vec{u}_{H1}=a\bullet\vec{u}_1+b_1\bullet\vec{u}_2+c_1\bullet\vec{u}_3$, $\vec{u}_{H2}=a\bullet\vec{u}_1+b_2\bullet\vec{u}_2+c_2\bullet\vec{u}_3$,

$$\vec{u}_{H3}=a\bullet\vec{u}_1+b_3\bullet\vec{u}_2+c_3\bullet\vec{u}_3,\text{ where } a=-\cos\alpha,$$

$$b_1=\sin\alpha\cos\varphi, b_2=\sin\alpha\cos\left(\varphi+\frac{2\pi}{3}\right), b_3=\sin\alpha\cos\left(\varphi+\frac{4\pi}{3}\right)\text{ and}$$

$$c_1=\sin\alpha\sin\varphi, c_2=\sin\alpha\sin\left(\varphi+\frac{2\pi}{3}\right), c_3=\sin\alpha\sin\left(\varphi+\frac{4\pi}{3}\right).$$

In summary, this configuration requires the following parameters: A_0, A_1, B_1, d_H, a, b, c.

2.7 2H-planar configuration

An example is the Hε21 and Hε22 atoms bound to Nε2 of glutamine (Fig. 1F). Each of the two H-atoms forms an ideal angle α with A_0-A_1 and they are located in the plane formed by the neighboring non-H atoms. One of the H-atoms, H_1, forms the dihedral angle φ equal to 0 between the planes H_1-A_0-A_1 and A_0-A_1-B_1. For the second atom, H_2, this angle is equal to π.

To describe the chain of transformations we reuse the algorithm for the free-rotation configuration (Section 2.5). The parameters are determined for the H_1 atom with the ideal dihedral angle $\varphi = 0$. Step (f) for calculating the H atom coordinates becomes:

(f) $\vec{u}_{H1} = a\bullet\vec{u}_1 + b_1\bullet\vec{u}_2 + c_1\bullet\vec{u}_3$, $\vec{u}_{H2} = a\bullet\vec{u}_1 + b_2\bullet\vec{u}_2 + c_2\bullet\vec{u}_3$, where

$$a = -\cos\alpha, b_1 = \sin\alpha\cos\varphi, b_2 = \sin\alpha\cos(\varphi+\pi), c_1 = \sin\alpha\sin\varphi, c_2 = \sin\alpha\sin(\varphi+\pi)$$

If the angle φ is equal to zero, the coefficients can be simplified: $b_1 = \sin\alpha$, $b_2 = -\sin\alpha$ and $c_1 = c_2 = 0$. However, to keep this configuration versatile to allow for rotations of the H atom group around A_0-A_1, the general expression with φ is used.

In summary, this configuration requires the following parameters: A_0, A_1, B_1, d_H, a, b, c.

2.8 H atoms with a single parent

Some hydrogen atoms have a parent atom that has no further non-hydrogen neighbors covalently bound to it. Water molecules represent an example of this situation, as each H atom has only one non-H neighbor. It follows that the H atom positions cannot be expressed from connected non-H atoms, although it may be sometimes possible to parameterize the H atom using other interactions, such as hydrogen bonds and packing.

Therefore, the approach discussed in Sections 2.1–2.7, where the hydrogen position are redefined from information about covalent bonds and angles, cannot be used for H atoms that possess only a parent or a parent and only one neighbor.

This leaves two options. The first possibility is to refine such hydrogens as free (restrained) atoms leading to an increased number of parameters: at least three coordinates per hydrogen, e.g., by six for a water molecule. Another option is to introduce constraints and refine such groups as rigid using relevant *Phenix* options (Afonine, Grosse-Kunstleve, Urzhumtsev, & Adams, 2009). This increases the number of parameters by three for each rigid group regardless of the number of H atoms. In this work we focus on the parameterization of the non-water H atoms and do not discuss refinement of H atoms with a single parent here.

3. Riding H: Refinement targets and their gradients

The aim of refinement is to change the parameters of a model (such as the atomic positions, $\vec{r}_0, \vec{r}_1, \vec{r}_2, \vec{r}_3, \ldots, \vec{r}_N$) to optimize a function of a set of observations that may be experimental data, some statistical information, stereochemical prior knowledge, etc. The fit between model parameters and the observations is commonly expressed by a target function. The lower the value of this function, the better the fit, usually resulting in an improved model. Explicit modeling of hydrogen positions $\vec{r}_{H_0}, \vec{r}_{H_1}, \vec{r}_{H_2}, \ldots, \vec{r}_{H_M}$, may improve the macromolecular model and reduce the value of a target expressed through coordinates of the whole set of atoms:

$$R(\vec{r}_0, \vec{r}_1, \vec{r}_2, \vec{r}_3, \ldots, \vec{r}_N; \vec{r}_{H_0}, \vec{r}_{H_1}, \ldots, \vec{r}_{H_M}; \text{other parameters}) \rightarrow \min \quad (2)$$

For simplicity, in what follows we consider only atomic coordinates. Refining the coordinates of H atoms independently increases the number of parameters, which can compromise the optimization procedure, especially when the number of experimental observations is low. Therefore, constraints can be applied in the form of riding hydrogens. For example, considering hydrogen H_0 at position $\vec{r}_{H_0}$ in formula (2) in a coplanar configuration defined by atoms A_0, A_1 and A_2 in the positions $\vec{r}_0$, $\vec{r}_1$ and $\vec{r}_2$ (Fig. 1), and doing similarly for other H atoms, reduces the target (2) to a function of coordinates of only non-hydrogen atoms

$$R_{\text{riding}}(\vec{r}_0, \vec{r}_1, \vec{r}_2, \vec{r}_3, \ldots, \vec{r}_N) = R(\vec{r}_0, \vec{r}_1, \vec{r}_2, \vec{r}_3, \ldots, \vec{r}_N; \vec{r}_{H_0}(\vec{r}_0, \vec{r}_1, \vec{r}_2), \vec{r}_{H_1}(\ldots) \ldots, \vec{r}_{H_M}(\ldots)) \quad (3)$$

In practice, target (3) is calculated in two steps:

(a) Determine the coordinates of each hydrogen atom through the coordinates of neighboring non-H atoms using one of the algorithms described above, e.g., that in Section 2.2:

$$\vec{r}_{H_0} = \vec{R}_{coplanar}(\vec{r}_0, \vec{r}_1, \vec{r}_2) = (x_{H_0}(\vec{r}_0, \vec{r}_1, \vec{r}_2); y_{H_0}(\vec{r}_0, \vec{r}_1, \vec{r}_2); z_{H_0}(\vec{r}_0, \vec{r}_1, \vec{r}_2)) \quad (4)$$

(b) Calculate target (2) using a full set of atoms, both non-H and H.

Crystallographic programs minimize the target (3) by applying gradient methods (Booth, 1947; Cauchy, 1847). To calculate the gradient of the target efficiently, a two-step procedure is used that applies the chain rule and inverts the chain of transformations described in Section 2 (Lunin & Urzhumtsev, 1985). First, the gradient of the target is calculated with respect to all coordinates, as if all atoms (H and non-H) were independent:

$$\frac{\partial R}{\partial x_0}, \frac{\partial R}{\partial y_0}, \frac{\partial R}{\partial z_0}, \frac{\partial R}{\partial x_1}, \frac{\partial R}{\partial y_1}, \frac{\partial R}{\partial z_1}, \ldots \frac{\partial R}{\partial z_N}; \frac{\partial R}{\partial x_{H_0}}, \frac{\partial R}{\partial y_{H_0}}, \frac{\partial R}{\partial z_{H_0}}, \ldots, \frac{\partial R}{\partial x_{H_M}}, \frac{\partial R}{\partial y_{H_M}}, \frac{\partial R}{\partial z_{H_M}} \quad (5)$$

Then the gradient is recalculated with respect to the independent parameters, for example, in the case of atom H_0, expressed as in Eq. (4):

$$\frac{\partial R_{\text{riding}}}{\partial x_0} = \frac{\partial R}{\partial x_0} + \frac{\partial R}{\partial x_{H_0}}\frac{\partial x_{H_0}}{\partial x_0} + \frac{\partial R}{\partial y_{H_0}}\frac{\partial y_{H_0}}{\partial x_0} + \frac{\partial R}{\partial z_{H_0}}\frac{\partial z_{H_0}}{\partial x_0} \quad (6)$$

and similarly for all other coordinates. Each H atom contributes to the gradients of the non-H atoms that are used to describe its position. If the hydrogen group contains several H atoms, for example, as described in Sections 2.6 and 2.7, the derivatives of R_{riding} contain similar contributions from all of them:

$$\begin{aligned}\frac{\partial R_{\text{riding}}}{\partial x_0} = {} & \frac{\partial R}{\partial x_0} + \frac{\partial R}{\partial x_{H1}}\frac{\partial x_{H1}}{\partial x_0} + \frac{\partial R}{\partial y_{H1}}\frac{\partial y_{H1}}{\partial x_0} + \frac{\partial R}{\partial z_{H1}}\frac{\partial z_{H1}}{\partial x_0} \\ & + \frac{\partial R}{\partial x_{H2}}\frac{\partial x_{H2}}{\partial x_0} + \frac{\partial R}{\partial y_{H2}}\frac{\partial y_{H2}}{\partial x_0} + \frac{\partial R}{\partial z_{H2}}\frac{\partial z_{H2}}{\partial x_0} + \ldots \end{aligned} \quad (7)$$

Also, a non-H atom may be used to parameterize several hydrogen groups; then the gradient with respect to its coordinates is a sum of contributions from all these groups. The coordinates of the non-H atoms are refined; at each refinement iteration, the H atom position is updated from these new values of x_0, y_0, z_0, etc.

To simplify the calculation, approximations such as $\frac{\partial x_H}{\partial x_0} = \frac{\partial y_H}{\partial x_0} = \frac{\partial z_H}{\partial x_0} = 1$ and $\frac{\partial x_H}{\partial x_1} = \frac{\partial y_H}{\partial x_1} = \frac{\partial z_H}{\partial x_1} = 0$, can be made (Bourhis, Dolomanov, Gildea, Howard, & Puschmann, 2015). In our work, such approximations are not used and actually are not required. The transformation of the gradients based on the different H atom configurations is summarized in the Appendix.

4. Constructing the riding H model in *CCTBX*

4.1 Internal data structure for model and connectivity information

The *CCTBX* contains several entities describing an atomic model. One is the *hierarchy* (iotbx.pdb.hierarchy), which is a nested data structure describing the macromolecule at the model, chain, residue and atom level. The *hierarchy* is a container of atomic model information that makes it possible to access the atomic attributes of a model from files (such as mmCIF; Adams et al., 2019), such as chain identifiers, residue names, atom labels, coordinates and so on (Grosse-Kunstleve & Adams, 2010). While the *hierarchy* contains as much information as is available in the corresponding model file, it lacks any notion of atomic connectivity and covalent geometry. This information is available in another representation of the atomic model: the *cctbx.geometry_restraints* class (Grosse-Kunstleve, Afonine, & Adams, 2004). An object of this class is built from the *hierarchy* and from the information available in stereochemical libraries such as the Monomer Library (Engh & Huber, 1991; Vagin et al., 2004) or GeoStd (Moriarty & Adams, n.d., http://sourceforge.net/projects/geostd). The *cctbx.geometry_restraints* class is aware of atomic connectivity, it stores ideal and actual values of geometric parameters of the model (such as bonds, angles, torsions, planes, chiral volumes, and non-bonded interactions) and calculates functions to support refinement, such as the restraints target function and its derivatives.

Together, the coordinates of non-hydrogen atoms as well as the *hierarchy* and *cctbx.geometry_restraints* objects can be used to describe the configuration of hydrogen atoms in the model and calculate their riding model parameters (Sections 2.2–2.8) using ideal bond lengths and angles.

4.2 Building riding H from known connectivity

The riding H algorithm operates in two stages. First, the procedure determines the connectivity of H atoms in the input model. This means that for each H atom the information about atoms linked to each other (as well as corresponding bond lengths and angles) is accumulated for all neighbors defining its position. In the second step, this information is used to parameterize each H atom. This involves computing the coefficients described in Section 2 and checking if the input geometry is consistent with the assigned configuration. For example, if one H atom is missing in a propeller group, the information in the connectivity will be the same as for a 2H-planar group (such as in the ARG head group). Without checks, the riding H procedure would assume that the two H atoms are in a 2H-planar configuration. Such instances are therefore not parameterized as riding H because it is difficult to determine the actual configuration in this case.

When multiple conformations are present, determining the connectivity becomes complex. An H atom may then have several parents and neighbors, leading to a complicated network of connections. An example is shown in Fig. 6; residue i (right) adopts a double conformation and residue i + 1 (left) has a single conformation. To maintain planarity of the peptide unit that connects the two residues, the amide H atoms are also in a double conformation while the amide N atom is in a single conformation. For example, the H atom in conformation A has the parent atom N, which in turn is connected to the H atom in conformation B, the C atoms of both conformations A and B, and atom C_α in residue i + 1. To parameterize this H atom in a coplanar configuration, the correct subset of connections is needed. In this example, the atoms in conformation B are irrelevant for the atom H in conformation A. The assignment of parent and neighbor atoms in

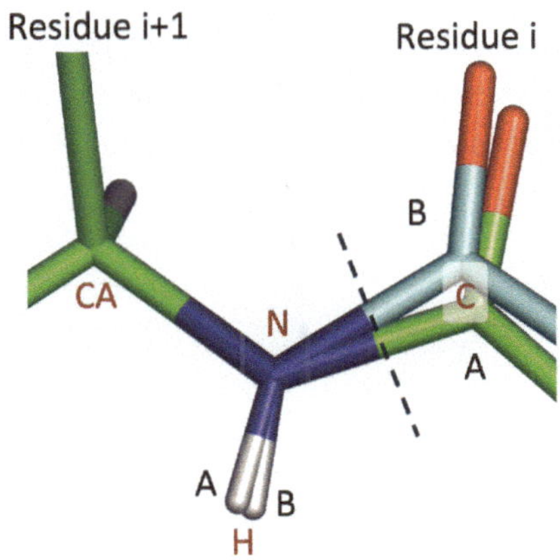

Fig. 6 Illustration of connectivity when the H atom adopts a double conformation. Atom name labels are in orange, alternative conformation labels are in black.

the presence of multiple conformations is achieved by exploiting the *hierarchy* object, which interprets the atomic model in terms of alternates (Grosse-Kunstleve & Adams, 2010).

The algorithms that build the connectivity of H atoms and determine the parameterization are written in Python. This step typically takes 10% of the computing time necessary to generate the *hierarchy* and *cctbx.geometry_restraints* objects. The algorithms that rebuild H atoms and calculate the gradients are written in C++. They are typically an order of magnitude faster than calculating the target function (less than 0.1 s in most cases).

5. Summary

This work outlines future directions of the synergetic use of neutron and X-ray diffraction data and describes the implementation of the riding hydrogen model as a fundamental prerequisite for it. We argue that the traditional joint X-ray and neutron refinement approach that yields one atomic model meant to fit both datasets is inadequate. Instead, we suggest refining two atomic models, each one fitting the respective dataset, while also benefitting from the expectation that these models should be similar.

Riding H atoms can be found in several different configurations that are not specific to protein residues. For each configuration, the H atom position is parameterized using coordinates from a set of non-H neighbors and ideal bond lengths and angles. We successfully implemented the riding hydrogen model for typical H atom geometries. The parameterization and the expressions for the gradients of refinement targets with respect to the independent parameters are described for each configuration.

The riding H option is available in *CCTBX* and *Phenix* since release 1.15.

Acknowledgments

This work was supported by the US National Institutes of Health (NIH) (grant P01GM063210), the Phenix Industrial Consortium and the NIH-funded (grant R01GM071939) Macromolecular Neutron Consortium between Oak Ridge National Laboratory and Lawrence Berkeley National Laboratory. This work was supported in part by the US Department of Energy under Contract No. DE-AC02-05CH1123. A.U. acknowledges the support and the use of resources of the French Infrastructure for Integrated Structural Biology FRISBI ANR-10-INBS-05 and of Instruct-ERIC.

Appendix. Gradient calculation for riding-H atoms

A.1 Chain rule and common steps

A.1.1 Gradient components

Each riding H group consists of one or several H atoms at the positions $\vec{r}_{Hm} = (x_{Hm}, y_{Hm}, z_{Hm}), m = 0, 1, \ldots, M$, and several non-hydrogen atoms at the positions $\vec{r}_k = (x_k, y_k, z_k), k = 0, 1, \ldots, K$ including the 'parent' atom with $k=0$ to which the hydrogen atoms of the group are bound. We suppose that we know the components of the gradient of the target R with respect to all atoms:

$$\vec{\nabla}_{Hm}R = \left(\frac{\partial R}{\partial x_{Hm}}, \frac{\partial R}{\partial y_{Hm}}, \frac{\partial R}{\partial z_{Hm}}\right); \vec{\nabla}_{rk}R = \left(\frac{\partial R}{\partial x_k}, \frac{\partial R}{\partial y_k}, \frac{\partial R}{\partial z_k}\right)$$

The gradient $\vec{\nabla}_{rk}R_{riding}$ of the target with respect to independent parameters (coordinates of non-hydrogen atoms) is a sum of $\vec{\nabla}_{rk}R$ with the contributions $\vec{\nabla}_{rkH}R$, recalculated from $\vec{\nabla}_{Hm}R$, possibly from several H atoms and several H groups.

Calculating these contributions $\vec{\nabla}_{rkH}R$ requires inversion of the formulae used to express the riding H atoms. Below we present calculation steps for each H atom configuration. There are some common expressions, so they are elaborated once and later only referred to.

A.1.2 Riding hydrogen position

The last step for all configurations consists in calculating the hydrogen position at the distance d_H along the unit vector $\vec{u}_{H0} = (x_{uH0}, y_{uH0}, z_{uH0})$ that starts from the parent non-H atom A_0 with coordinates $\vec{r}_0 = (x_0, y_0, z_0)$:

$$\vec{r}_H = \vec{r}_0 + d_H\vec{u}_{H0}$$

For the gradient calculation, we start from the gradient components $\vec{\nabla}_H R$ supposed to be known. Then

$$\vec{\nabla}_{uH0}R = d_H\vec{\nabla}_H R$$
$$\vec{\nabla}_{r0H}R = \vec{\nabla}_H R$$

In the second expression, the subscript $r0H$ is used to indicate the contribution to the coordinates of the 'parent' atom A_0 from the riding H; we shall distinguish this from the contribution $\vec{\nabla}_{r0}R$ calculated directly from the target as explained above.

A.1.3 Vector normalization

Normalization of a vector $\vec{r} = (x, y, z)$ gives a new vector

$$\vec{u} = (x_u(x, y, z), y_u(x, y, z), z_u(x, y, z)) = \|.\|^{-1}(x, y, z)$$

where $\|.\| = (x^2 + y^2 + z^2)^{1/2}$. Let $\vec{\nabla}_u R = \left(\frac{\partial R}{\partial x_u}, \frac{\partial R}{\partial y_u}, \frac{\partial R}{\partial z_u}\right)$ be known. Then the gradient $\vec{\nabla}_r R$ with respect to (x, y, z) has the x-component

$$\frac{\partial R}{\partial x} = \frac{\partial R}{\partial x_u}\frac{\partial x_u}{\partial x} + \frac{\partial R}{\partial y_u}\frac{\partial y_u}{\partial x} + \frac{\partial R}{\partial z_u}\frac{\partial y_u}{\partial x} = \|.\|^{-1}\frac{\partial R}{\partial x_u} - x\|.\|^{-3}\left(\vec{r}\bullet\vec{\nabla}_u R\right)$$

and similar expressions for $\frac{\partial R}{\partial y}$ and $\frac{\partial R}{\partial z}$. Or in vector form:

$$\vec{\nabla}_r R = \|.\|^{-1}\vec{\nabla}_u R - \|.\|^{-3}\left(\vec{r}\bullet\vec{\nabla}_u R\right)\vec{r}$$

A.1.4 Cross product

The vector product of $\vec{r}_1 = (x_1, y_1, z_1)$ and $\vec{r}_2 = (x_2, y_2, z_2)$ gives

$$\vec{v} = \vec{r}_1 \times \vec{r}_2 = (x_v, y_v, z_v) = (y_1 z_2 - z_1 y_2, z_1 x_2 - x_1 z_2, x_1 y_2 - y_1 x_2)$$

Let $\vec{\nabla}_v R = \left(\frac{\partial R}{\partial x_v}, \frac{\partial R}{\partial y_v}, \frac{\partial R}{\partial z_v}\right)$ be known. Then

$$\vec{\nabla}_{r1} R = \vec{r}_2 \times \vec{\nabla}_v R,$$
$$\vec{\nabla}_{r2} R = \vec{\nabla}_v R \times \vec{r}_1$$

A.2 Gradient for particular riding H configurations

The next sections explain how to calculate the gradient of the target function with respect to independent coordinates, starting from the values given in Section A.1.1. For each configuration, we explicitly list the steps to which we apply the chain rule. Generally, we invert the steps described for each geometry, starting from the last and moving backward.

A.2.1 Coplanar configuration

For definition of parameters, see Section 2.2.

We start from the last step (d) and move backward to (a) after getting $\vec{\nabla}_{uH0}R$ and $\vec{\nabla}_{r0H}R$ from $\vec{\nabla}_H R$ according to Section A.1.2.

(d) obtain $\vec{\nabla}_{rH0}R$ from $\vec{\nabla}_{uH0}R$ according to Section A.1.3

(c) $\vec{\nabla}_{u10}R = a\vec{\nabla}_{rH0}R$, $\vec{\nabla}_{u20}R = b\vec{\nabla}_{rH0}R$

(b) obtain $\vec{\nabla}_{r10}R$ from $\vec{\nabla}_{u10}R$ and $\vec{\nabla}_{r20}R$ from $\vec{\nabla}_{u20}R$ according to Section A.1.3

(a) $\vec{\nabla}_{r1}R_{riding} = \vec{\nabla}_{r10}R$, $\vec{\nabla}_{r2}R_{riding} = \vec{\nabla}_{r20}R$

$$\vec{\nabla}_{r0}R_{riding} = \vec{\nabla}_{r0H}R - \vec{\nabla}_{r10}R - \vec{\nabla}_{r20}R$$

A.2.2 2H-tetrahedral configuration

For definition of parameters, see Section 2.3.

We start from getting $\vec{\nabla}_{uH0}R$ and $\vec{\nabla}_{r0H}R$ from $\vec{\nabla}_H R$ according to Section A.1.2.

(g) $\vec{\nabla}_{d0}R = (\cos\delta)\vec{\nabla}_{d0}R$, $\vec{\nabla}_{v0}R = (\sin\delta)\vec{\nabla}_{d0}R$

(f) $\vec{\nabla}_d R$ from $\vec{\nabla}_{d0}R$ according to Section A.1.3

(e) $\vec{\nabla}^{(1)}_{u10}R = a\vec{\nabla}_d R$ and $\vec{\nabla}^{(1)}_{u20}R = b\vec{\nabla}_d R$

(d) obtain $\vec{\nabla}_v R$ from $\vec{\nabla}_{v0}R$ according to Section A.1.3

(c) obtain $\vec{\nabla}^{(2)}_{u10}R$ and $\vec{\nabla}^{(2)}_{u20}R$ from $\vec{\nabla}_v R$according to Section A.1.4 and then calculate

$$\vec{\nabla}_{u10}R = \vec{\nabla}^{(1)}_{u10}R + \vec{\nabla}^{(2)}_{u10}R \text{ and } \vec{\nabla}_{u20}R = \vec{\nabla}^{(1)}_{u20}R + \vec{\nabla}^{(2)}_{u20}R$$

(b) obtain $\vec{\nabla}_{r10}R$ from $\vec{\nabla}_{u10}R$ and $\vec{\nabla}_{r20}R$ from $\vec{\nabla}_{u20}R$ according to Section A.1.3

(a) $\vec{\nabla}_{r1}R_{riding} = \vec{\nabla}_{r10}R$, $\vec{\nabla}_{r2}R_{riding} = \vec{\nabla}_{r20}R$

$$\vec{\nabla}_{r0}R_{riding} = \vec{\nabla}_{r0H}R - \vec{\nabla}_{r10}R - \vec{\nabla}_{r20}R$$

A.2.3 1H-tetrahedral configuration

For definition of parameters, see Section 2.4.

We start from getting $\vec{\nabla}_{uH0}R$ and $\vec{\nabla}_{r0H}R$ from $\vec{\nabla}_H R$ according to Section A.1.2.

(d) obtain $\vec{\nabla}_{rH0}R$ from $\vec{\nabla}_{uH0}R$ according to Section A.1.3

(c) $\vec{\nabla}_{u10}R = a\vec{\nabla}_{rH0}R, \vec{\nabla}_{u20}R = b\vec{\nabla}_{rH0}R, \vec{\nabla}_{u30}R = c\vec{\nabla}_{rH0}R$

(b) obtain $\vec{\nabla}_{r10}R$ from $\vec{\nabla}_{u10}R$, $\vec{\nabla}_{r20}R$ from $\vec{\nabla}_{u20}R$, $\vec{\nabla}_{r30}R$ from $\vec{\nabla}_{u30}R$ according to Section A.1.3

(a) $\vec{\nabla}_{r1}R_{riding} = \vec{\nabla}_{r10}R, \vec{\nabla}_{r2}R_{riding} = \vec{\nabla}_{r20}R, \vec{\nabla}_{r3}R_{riding} = \vec{\nabla}_{r30}R,$
$\vec{\nabla}_{r0}R_{riding} = \vec{\nabla}_{r0H}R - \vec{\nabla}_{r10}R - \vec{\nabla}_{r20}R - \vec{\nabla}_{r30}R$

A.2.4 Single H with rotational degree of freedom

For definition of parameters, see Section 2.5.

We start from getting $\vec{\nabla}_{uH0}R$ and $\vec{\nabla}_{r0H}R$ from $\vec{\nabla}_{H}R$ according to Section A.1.2.

(f) $\vec{\nabla}^{(1)}_{u10}R = a\vec{\nabla}_{rH0}R, \vec{\nabla}^{(1)}_{u20}R = b\vec{\nabla}_{rH0}R, \vec{\nabla}_{u30}R = c\vec{\nabla}_{rH0}R$

(e) obtain $\vec{\nabla}^{(2)}_{u10}R$ and $\vec{\nabla}^{(2)}_{u20}R$ from $\vec{\nabla}_{u30}R$ according to Section A.1.4; superscript (2) is used to distinguish these derivatives from those calculated above $\vec{\nabla}^{(1)}_{u20}R$ and $\vec{\nabla}^{(1)}_{u10}R$; $\vec{\nabla}_{u2}R = \vec{\nabla}^{(1)}_{u20}R + \vec{\nabla}^{(2)}_{u20}R$;

(d) obtain $\vec{\nabla}_{v}R$ from $\vec{\nabla}_{u2}R$ according to Section A.1.2

(c) $\vec{\nabla}_{rBA}R = \vec{\nabla}_{v}R - \vec{u}_1\left(\vec{u}_1\bullet\vec{\nabla}_{v}R\right);\qquad \vec{\nabla}_{u1v}R = -\vec{\nabla}_{v}R(\vec{r}_{B1A1}\bullet\vec{u}_1) - \vec{r}_{B1A1}$
$\left(\vec{u}_1\bullet\vec{\nabla}_{v}R\right)$; $\vec{\nabla}_{u1}R = \vec{\nabla}^{(1)}_{u10}R + \vec{\nabla}^{(2)}_{u10}R + \vec{\nabla}_{u1v}R$

(b) obtain $\vec{\nabla}_{r01}R$ from $\vec{\nabla}_{u1}R$ according to Section A.1.2

(a) $\vec{\nabla}_{r1}R_{riding} = -\vec{\nabla}_{r01}R - \vec{\nabla}_{rBA}R, \vec{\nabla}_{rB1}R_{riding} = \vec{\nabla}_{rB1}R,$
$\vec{\nabla}_{r0}R_{riding} = \vec{\nabla}_{r0H}R + \vec{\nabla}_{r01}R$

A.2.5 3H-propeller

For definition of parameters, see Section 2.6.

We start from getting $\vec{\nabla}_{uH1}R$, $\vec{\nabla}_{r0H1}R$, $\vec{\nabla}_{uH2}R$, $\vec{\nabla}_{r0H2}R$ and $\vec{\nabla}_{uH3}R$, $\vec{\nabla}_{r0H3}R$ from $\vec{\nabla}_{H1}R$, $\vec{\nabla}_{H2}R$, $\vec{\nabla}_{H3}R$, respectively, according to Section A.1.2. Then

(f) $\vec{\nabla}_{u10}R = a\sum_{m=1}^{3}\vec{\nabla}_{r0Hm}R, \vec{\nabla}_{u20}R = \sum_{m=1}^{3} b_m\vec{\nabla}_{r0Hm}R, \vec{\nabla}_{u30}R$

$$= \sum_{m=1}^{3} c_m\vec{\nabla}_{r0Hm}R$$

Steps (e)–(a) repeat the steps from Section A.2.4 except the last expression for $\vec{\nabla}_{r0} R_{riding}$ that becomes

$$\vec{\nabla}_{r0} R_{riding} = \vec{\nabla}_{r01} R + \sum_{m=1}^{3} \vec{\nabla}_{r0Hm} R$$

A.2.6 2H-planar

For definition of parameters, see Section 2.7.

We start from getting $\vec{\nabla}_{uH1} R$, $\vec{\nabla}_{r0H1} R$ and $\vec{\nabla}_{uH2} R$, $\vec{\nabla}_{r0H2} R$ from $\vec{\nabla}_{H1} R$ and $\vec{\nabla}_{H2} R$, respectively, according to Section A.1.2. Then

$$\vec{\nabla}_{u10} R = a\left(\vec{\nabla}_{r0H1} R + \vec{\nabla}_{r0H2} R\right),$$

(f) $$\vec{\nabla}_{u20} R = b\left(\vec{\nabla}_{r0H1} R - \vec{\nabla}_{r0H2} R\right),$$

$$\vec{\nabla}_{u30} R = c\left(\vec{\nabla}_{r0H1} R - \vec{\nabla}_{r0H2} R\right)$$

Steps (e)–(a) repeat the steps from Section A.2.4 except the last expression for $\vec{\nabla}_{r0} R_{riding}$ that becomes

$$\vec{\nabla}_{r0} R_{riding} = \vec{\nabla}_{r01} R + \vec{\nabla}_{r0H1} R + \vec{\nabla}_{r0H2} R$$

References

Adams, P. D., Afonine, P. V., Baskaran, K., Berman, H. M., Berrisford, J., Bricogne, G., et al. (2019). Announcing mandatory submission of PDBx/mmCIF format files for crystallographic depositions to the Protein Data Bank (PDB). *Acta Crystallographica Section D: Structural Biology*, *75*, 451–454.

Adams, P. D., Mustyakimov, M., Afonine, P. V., & Langan, P. (2009). Generalized X-ray and neutron crystallographic analysis: More accurate and complete structures for biological macromolecules. *Acta Crystallographica Section D: Structural Biology*, *65*, 567–573.

Afonine, P. V., Grosse-Kunstleve, R. W., Urzhumtsev, A., & Adams, P. D. (2009). Automatic multiple-zone rigid-body refinement with a large convergence radius. *Journal of Applied Crystallography*, *42*, 607–615.

Afonine, P. V., Mustyakimov, M., Grosse-Kunstleve, R. W., Moriarty, N. W., Langan, P., & Adams, P. D. (2010). Joint X-ray and neutron refinement with phenix.refine *Acta Crystallographica Section D: Structural Biology*, *66*, 1153–1163.

Allen, F. H. (1986). A systematic pairwise comparison of geometric parameters obtained by X-ray and neutron diffraction. *Acta Crystallographica Section B: Structural Science, Crystal Engineering and Materials*, *42*, 515–522.

Allen, F. H., & Bruno, I. J. (2010). Bond lengths in organic and metal-organic compounds revisited: X-H bond lengths from neutron diffraction data. *Acta Crystallographica Section B: Structural Science, Crystal Engineering and Materials*, *66*, 380–386.

Booth, A. D. (1947). Application of the method of steepest descents to X-ray structure analysis. *Nature*, *160*, 196.

Bourhis, L. J., Dolomanov, O. V., Gildea, R. J., Howard, J. A. K., & Puschmann, H. (2015). The anatomy of a comprehensive constrained, restrained refinement program for the modern computing environment - Olex2 dissected. *Acta Crystallographica Section A: Foundations of Crystallography*, *71*, 59–75.
Busing, W. R., & Levy, H. A. (1964). The effect of thermal motion on the estimation of bond lengths from diffraction measurements. *Acta Crystallographica*, *17*, 142–146.
Cauchy, A. L. (1847). Méthode générale pour la résolution des systèmes d'équations simultanées. *Comptes Rendus de l'Academie des Sciences*, *25*, 536–538.
Chen, J. C.-H., & Unkefer, C. J. (2017). Fifteen years of the protein crystallography station: The coming of age of macromolecular neutron crystallography. *IUCrJ*, *4*, 72–86.
Coates, L., Cuneo, M. J., Frost, M. J., He, J., Weiss, K. L., Tomanicek, S. J., et al. (2015). The macromolecular neutron diffractometer MaNDi at the spallation neutron source. *Journal of Applied Crystallography*, *48*, 1302–1306.
Coppens, P. (1967). Comparative X-ray and neutron diffraction study of bonding effects in s-Triazine. *Science*, *158*, 1577–1579.
Cramer, G. (1750). *Introduction à l'Analyse des lignes Courbes algébriques*. Geneva: Europeana.
Dauter, M., & Dauter, Z. (2017). Many ways to derivatize macromolecules and their crystals for phasing. *Methods in Molecular Biology*, *1607*, 349–356.
Engh, R. A., & Huber, R. (1991). Accurate bond and angle parameters for X-ray protein structure refinement. *Acta Crystallographica Section A: Foundations and Advances*, *47*, 392–400.
Fraser, J. S., van den Bedem, H., Samelson, A. J., Lang, P. T., Holton, J. M., Echols, N., et al. (2011). Accessing protein conformational ensembles using room-temperature X-ray crystallography. *Proceedings of the National Academy of Sciences of the United States of America*, *108*, 16247–16252.
Grosse-Kunstleve, R. W., & Adams, P. D. (2010). cctbx PDB handling tools. *Computational Crystallography Newsletter*, *1*, 4–11.
Grosse-Kunstleve, R. W., Afonine, P. V., & Adams, P. D. (2004). cctbx news. *IUCr Computer Communication Newsletter*, *4*, 19–36.
Grosse-Kunstleve, R. W., Sauter, N. K., Moriarty, N. W., & Adams, P. D. (2002). The computational crystallography toolbox: Crystallographic algorithms in a reusable software framework. *Journal of Applied Crystallography*, *35*, 126–136.
Gruene, T., Hahn, H. W., Luebben, A. V., Meilleur, F., & Sheldrick, G. M. (2014). Refinement of macromolecular structures against neutron data with SHELXL2013. *Journal of Applied Crystallography*, *47*, 462–466.
Kosinski, A. A. (2001). Cramer's rule is due to Cramer. *Mathematics Magazine*, *74*, 310–312.
Liebschner, D., Afonine, P. V., Baker, M. L., Bunkóczi, G., Chen, V. B., Croll, T. I., et al. (2019). Macromolecular structure determination using X-rays, neutrons and electrons: Recent developments in Phenix. *Acta Crystallographica Section D: Structural Biology*, *75*, 861–877.
Liebschner, D., Afonine, P. V., Moriarty, N. W., Langan, P., & Adams, P. D. (2018). Evaluation of models determined by neutron diffraction and proposed improvements to their validation and deposition. *Acta Crystallographica Section D: Structural Biology*, *74*, 800–813.
Lunin, V. Y., & Urzhumtsev, A. G. (1985). Program construction for macromolecule atomic model refinement based on the fast Fourier transform and fast differentiation algorithms. *Acta Crystallographica Section A: Foundations of Crystallography*, *41*, 327–333.
Moriarty, N. W. & Adams, P. D. n.d. GeoStd https://sourceforge.net/projects/geostd/
Ostermann, A., Tanaka, I., Engler, N., Niimura, N., & Parak, F. G. (2002). Hydrogen and deuterium in myoglobin as seen by a neutron structure determination at 1.5 Å resolution. *Biophysical Chemistry*, *95*, 183–193.

Petrova, T., & Podjarny, A. (2004). Protein crystallography at subatomic resolution. *Reports on Progress in Physics*, *67*, 1565–1605.
Sheldrick, G. M., & Schneider, T. R. (1997). SHELXL: High-resolution refinement. *Methods in Enzymology*, *277*, 319–343.
Shu, F., Ramakrishnan, V., & Schoenborn, B. P. (2000). Enhanced visibility of hydrogen atoms by neutron crystallography on fully deuterated myoglobin. *Proceedings of the National Academy of Sciences of the United States of America*, *97*, 3872–3877.
Turkenburg, J. P., & McAuley, K. E. (2013). Data collection and processing. *Acta Crystallographica Section D: Biological Crystallography*, *69*, 1193–1194.
Vagin, A. A., Steiner, R. A., Lebedev, A. A., Potterton, L., McNicholas, S., Long, F., et al. (2004). REFMAC5 dictionary: Organization of prior chemical knowledge and guidelines for its use. *Acta Crystallographica Section D: Biological Crystallography*, *60*, 2184–2195.
Wlodawer, A. (1980). Studies of ribonuclease-A by X-ray and neutron diffraction. *Acta Crystallographica Section B: Structural Science, Crystal Engineering and Materials*, *36*, 1826–1831.
Wlodawer, A., & Hendrickson, W. A. (1982). A procedure for joint refinement of macromolecular structures with X-ray and neutron diffraction data from single crystals. *Acta Crystallographica Section A: Foundations and Advances*, *38*, 239–247.
Wlodawer, A., Miller, M., & Sjölin, L. (1983). Active site of RNase: Neutron diffraction study of a complex with uridine vanadate, a transition-state analog. *Proceedings of the National Academy of Sciences of the United States of America*, *80*, 3628–3631.

Further reading

Blakeley, M. P., Ruiz, F., Cachau, R., Hazemann, I., Meilleur, F., Mitschler, A., et al. (2008). Quantum model of catalysis based on a mobile proton revealed by subatomic x-ray and neutron diffraction studies of h-aldose reductase. *Proceedings of the National Academy of Sciences of the United States of America*, *105*, 1844–1848.
Fisher, S. J., Wilkinson, J., Henchman, R. H., & Helliwell, J. R. (2009). An evaluation review of the prediction of protonation states in proteins versus crystallographic experiment. *Crystallography Reviews*, *15*, 231–259.
Howard, E. I., Blakeley, M. P., Haertlein, M., Haertlein, I. P., Mitschler, A., Fisher, S. J., et al. (2011). Neutron structure of type-III antifreeze protein allows the reconstruction of AFP–ice interface. *Journal of Molecular Recognition*, *24*, 724–732.
Howard, E. I., Guillot, B., Blakeley, M. P., Haertlein, M., Moulin, M., Mitschler, A., et al. (2016). High-resolution neutron and X-ray diffraction room-temperature studies of an H-FABP–oleic acid complex: Study of the internal water cluster and ligand binding by a transferred multipolar electron-density distribution. *IUCrJ*, *3*, 115–126.

CHAPTER TEN

Interactive model building in neutron macromolecular crystallography

Derek T. Logan*

Section for Biochemistry and Structural Biology, Department of Chemistry, Lund University, Lund, Sweden
*Corresponding author: e-mail address: derek.logan@biochemistry.lu.se

Contents

Abstract

This chapter aims to give an overview of the process of interactive model building in macromolecular neutron crystallography for the researcher transitioning from X-ray crystallography alone. The two most popular programs for refinement and model building, *phenix.refine* and Coot, respectively, are used as examples, and familiarity with the programs is assumed. Some work-arounds currently required for proper communication between the programs are described. We also discuss the appearance of nuclear density maps and how this differs from that of electron density maps. Advice is given to facilitate deposition of jointly refined neutron/X-ray structures in the Protein Data Bank.

Methods in Enzymology, Volume 634
ISSN 0076-6879
https://doi.org/10.1016/bs.mie.2019.11.017

1. Introduction

Model building and refinement, with concomitant use of modern validation tools to ensure model quality, are the final, and often the most enjoyable, phases of a structure determination project, and the last ones before deposition and publication. A plethora of tools have been developed over the past 10–15 years that provide an increasingly integrated approach to refinement, validation and deposition of X-ray crystal structures. This chapter will describe the special considerations in interactive model building of a protein structure derived from neutron crystallography. We will assume that the reader is familiar with protein X-ray crystallography, including a working knowledge of the most widely-used graphics program for model building, Coot (Emsley, Lohkamp, Scott, & Cowtan, 2010). Furthermore, we will assume that the reader is using the program *phenix.refine* (Afonine, Ralf, Headd, & Thomas, 2012) for reciprocal-space refinement of a structure for which both neutron and X-ray data are available. Another less commonly used package is nCNS (Adams, Mustyakimov, Afonine, & Langan, 2009). Development is underway for joint neutron X-ray refinement in other programs such as Refmac5 (Murshudov et al., 2011) and neutron-only refinement is possible in SHELXL (Gruene, Hahn, Luebben, Meilleur, & Sheldrick, 2014). Nevertheless, *phenix.refine* is currently the most commonly-used program for refinement of neutron crystal structures. We will also assume the realistic scenario that the model has been built almost to completion using traditional X-ray crystallography and that the neutron-specific model building is oriented toward correct placement of hydrogen/deuterium atoms and minor adjustments of the model.

We will focus on the idiosyncrasies of nuclear density maps and how to use them optimally in conjunction with electron density maps to build the best model. There are also some challenges in the use of PDB files with deuterium atoms in Coot, as well as aspects to take special consideration of when depositing jointly refined structures in the Protein Data Bank. All programs are naturally under development, and given the increasing number of neutron crystal structures being determined, there is hope that more and more tools relevant to neutron crystallography will be integrated into existing workflows. The observations made in this chapter refer to the program versions available in October 2019. The bibliography is of necessity selective, and the reader is referred to the many other articles cited in the publications mentioned here.

2. Getting started with model building

In X-ray crystallography, after a given cycle of refinement one will have an atomic model in PDB or mmCIF format and an MTZ file (Winn et al., 2011) containing the original observed structure factors, calculated structure factors and phases for the current model, an R-free flag and coefficients for the calculation of electron density maps by Fourier transformation in Coot. For the purposes of this chapter, we will assume that the coordinates are maintained in PDB format until just before deposition. For X-ray-only structures, the map coefficients are easily read into Coot using the "Auto Open MTZ" command from the "File" menu. This will automatically generate electron density maps with coefficients 2m|F_o|-D|F_c| and m|F_o|-D|F_c|[a] from the pairs of columns `2FOFCWT/PH2FOFCWT` and `FOFCWT/PHFOFCWT`, respectively, in the MTZ file. However, for joint neutron/X-ray refinement the situation is not so simple due to the plethora of columns in the output file. From *phenix.refine* these columns are:

```
2FOFCWT_xray                 PH2FOFCWT_xray
2FOFCWT_filled_xray          PH2FOFCWT_filled_xray
FOFCWT_xray                  PHFOFCWT_xray

2FOFCWT_neutron              PH2FOFCWT_neutron
2FOFCWT_filled_neutron       PH2FOFCWT_filled_neutron
FOFCWT_neutron               PHFOFCWT_neutron
```

i.e., three pairs of structure factor amplitudes and phases for each of the X-ray and neutron datasets. The coefficients containing the word "filled" have F_o values missing in the observed dataset substituted by D|F_{model}|. For X-ray datasets of normal completeness, there is not normally much difference between maps calculated with and without such "filling," but given that neutron datasets can be significantly less complete (often 80–85%), it can be worth checking both types of map, as "filled" maps can be severely model-biased in case of poorly complete |F_o|.

2.1 Visualizing nuclear and electron density maps in Coot

The easiest way to use Coot to visualize the maps is to open the program through the Phenix graphical user interface (Liebschner et al., 2019). When a *phenix.refine* job is started, a Coot window will automatically open

[a] $\mathbf{F}_c$ are in reality $\mathbf{F}_{model}$, see Afonine et al. (2012).

and the maps and the current model will be shown. The X-ray maps are continuously updated after each cycle of refinement, but the neutron map will only appear after the final cycle. This is to reduce the running time for a given refinement job, as map calculation adds running time to each cycle. If this window is not already open, one can press the button "Open in Coot" on the result tab of any joint X-ray/neutron refinement. Note that the same function does not currently exist for PyMOL.[b]

If you have run *phenix.refine* on the command line, for example as part of a scripted testing of different weights for the X-ray and neutron terms within the total energy function (Afonine et al., 2010), you will want to open Coot in a stand-alone manner. Coot does not recognize these map coefficients automatically when using the "Auto Open MTZ" command from the pull-down menus. Instead, you can install a short script (kindly provided by Paul Emsley):

```
;; return a list of map numbers
(define (x-n-open mtz-file-name)

  (map (lambda (col-pair)
        (make-and-draw-map mtz-file-name (car col-pair) (cadr col-pair)
"" 0 (caddr col-pair)))
        (list
           (list "2FOFCWT_xray" "PH2FOFCWT_xray" 0)
           (list "FOFCWT_xray"  "PHFOFCWT_xray" 1)
           (list "2FOFCWT_neutron" "PH2FOFCWT_neutron" 0)
           (list "FOFCWT_neutron" "PHFOFCWT_neutron" 1))))

(if (defined? 'coot-main-menubar)
    (let ((menu (coot-menubar-menu "Neutron")))
      (add-simple-coot-menu-menuitem
       menu "Quick-load MTZ"
       (lambda () (x-n-open (most-recently-created-file "*.mtz" "."))))

      (add-simple-coot-menu-menuitem
        menu "Open MTZ"
        (lambda ()
           (let* ((fs-window (gtk-file-selection-new "file
selection")))
                (gtk-signal-connect
                     (gtk-file-selection-ok-button fs-window)
                     "clicked"
```

[b] http://www.pymol.org

```
                    (lambda ()
                      (let ((t (gtk-file-selection-get-filename
  fs-window)))
                        (x-n-open t)
                        (gtk-widget-destroy fs-window))))
               (gtk-widget-show fs-window)))))
```

Save this script as e.g. `xray-neutron.scm` to your `.coot-preferences` directory and restart Coot. A button will appear in the menu bar labeled "Neutron" in which there will be a menu item "Quick Open MTZ." This will open the most recently created MTZ file in the directory that you started Coot in and display all four maps, $2m|F_o|\text{-}D|F_c|$ and $m|F_o|\text{-}D|F_c|$ for both neutrons and X-rays. The second option will be "Open MTZ," which will enable you to open any output MTZ file from *phenix.refine* with these column labels, through a pop-up file selection window. By the time of publication, this script may have been included in Curlew, Coot's Extension Wrangler, which will make its implementation easier.

For reasons that will be described below, you may also have to modify the hydrogen atom names in your PDB file from *phenix.refine* for convenience of model building in Coot.

3. Appearance of nuclear density maps

Atoms scatter neutrons through their nuclei, in contrast to X-rays which are scattered by the electrons. Thus peaks in nuclear density maps show the true positions of atomic nuclei rather than the centers of electronic distributions. The principle use of neutron macromolecular crystallography (NMX) is to locate hydrogen atoms. Even at atomic resolution, where H atoms become visible in electron density maps, the peaks used to interpret H atom positions are shifted toward the center of the bond to the heavy atom to which they are attached. In NMX, H atoms are exchanged to D to increase their visibility, either by soaking hydrogenous proteins in solutions containing D_2O to replace polar H, or by perdeuterating the protein and replacing even the aliphatic H by D. The appreciable scattering of deuterium nuclei makes them just as visible as those of C, N or O. In contrast, the most common isotope of hydrogen, ^1H, scatters neutrons negatively. More precisely, the coherent neutron scattering length of deuterium (^2H or D, 6.67 fm) is comparable to that of these heavier atoms (e.g., 6.65 fm for C), whereas the scattering length of the ^1H isotope is −3.74 fm. Both of these factors have an effect on the appearance of nuclear density maps, which can take the first-time neutron crystallographer with a background in X-rays some time to get used to. This is particularly true for nuclear density maps from hydrogenous proteins whose polar hydrogen atoms have been exchanged for deuterium by soaking in D_2O but whose aliphatic hydrogens are intact. Such maps can be perceived as "ugly" by the X-ray crystallographer because the hydrogen atoms' negative scattering has a resolution-dependent cancelation effect on the heavier atoms to which they are bonded. Furthermore, cancelation effects shift the apparent negative peak for the H atom along the bond vector further away from its correct position and diminish the density for the corresponding heavier atoms (Afonine et al., 2010). Nuclear density maps from proteins that have been fully perdeuterated to improve data quality (Shu, Ramakrishnan, & Schoenborn, 2000) are less problematic, but it is not always possible to obtain large enough crystals of the perdeuterated protein. In any case, the important information on the polar hydrogen atoms should be clear in both types of map, assuming a sufficiently high degree of exchange of polar H to D. At an H:D ratio of around 2:1, the scattering from the two isotopes cancels completely.

3.1 Hydrogenous, deuterium-exchanged proteins

Fig. 1A shows an example from the crystal structure determined at 1.9 Å resolution of urate oxidase from *Aspergillus flavus* (PDB ID 4N3M), a non-perdeuterated, deuterium exchanged protein (Oksanen, Blakeley, El-Hajji, Ryde, & Budayova-Spano, 2014) in complex with the ligand 8-azaxanthine. For this figure, the model has been refined without H or D atoms on Arg176 and Gln228 to generate "omit maps." The aliphatic part of the well-ordered Arg176 side chain appears "thinner" in the $2m|F_o|\text{-}D|F_c|$ map than in a corresponding X-ray map due to the cancelation effect of the native hydrogen atoms, which appear as negative peaks in the omit map (and indeed appear at a negative level even in the $2m|F_o|\text{-}D|F_c|$ map). In contrast, the exchanged deuterium atoms of at least one of the terminal NH_2 groups of Arg176 appear as positive peaks and the $2m|F_o|\text{-}D|F_c|$ maps have "bulges" where the D atoms should be. The same is true of the NH group of the guanidinium moiety and the protonated N1 atom of the ligand. It is noteworthy that a few deuterium atoms that we know to be present are not clearly visible in the omit maps while they are quite visible in the $2m|F_o|\text{-}D|F_c|$ maps, e.g. one of the deuteriums of the right-hand terminal amino group on the arginine in Fig. 1A and another in the amide group of the glutamine side chain. This emphasizes the importance of placing as many hydrogen or deuterium atoms as possible based on unambiguous geometrical considerations in order to focus on the significant ones.

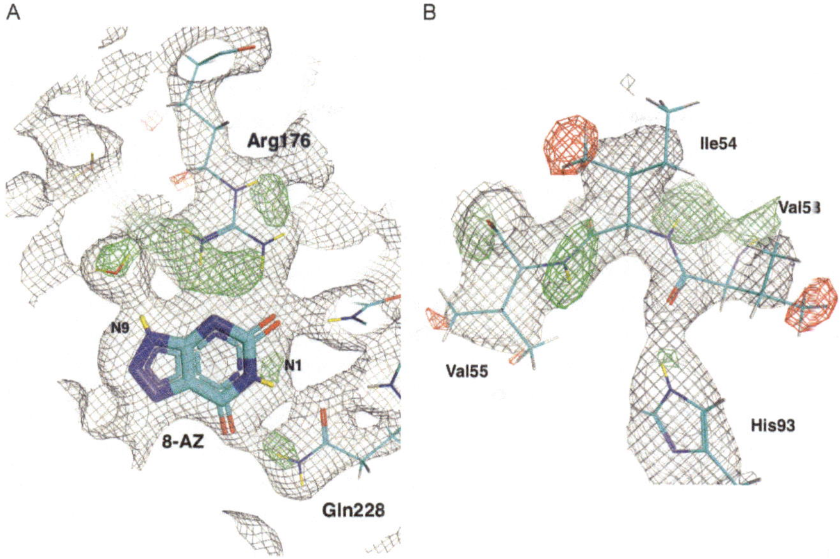

Fig. 1 Example of a nuclear density map from a non-perdeuterated protein, urate oxidase from *Aspergillus flavus* in complex with 8-azaxanthine (PDB ID 4N3M). The blue mesh represents $2m|F_o|$-$D|F_c|$ nuclear density contoured at 1.0σ; the green and red meshes represent $m|F_o|$-$D|F_c|$ omit maps contoured at $+2.5\sigma$ and -2.5σ, respectively. (A) Arg176 and Gln228 coordinating the substrate in the active site. (B) The region around Ile54, including an H-bond to His93.

Fig. 1B shows another example from urate oxidase, the area around residue Ile54. There are clear positive difference peaks in the nuclear density map for the main chain amide deuterium atoms and also for the deuteron in a hydrogen bond between a His side chain and a main chain carbonyl moiety. In contrast, a strong negative difference peak characterizes one of the methyl groups of the Ile side chain.

Direct identification of hydrogen atoms in non-perdeuterated proteins at resolutions poorer than 2.0 Å is difficult, though it of course depends on how well-ordered the residues are. This was first identified in the early 1980s (Kossiakoff & Spencer, 1980) and again discussed more recently (Shu et al., 2000), both studies using myoglobin as an example. These authors demonstrated that the hydrogen atoms are poorly visible even in nuclear density maps at 2.0 Å resolution calculated from model structure factors $|F_c|$ (that is to say, having ideal amplitudes and phases). The conclusion was that, not only do the hydrogen atoms have an effect on the visibility of the heavier atoms, but the positive scattering of the heavier atoms has a detrimental effect on the visibility of the hydrogen atoms at negative contour levels. In a follow-up study where myoglobin data to a higher resolution of 1.5 Å were used, the problem persisted, even for ideal, calculated maps (Ostermann, Tanaka, Engler, Niimura, & Parak, 2002). The authors proposed a solution involving subtraction of a normalized map calculated without hydrogen atoms from an equivalent map calculated with hydrogen atoms, which significantly improved hydrogen visibility in myoglobin. This correction does not affect refinement, only map interpretation. However, the proposed solution has not yet been implemented in any widely available program, so we will not go into it further here and refer the reader to the cited literature for details.

As noted above, in nuclear density maps for H/D exchanged proteins, the relative scattering lengths of H and D will result in a complete cancelation of the nuclear density when the ratio of H to D is around 2:1, and there will be a reduction in the density at even lower values.

3.2 Effect of resolution

Even large crystals almost always diffract neutrons to lower resolution than they do X-rays, thus it is important to consider the effect of resolution on the visibility of hydrogen and/or deuterium atoms and their contribution to cancelation effects. This was already examined in the studies mentioned above. In 2012, Chen et al. published a new study in which they obtained data to the exceptional resolution of 1.1 Å from crystals of the small 46-amino acid protein crambin measuring 4 mm^3 (Chen, Hanson, Fisher, Langan, & Kovalevsky, 2012). A neutron-only refinement was carried out using nCNS and SHELXL-97 (Langan et al., 2008; Sheldrick & Schneider, 1997). Truncation of the experimental maps at different resolutions confirmed that the cancelation effect of hydrogen atoms on the nuclear density, particularly of side chains, was much more pronounced at lower resolutions, especially below 2.0 Å, where it begins to have a serious effect on the continuity of the nuclear density between the heavier atoms (Fig. 2A). This can be a particular problem for more poorly ordered, surface-exposed side chains such as lysine. Regarding the direct visibility of non-exchanged hydrogen atoms in $2m|F_o|$-$D|F_c|$ maps, a well-ordered tyrosine side chain was used as an illustration (Fig. 2C), and it could be concluded that direct, unambiguous identification of H atoms in hydrogenous proteins is limited to resolutions better than 1.5 Å. A similar analysis of H/D atom visibility by Afonine et al. also included a consideration of how partial D atom

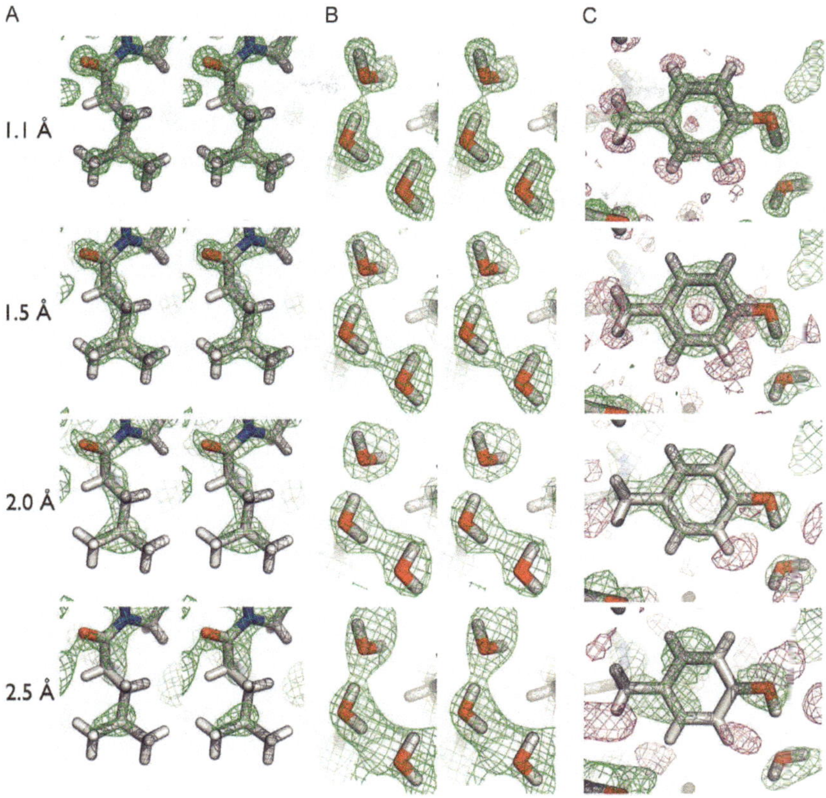

Fig. 2 Effect of resolution on visibility of atoms in an H/D-exchanged protein structure. (A) 2m|F_o|-D|F_c| nuclear density for Leu18 in crambin at 1.1, 1.5, 2.0, and 2.5 Å resolution. (B) Nuclear density for D_2O molecules. (C) Nuclear density for Tyr44. Maps are contoured at 1.5σ, with positive $2F_o$-F_c density in green and negative 2m|F_o|-D|F_c| density in violet. *Reproduced with permission from Chen, J. C. H., Hanson, B. L., Fisher, S. Z., Langan, P., & Kovalevsky, A. Y. (2012). Direct observation of hydrogen atom dynamics and interactions by ultrahigh resolution neutron protein crystallography.* Proceedings of the National Academy of Sciences of the United States of America, 109 *(38), 15301–15306. doi: 10.1073/pnas.1208341109.*

occupancy might affect visibility (Afonine et al., 2010). Luckily, for most aliphatic and aromatic hydrogen atoms, the placement is unambiguous for stereochemical reasons.

3.3 Perdeuterated proteins

In contrast to the case for H/D-exchanged proteins, nuclear density maps from perdeuterated proteins are more pleasing to the eye of the X-ray crystallographer, as they are more continuous. The exchangeable protons should be clearly visible, as for the non-perdeuterated proteins, but in addition all aliphatic deuterium atoms are seen at approximately the same level as C, N, or O, resulting in "bulges" on almost all atoms bonded to D. There are no cancelation effects on the scattering of either the heavy or light atoms. Given the reduction in background from incoherent scattering afforded by replacing all H by D, which often also allows the use of smaller crystals for data collection, it is difficult to see any disadvantages with using perdeuterated protein other than expense and the potential difficulty of growing larger crystals. An example of nuclear scattering density from the 1.7 Å crystal structure of perdeuterated galectin-3C in complex with lactose (Manzoni et al., 2016; Manzoni et al., 2018) is shown in Fig. 3. Note that the lactose ligand is not deuterated and therefore suffers from cancelation effects around the carbon atoms, particularly affecting the visibility of the right hand glucose ring, which is more flexible than the left hand galactose ring (Fig. 3), but which is clearly visible in electron density maps from the same crystal (not shown). An "omit" nuclear density map in which D atoms have been omitted from the histidine and the ligand clearly shows the

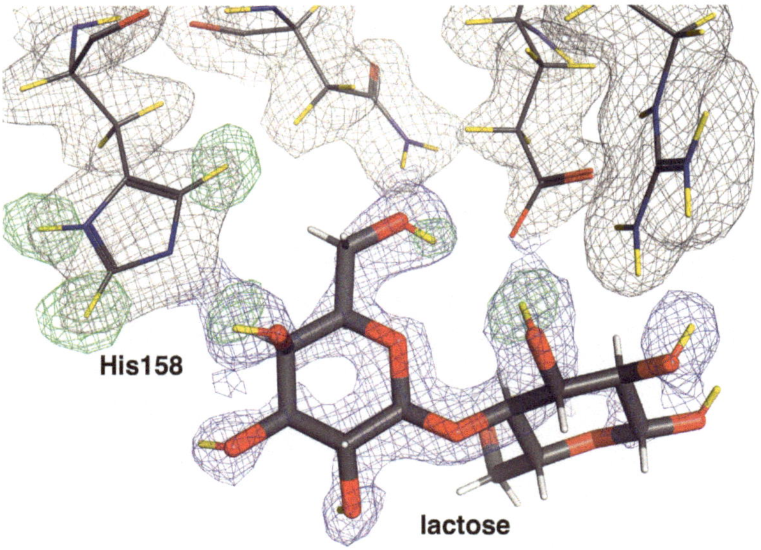

Fig. 3 Example of nuclear density from a fully deuterated protein, galectin-3C in complex with non-deuterated lactose (Manzoni et al., 2018). The $2m|F_o|$-$D|F_c|$ nuclear density around lactose is contoured at 0.6σ and colored blue; around protein residues it is contoured at 0.9σ and colored gray. An "omit" map where deuterium atoms have been omitted from lactose and the histidine side chain is contoured at 4σ and colored green. Hydrogen atoms are drawn in white and deuterium atoms in yellow.

aliphatic deuterium atoms and also the unusual protonation of the histidine on the ND1 atom. The positions of all polar D atoms on the inside of the lactose are also very well defined in omit maps at a high contour level.

3.4 Appearance of sulfur atoms

Sulfur scatters neutrons weakly (coherent scattering length 2.8 fm), thus the S atoms in Cys and Met are poorly visible in nuclear density maps. However, this is no problem for model building, as sulfur, being electron-rich, is usually very easily distinguished in the electron density maps (Fig. 4).

3.5 Appearance of water molecules

No matter whether the protein is hydrogenous or perdeuterated, the water molecules in a neutron crystal structure should always be D_2O, as the crystals are invariably solvent exchanged, both to increase the visibility of polar hydrogen atoms and to reduce overall background in the diffraction experiment from the incoherent scattering of hydrogen. As previously described in several studies, D_2O molecules have three characteristic appearances in nuclear density maps that depend on the degree to which the molecules are rotationally ordered and thus to the number of hydrogen bonds that they make (Fig. 5).

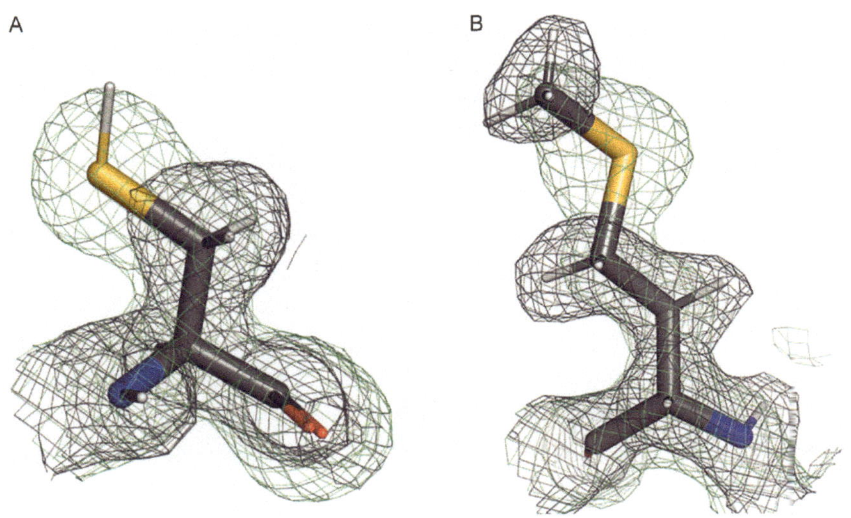

Fig. 4 Examples of nuclear and electron density from a fully deuterated protein, galectin-3C. (A) The area around Cys173 and (B) the area around Met130. The nuclear density is contoured at 1.0σ and colored gray. The electron density is also contoured at 1σ and is colored pale green.

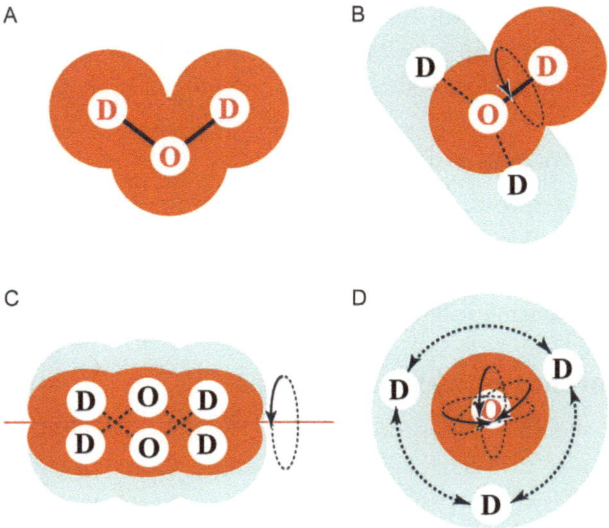

Fig. 5 Schematic diagram showing the order or disorder of D_2O molecules corresponding to their individual shapes: (A) triangular, (B) short ellipsoidal, (C) long ellipsoidal and (D) spherical. The approximate peak shapes in nuclear density maps are shown in orange areas (high positive contours) and cyan regions correspond to low diffuse contours. The interpretations of these shapes are indicated by the "O" and "D" labels: ordered O and D atoms are shown in red, while disordered atoms are shown with black letters. *Reproduced from Chatake, T., Kurihara, K., Tanaka, I., Tsyba, I., Bau, R., Jenney, F. E., … Niimura, N. (2004). A neutron crystallographic analysis of a rubredoxin mutant at 1.6 Å resolution.* Acta Crystallographica Section D: Biological Crystallography, 60, *1364–1373. doi: 10.1107/S090744490401176X by permission of the International Union of Crystallography.*

At higher resolutions, water molecules with two ordered protons have a characteristic "boomerang" shape due to the equal visibility of the oxygen and deuterium atoms (Fig. 5A). Where neither of the deuterons is well ordered, the density will be spherical (Fig. 5D). At higher resolution one can distinguish two types of ellipsoidal nuclear density, "short" and "long" (Fig. 5B and C). The former may indicate that one of the protons is ordered and the other rotationally disordered around the O-D axis of the first; the second may indicate that two water molecules with ordered protons are disordered around the longitudinal axis of the density (Chatake et al., 2004).

In cases where there is ambiguity in the orientation, it is very useful to combine information from the nuclear and electron density maps. The electron density map will report only on the position of the oxygen atom, while the nuclear density map will report on the deuterium atom(s) as well.

This can help resolve which of the often equally large peaks in the nuclear density map comes from the oxygen atom and thus which are from the deuterium atoms. In 2003, Chatake et al. made a systematic study of the water molecules in neutron crystal structures of myoglobin, rubredoxin, and a rubredoxin mutant at 1.5–1.6Å resolution (Chatake, Ostermann, Kurihara, Parak, & Niimura, 2003). In these three proteins, the average populations of the "boomerang," ellipsoidal, and spherical nuclear density forms were 29%, 16%, and 55%, though it is not always possible to assign a water molecule unambiguously to one of the categories.

As well as the electron density map, hydrogen-bonding considerations with the protein and other water molecules can also be useful in deciding how to model a particular water molecule. Chatake et al. observed that "boomerang"-shaped water molecules generally have the lowest B-factors and the largest number of anchors to the protein (two or three of their atoms H-bonded; Fig. 6A), while the molecules with elongated density usually have one or two anchor points (Fig. 6B and C) and the ones with spherically-shaped waters have only one anchor, or none at all (Fig. 6D). Furthermore, it is unusual to find a triangular-shaped nuclear density for a water molecule outside the first hydration shell, and if such a molecule is found, it will be connected to the surface of the protein through another triangular water molecule, not through an elongated or spherical one (Chatake et al., 2003).

One should also be aware that the visibility of water molecules in nuclear density maps is more sensitive to disorder than in X-ray maps. It is frequently not possible to identify water molecules in the neutron maps that are clearly visible in the X-ray maps even if the resolution is similar in both datasets. For example, in the 2004 study by Chatake et al., only 31 of the 53 water molecules identified in the 1.5Å X-ray structure had clear difference Fourier peaks above 3σ in the neutron difference maps at 1.6Å (Chatake et al., 2004). In our study of galectin-3C, crystals diffracted X-rays to almost 1.0Å at room temperature while neutron diffraction from the same crystals was limited to 1.6–1.8Å, depending on crystal size. All X-ray data were used for refinement, as this allowed the used of anisotropic B-factors that also improved the fit of the model to the neutron data, but since only water molecules present in the nuclear density maps were included in the final model, this led to a situation where the model did not completely explain the X-ray data. In such cases, it may be better to deposit a separate X-ray model for the high-resolution X-ray data from the same crystal.

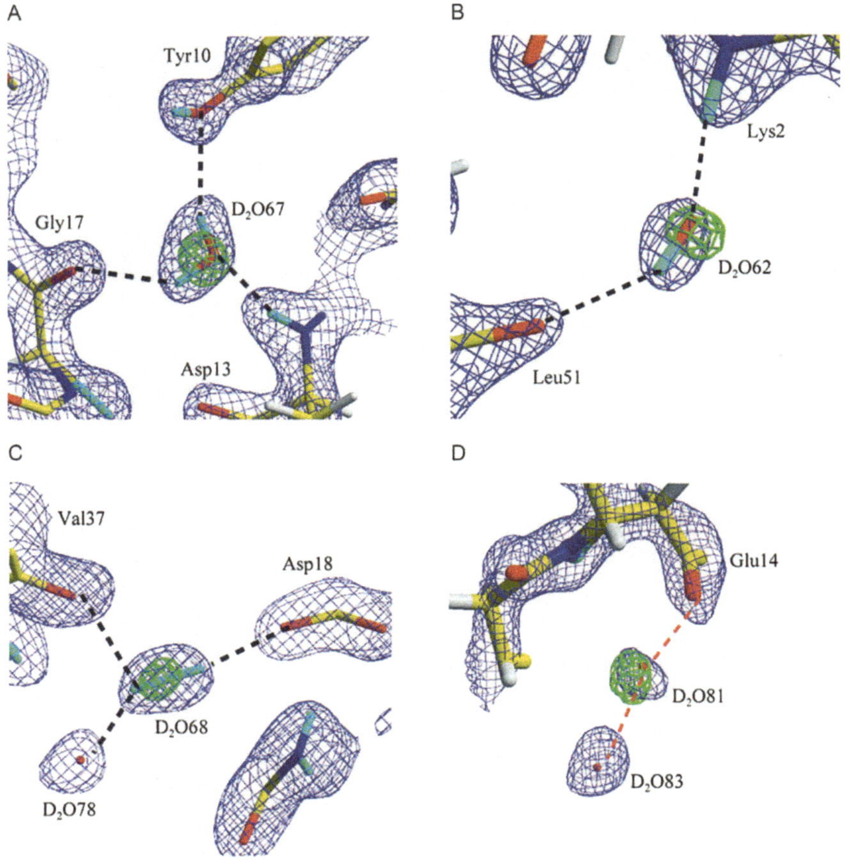

Fig. 6 2m|F_o|-D|F_c| maps of water molecules of hydration for a rubredoxin triple mutant observed by neutron crystallography. (A) triangular shape, (B) short ellipsoidal shape, (C) long ellipsoidal shape and (D) spherical shape. The blue mesh corresponds to the nuclear density map, while the green mesh corresponds to oxygen atom peaks from X-ray data. Atoms identifiable from the neutron data are shown as sticks, colored as follows: C, yellow; N, dark blue/violet; O, red; D, light blue/cyan. In (A) all atoms of the central D_2O molecule are visible (cyan/red/cyan), whereas in the other panels only some of the solvent atoms can be located: O and D in (B), two D in (C) and O only in (D). *Reproduced from Chatake, T., Kurihara, K., Tanaka, I., Tsyba, I., Bau, R., Jenney, F. E., ... Niimura, N. (2004). A neutron crystallographic analysis of a rubredoxin mutant at 1.6 Å resolution.* Acta Crystallographica Section D: Biological Crystallography, 60, *1364–1373. doi: 10.1107/S090744490401176X by permission of the International Union of Crystallography.*

In 2012, the ultra-high resolution neutron data obtained for crambin by Chen et al. enabled an examination of the effects of resolution on the appearance of water molecules (Chen et al., 2012). The results, summarized in Fig. 2B, once again lead to the conclusion that resolutions better than 2.0 Å are preferable for unambiguous identification of the shape and H-bonding patterns of water molecules in neutron diffraction experiments.

3.6 Distinguishing ions from water molecules

In principle, neutron crystallography can also be useful for distinguishing metal or other ions from water molecules in nuclear density maps. Again, this is due to the non-linear relationship of neutron scattering length to atomic number. The value for O is 5.8 fm while for Cl it is 9.6 fm. Li has a significant negative scattering (−1.9 fm), Na has 3.6 fm and Mg 5.4 fm. However, given the relative similarity of these values and that both water molecules and ions can have partial occupancy, the usefulness of neutrons in this context (except for Li) is relatively limited.

3.7 Hydronium ions and protons

Under certain circumstances, neutron crystallography can identify more unusual species such as hydronium ions, protons and hydroxide ions (Cuypers et al., 2013; Kovalevsky et al., 2011). The deuterated hydronium ion D_3O^+ is distinguished by a trigonal pyramidal shape in a nuclear density map (Fig. 7A). Special care must be taken to ensure that the density cannot be equally well explained by two partially occupied water molecules, i.e. the occupancies of all three deuterons should be the same as the occupancy of the O atom. For an isolated deuteron, it should be at hydrogen bonding distance from one or (preferably) more H-bond acceptors (Fig. 7B) and, importantly, should be absent from an electron density map from the same crystal.

4. Interactive model building of neutron crystal structures in Coot

As mentioned in Section 1, we will assume that the protein model is almost completely built into conventional X-ray maps, that most of the hydrogen atoms have been added and using e.g. phenix.ready_set or *REDUCE* (Word et al., 1999) and refined using *phenix.refine*. The concern of the following section is how to make the final adjustments to the model.

4.1 Modification of PDB files to allow real space refinement and geometry optimization

The ability to carry out real space refinement (RSR) and geometry optimization in Coot is essential for model building. However, the mutually

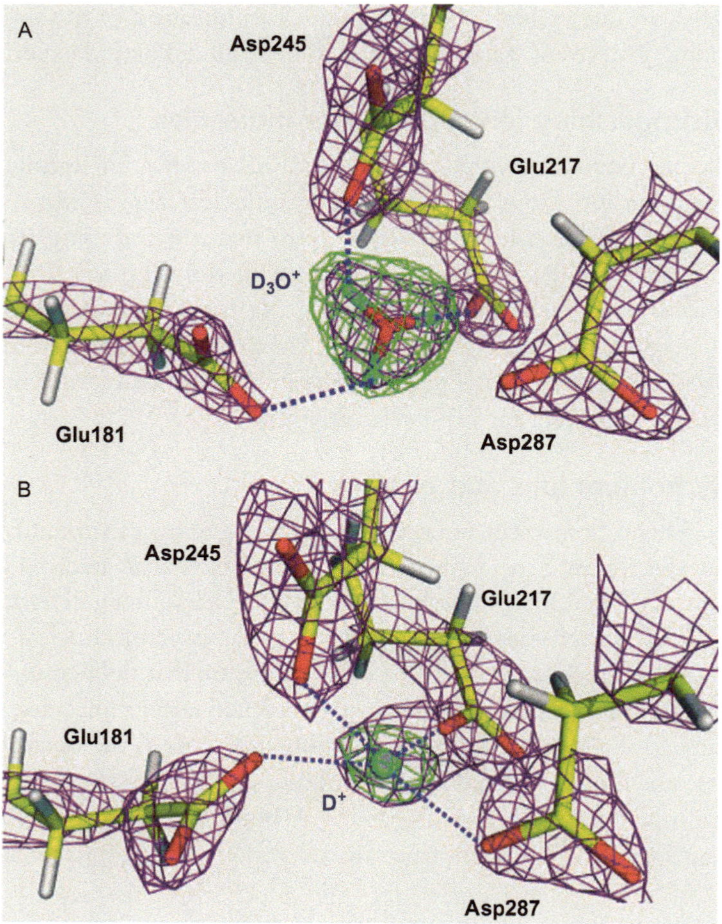

Fig. 7 The appearance of hydronium ions and protons (as D_3O^+ and D^+) in nuclear density maps. The M1 metal-binding site of the D-xylose isomerase active site: $2m|F_o|-D|F_c|$ (violet) and omit $m|F_o|-D|F_c|$ (green) positive neutron scattering density maps of the active site of D-xylose isomerase showing (A) D_3O^+ at pH 7.7 (1.8 Å resolution, PDB 3KCJ) and (B) D^+ at pH 5.9 (2.0 Å resolution, PDB 3QZA). Hydrogen bonds are shown as blue dotted lines. *Reproduced with permission from Kovalevsky, A. Y., Hanson, B. L., Mason, S. A., Yoshida, T., Fisher, S. Z., Mustyakimov, M., … Langan, P. (2011). Identification of the elusive hydronium ion exchanging roles with a proton in an enzyme at lower pH values.* Angewandte Chemie, International Edition, 50 *(33), 7520–7523. doi: 10.1002/anie.201101753. Copyright Wiley-VCH Verlag GmbH & Co. KGaA.*

incompatible atomic naming conventions adopted by Coot and *phenix.refine* can make this difficult. For example, when flipping an asparagine side chain or attempting to rotate a water molecule as a rigid body, the hydrogen atoms will not follow the heavy atoms. Attempts to optimize geometry will result in "exploding" models with deuterium atoms that move to distances far from the heavy atoms. In the early stages of model building and refinement, one can in principle overcome this by regenerating all hydrogen/deuterium positions afresh in each refinement cycle, but this is inappropriate in the later cycles, where one may have invested considerable time in optimizing rotatable, polar atom positions, particularly for water molecules.

In more detail, the atomic naming convention used by phenix.refine is not completely compatible with the geometry restraints libraries used by Coot. This is because Coot expects hydrogen atoms to be called H*, e.g. HA for the hydrogen atom on CA, HB1, HB2 and HB3 for the methyl hydrogen atoms of alanine, and so on. The actual element type is specified in column 78 of the PDB file. Thus, one solution to the problem for fully deuterated proteins or proteins where all exchangeable hydrogen atoms are replaced by deuterium is to globally rename all deuteriums to hydrogens then convert them back again after work in Coot is complete. This can be done through a simple `sed` script:

```
s/ DA/ HA/g
s/ DB/ HB/g
s/ DG/ HG/g
s/ DD/ HD/g
s/ DE/ HE/g
s/ DH/ HH/g
s/ DZ/ HZ/g
s/ D / H /g
s/ D1 / H1 /g
s/ D2 / H2 /g
```

This can be saved as a file and run using the command

```
sed -f sed.script input_withD.pdb > output_withH.pdb
```

The three spaces after "D" and "H" on line 8 are important, as with only two spaces, there is a risk that all element labels in column 78 are also replaced by "H". The first seven lines convert the names of all possible deuterium atoms in the side chains, the eighth line deals with amide deuterium and the last two lines take care of the deuterium on water molecules.

Real space refinement against neutron maps in Coot is of limited utility for fitting H/D atoms, as the selected atoms are refined using the sum of the stereochemical restraints and an atomic number-weighted sum of the *electron*-density values over all the atomic centers. Coot does not use neutron scattering tables. However, RSR can be useful for adjusting the positions of the heavier atoms, especially using the electron density maps.

Note that hydrogen/deuterium bond lengths in Coot will be optimized to the values in Coot's geometry libraries, which are based on X-ray crystallography. These tend to be systematically shorter than X—H bond lengths determined by neutron crystallography, as the atomic positions are biased toward the center of the electron density distribution, which is in the X—H bond. However, this will only affect the small subset of residues being remodeled, and the minor distortions will easily be taken care of in the following cycles of joint neutron/X-ray refinement in *phenix.refine*, where bond lengths based on neutron crystal structures can be chosen.

After model building in Coot, the coordinates can be converted back to an acceptable format for phenix.refine using the inverse sed script:

```
s/ HA/ DA/g
s/ HB/ DB/g
s/ HG/ DG/g
s/ HD/ DD/g
s/ HE/ DE/g
s/ HH/ DH/g
s/ HZ/ DZ/g
s/ H / D /g
s/ H1 / D1 /g
s/ H2 / D2 /g
sed -f sed.script input_withD.pdb > output_withH.pdb
```

When dealing with PDB files that have only the polar H atoms exchanged to D, or which have mixed H/D atoms for the polar hydrogens, the first script will also work to make the file suitable for Coot. However, the situation is more complex on converting the file back again, to avoid changing the names of all H atoms, including the non-exchanges ones, to D. A more specific set of `sed` commands must be given, which I must leave up to the reader.

An alternative method to overcome the nomenclature discrepancy is to generate a full set of restraint libraries for each amino acid with D instead of H, which has been applied by several researchers (personal communication). However, these libraries are not publicly available, thus the above manipulation of the coordinate file appears to be the most convenient solution.

4.2 Modeling protein residues in Coot

As long as the atom names have been changed to H then manipulation of the existing coordinates in Coot should pose no problems. The H atoms will move along with the heavy atoms to which they are attached during geometry optimization, real space refinement and when side chains are rotated or flipped manually. If adding an alternate conformation to a residue that is already hydrogenated, the alternate conformation will also have hydrogens. However, extra steps are required if adding a completely new residue, e.g. at the N- or C-terminus or in a loop. This is done using the "Add Residue" and "Mutate and AutoFit" commands. The residue introduced will have no hydrogen atoms. These can be added after saving the coordinates using, e.g.

```
phenix.reduce --no-build my_protein.pdb > my_protein_H.pdb
```

The `-no_build` flag ensures that existing H atoms are not disturbed. However, this command will hydrogenate every residue where H atoms are perceived to be missing. Since you may have put considerable effort into modeling unusual protonation states of protein residues and the form of water molecules (i.e. having removed protons that you deem not to be visible), you will not necessarily want to have hydrogens added automatically again. One solution is to excise the coordinates for the hydrogenated residue you just created with *phenix.reduce* using a text editor and add these to your master coordinate file. It is worth noting that, while *phenix.reduce* does not add hydrogens to water molecules, *phenix.ready_set* does.

4.3 Placement of water molecules in Coot

How to place a new water molecule including deuterium atoms? Simply using the "Place atom at pointer" or "Find waters" tool in Coot will only place the oxygen atom. Thus, the best solution is to move the pointer to each nuclear density peak in turn then using "Get Monomer" from the "File" menu. This will import a water molecule with hydrogen atoms and place it at the pointer. It can now be real-space refined and manually reoriented using the "Rotate/Translate Zone" menu item. When water placement is done, these separate molecules should be merged into the main molecule containing the protein structure. The second `sed` script described above will then convert the H1 and H2 atom names to D1 and D2 along with those of the rest of the protein, for refinement in *phenix.refine*. However, the atom type in column 78 of the PDB file needs to be changed manually from H to D using a text editor.

For partially exchanged H/D sites, one option in phenix.refine is to refine both H and D with partial occupancies that are forced to sum to 1.0. However, the presence of H and D at the same position may also cause problems in Coot. The author's preferred solution would be to refine using deuterium only and then work out the degree of exchange from the relative neutron scattering lengths of H and D.

5. Depositing a joint X-ray/neutron structure in the Protein Data Bank

Deposition of three-dimensional macromolecular structures is done through the PDB interface at deposit.wwpdb.org, which will direct the user to the appropriate regional server. The deposition tools provided by the Protein Data Bank are currently not optimized for structures refined jointly using X-ray and neutron data, due to the historically low number of structures deposited. Often a significant amount of manual intervention from wwPDB staff is still necessary, and it is necessary to maintain a dialog with them. With the growth in popularity of neutron crystallography (half of the structures deposited as of October 2019 have been solved in the past five years), this situation may improve, but we describe here some tips that will facilitate deposition with the current interface.

Since July 01, 2019, the PDB requires coordinate deposition in mmCIF format (Adams et al., 2019), and suitable files are output by phenix.refine and Refmac5. Thus coordinate deposition is normally no more problematic for neutron structures than for X-ray ones. The wwPDB still allows deposition of merged diffraction data either in mmCIF format or in the older MTZ format. Nevertheless, for convenient deposition of the two separate datasets used for a joint neutron/X-ray diffraction, one should use mmCIF, which has the advantage of being in ASCII file format and thus modifiable using a text editor. For correct interpretation by PDB deposition tools, the two sets of diffraction data should be declared in separate data blocks. The latest version of *phenix.refine* will output two separate data blocks if the box "Write Fobs, Fmodel and R-free-flags into mmCIF file" is checked under the "Output" tab of the "Configure" menu. These blocks will be identifiable by the keywords "`data_your_project_name`" and "`data_your_project_name_neutron`" on the first line of the data block. Otherwise, a convenient way to prepare diffraction data files for deposition is to use the SF-TOOL web server: http://sf-tool.wwpdb.org/ either using the separate raw data files or the merged output file from *phenix.refine*, and make a separate mmCIF file for each of the datasets, using the semi-automatic method to ensure that the correct columns for each dataset are selected. If there is a single free R-factor column for both datasets, it should be included in the first dataset. Having created these two separate mmCIF files, use the Linux command `cat` or a text editor to combine them together into one file. It is important to ensure that the X-ray dataset is the first one, as this will enable the PDB validation to proceed using the X-ray data alone (there are at present no validation tools to check goodness of fit to nuclear density maps). The combined mmCIF file can then be uploaded to the deposition session.

The PDB does not calculate nuclear density maps in the way that is now routine for X-ray maps, thus it can be helpful to deposit a CIF file containing the pre-calculated map coefficients from *phenix.refine* that the PDB user can download. In this way the user will see the maps that you calculated.

6. Visualizing nuclear (and electron) density maps in PyMOL

Finally, you may also wish to display the nuclear and electron density maps in PyMOL, especially for publication-quality images. Recent versions of PyMOL have the ability to Fourier-transform MTZ files to density maps using the Schrödinger PrimeX module, in a similar way to Coot, which obviates the need to calculate maps separately. For the original figures in this chapter I have used MacPyMOL v2.3.3. MTZ files can be opened through the "Open" menu. The correct columns can then be selected, as well as an appropriate prefix for the map objects, as in Fig. 8.

This operation can be laborious if one wants to display both $2m|F_o|$-$D|F_c|$ and difference maps for both X-rays and neutrons. Thus, to avoid having to open such "non-standard" MTZ files multiple times, it may be advantageous to use simple scripts. Here is one example, which will display both neutron and X-ray maps. The input file names and object names in PyMOL can of course be changed to anything you like.

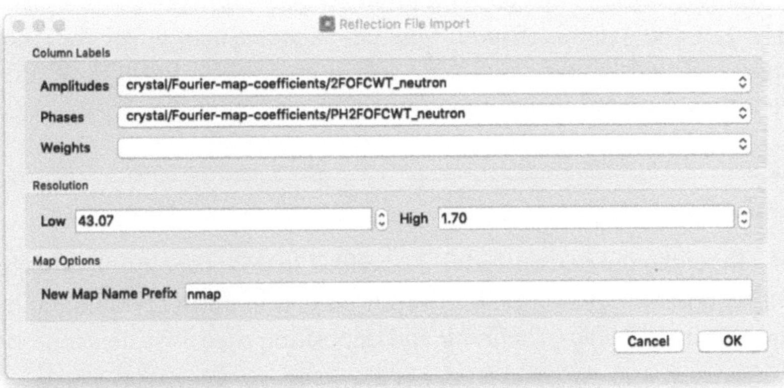

Fig. 8 The MTZ column selection menu from PyMOL.

```
set mesh_width, 0.3
# First open the nuclear density map - make sure that PyMOL reads
only the neutron columns
set default_phenix_names, 2FOFCWT_neutron PH2FOFCWT_neutron
FOFCWT_neutron PHFOFCWT_neutron
load 6EYM.mtz, nmap
load 6EYM.pdb, mol
isomesh 2fofc_n, nmap.2fofc, 1.0, mol, carve=5.0
colour purple, 2fofc_n
isomesh fofc+_n, nmap.fofc, 3.0, mol, carve=5.0
colour yellow, fofc+_n
isomesh fofc-_n, nmap.fofc, -3.0, mol, carve=5.0
colour orange, fofc-_n
# Now the X-ray map - change default map coefficients to the X-ray
ones
set default_phenix_names, 2FOFCWT_xray PH2FOFCWT_xray FOFCWT_xray
PHFOFCWT_xray
load 6EYM_001.mtz, xmap
isomesh 2fofc_x, xmap.2fofc, 1.0, mol, carve=5.0
colour white, 2fofc_x
isomesh fofc+_x, xmap.fofc, 3.0, mol, carve=5.0
colour green, fofc+_x
isomesh fofc-_x, xmap.fofc, -3.0, mol, carve=5.0
colour red, fofc+_x
show lines, mol
```

This script would contour the nuclear and electron density maps at a level of 1.0σ for the $2m|F_o|$-$D|F_c|$ map and ±3.0σ for the difference maps around the entire molecule with a cushion of 5.0 Å. The appearance will be similar to Figs. 1, 3, and 4 in this chapter. One can of course manipulate the amount of map displayed by judicious atom selection. See the PyMOL Wiki[c] for more information on choice of map coefficients.

7. Conclusion

Hopefully the practical advice provided in this chapter will prove useful to the relative newcomer to macromolecular neutron crystallography in the adventure of solving, refining and depositing their first neutron crystal

[c] https://pymol.org/dokuwiki/doku.php?id=media:edmaps

structure. It is obvious that the process currently involves more care and manual intervention than X-ray crystallography, but the outcome should be a just reward.

Acknowledgments

I would like to thank all colleagues who offered comments on this chapter, including Zoë Fisher, Esko Oksanen, Pavel Afonine and Dorothee Liebschner. Thanks also to Paul Emsley for providing the scheme script for opening joint neutron/X-ray MTZ files from *phenix.refine*.

References

Adams, P. D., Afonine, P. V., Baskaran, K., Berman, H. M., Berrisford, J., Bricogne, G., et al. (2019). Announcing mandatory submission of PDBx/mmCIF format files for crystallographic depositions to the Protein Data Bank (PDB). *Acta Crystallographica Section D: Structural Biology*, *75*, 451–454. https://doi.org/10.1107/S2059798319004522.

Adams, P. D., Mustyakimov, M., Afonine, P. V., & Langan, P. (2009). Generalized X-ray and neutron crystallographic analysis: More accurate and complete structures for biological macromolecules. *Acta Crystallographica Section D: Biological Crystallography*, *65*(Pt. 6), 567–573. https://doi.org/10.1107/S0907444909011548.

Afonine, P. V., Mustyakimov, M., Grosse-Kunstleve, R. W., Moriarty, N. W., Langan, P., & Adams, P. D. (2010). Joint X-ray and neutron refinement with phenix.refine. *Acta Crystallographica Section D: Biological Crystallography*, *66*(Pt. 11), 1153–1163. https://doi.org/10.1107/S0907444910026582.

Afonine, P. V., Ralf, W., Headd, J. J., & Thomas, C. (2012). Towards automated crystallographic structure refinement with phenix.refine. *Acta Crystallographica Section D: Biological Crystallography*, *68*, 352–367. https://doi.org/10.1107/S0907444912001308.

Chatake, T., Kurihara, K., Tanaka, I., Tsyba, I., Bau, R., Jenney, F. E., et al. (2004). A neutron crystallographic analysis of a rubredoxin mutant at 1.6 Å resolution. *Acta Crystallographica Section D: Biological Crystallography*, *60*, 1364–1373. https://doi.org/10.1107/S090744490401176X.

Chatake, T., Ostermann, A., Kurihara, K., Parak, F. G., & Niimura, N. (2003). Hydration in proteins observed by high-resolution neutron crystallography. *Proteins: Structure, Function and Genetics*, *50*(3), 516–523. https://doi.org/10.1002/prot.10303.

Chen, J. C. H., Hanson, B. L., Fisher, S. Z., Langan, P., & Kovalevsky, A. Y. (2012) Direct observation of hydrogen atom dynamics and interactions by ultrahigh resolution neutron protein crystallography. *Proceedings of the National Academy of Sciences of the United States of America*, *109*(38), 15301–15306. https://doi.org/10.1073/pnas.1208341109.

Cuypers, M. G., Mason, S. A., Blakeley, M. P., Mitchell, E. P., Haertlein, M., & Forsyth, V. T. (2013). Near-atomic resolution neutron crystallography on perdeuterated *Pyrococcus furiosus* rubredoxin: Implication of hydronium ions and protonation state equilibria in redox changes. *Angewandte Chemie, International Edition*, *52*(3), 1022–1025. https://doi.org/10.1002/anie.201207071.

Emsley, P., Lohkamp, B., Scott, W. G., & Cowtan, K. (2010). Features and development of Coot. *Acta Crystallographica Section D: Biological Crystallography*, *66*(Pt. 4), 486–501. https://doi.org/10.1107/S0907444910007493.

Gruene, T., Hahn, H. W., Luebben, A. V., Meilleur, F., & Sheldrick, G. M (2014). Refinement of macromolecular structures against neutron data with SHELXL2013. *Journal of Applied Crystallography*, *47*(1), 462–466. https://doi.org/10.1107/S1600576713027659.

Kossiakoff, A. A., & Spencer, S. A. (1980). Neutron diffraction identifies his 57 as the catalytic base in trypsin. *Nature, 288*(5789), 414–416. https://doi.org/10.1038/288414a0.

Kovalevsky, A. Y., Hanson, B. L., Mason, S. A., Yoshida, T., Fisher, S. Z., Mustyakimov, M., et al. (2011). Identification of the elusive hydronium ion exchanging roles with a proton in an enzyme at lower pH values. *Angewandte Chemie, International Edition, 50*(33), 7520–7523. https://doi.org/10.1002/anie.201101753.

Langan, P., Fisher, Z., Kovalevsky, A., Mustyakimov, M., Sutcliffe Valone, A., Unkefer, C., et al. (2008). Protein structures by spallation neutron crystallography. *Journal of Synchrotron Radiation, 15*(Pt. 3), 215–218. https://doi.org/10.1107/S0909049508000824.

Liebschner, D., Afonine, P. V., Baker, M. L., Bunkóczi, G., Chen, V. B., Croll, T. I., et al. (2019). Macromolecular structure determination using X-rays, neutrons and electrons: Recent developments in Phenix. *Acta Crystallographica Section D: Structural Biology, 75*(10), 861–877. https://doi.org/10.1107/s2059798319011471.

Manzoni, F., Saraboji, K., Sprenger, J., Kumar, R., Noresson, A. L., Nilsson, U. J., et al. (2016). Perdeuteration, crystallization, data collection and comparison of five neutron diffraction data sets of complexes of human galectin-3C. *Acta Crystallographica Section D: Structural Biology, 72*(11), 1194–1202. https://doi.org/10.1107/S2059798316015540.

Manzoni, F., Wallerstein, J., Schrader, T. E., Ostermann, A., Coates, L., Akke, M., et al. (2018). Elucidation of hydrogen bonding patterns in ligand-free, lactose- and glycerol-bound galectin-3C by neutron crystallography to guide drug design. *Journal of Medicinal Chemistry, 61*(10), 4412–4420. https://doi.org/10.1021/acs.jmedchem.8b00081.

Murshudov, G. N., Skubak, P., Lebedev, A. A., Pannu, N. S., Steiner, R. A., Nicholls, R. A., et al. (2011). REFMAC5 for the refinement of macromolecular crystal structures. *Acta Crystallographica Section D: Biological Crystallography, 67*(Pt. 4), 355–367. https://doi.org/10.1107/S0907444911001314.

Oksanen, E., Blakeley, M. P., El-Hajji, M., Ryde, U., & Budayova-Spano, M. (2014). The neutron structure of urate oxidase resolves a long-standing mechanistic conundrum and reveals unexpected changes in protonation. *PLoS One, 9*(1), e86651. https://doi.org/10.1371/journal.pone.0086651.

Ostermann, A., Tanaka, I., Engler, N., Niimura, N., & Parak, F. G. (2002). Hydrogen and deuterium in myoglobin as seen by a neutron structure determination at 1.5 Å resolution. *Biophysical Chemistry, 95*(3), 183–193. https://doi.org/10.1016/S0301-4622(01)00255-1.

Sheldrick, G. M., & Schneider, T. R. (1997). SHELXL: High-resolution refinement. *Methods in Enzymology, 277*, 319–343. https://doi.org/10.1016/S0076-6879(97)77018-6.

Shu, F., Ramakrishnan, V., & Schoenborn, B. P. (2000). Enhanced visibility of hydrogen atoms by neutron crystallography on fully deuterated myoglobin. *Proceedings of the National Academy of Sciences of the United States of America, 97*(8), 3872–3877. https://doi.org/10.1073/pnas.060024697.

Winn, M. D., Ballard, C. C., Cowtan, K. D., Dodson, E. J., Emsley, P., Evans, P. R., et al. (2011). Overview of the CCP4 suite and current developments. *Acta Crystallographica. Section D, Biological Crystallography, 67*(Pt. 4), 235–242. https://doi.org/10.1107/S0907444910045749.

Word, J. M., Lovell, S. C., LaBean, T. H., Taylor, H. C., Zalis, M. E., Presley, B. K., et al. (1999). Visualizing and quantifying molecular goodness-of-fit: Small-probe contact dots with explicit hydrogen atoms. *Journal of Molecular Biology, 285*(4), 1711–1733. https://doi.org/10.1006/jmbi.1998.2400.

CHAPTER ELEVEN

What are the current limits on determination of protonation state using neutron macromolecular crystallography?

Dorothee Liebschner[a,*], Pavel V. Afonine[a], Nigel W. Moriarty[a], Paul D. Adams[a,b]

[a]Molecular Biophysics and Integrated Bioimaging Division, Lawrence Berkeley National Laboratory (LBNL), Berkeley, CA, United States
[b]Department of Bioengineering, University of California Berkeley, Berkeley, CA, United States
*Corresponding author: e-mail address: dcliebschner@lbl.gov

Contents

Methods in Enzymology, Volume 634
ISSN 0076-6879
https://doi.org/10.1016/bs.mie.2020.01.008

Abstract

The rate of deposition of models determined by neutron diffraction, or a hybrid approach that combines X-ray and neutron diffraction, has increased in recent years. The benefit of neutron diffraction is that hydrogen atom (H) positions are detectable, allowing for the determination of protonation state and water molecule orientation. This study analyses all neutron models deposited in the Protein Data Bank to date, focusing on protonation state and properties of H (or deuterium, D) atoms as well as the details of water molecules. In particular, clashes and hydrogen bonds involving H or D atoms are investigated. As water molecules are typically the least reproducible part of a structural model, their positions in neutron models were compared to those in homologous high-resolution X-ray structures. For models determined by joint refinement against X-ray and neutron data, the water structure comparison was also carried out for models re-refined against the X-ray data alone. The homologues have generally fewer conserved water molecules where X-ray only was used and the positions of equivalent waters vary more than in the case of the hybrid X-ray model. As neutron diffraction data are generally less complete than X-ray data, the influence of neutron data completeness on nuclear density maps was also analyzed. We observe and discuss systematic map quality deterioration as result of data incompleteness.

1. Introduction

X-ray crystallography is the predominant method of determining the three-dimensional structure of macromolecules, accounting for 89% of all models deposited in the PDB (Berman, Henrick, & Nakamura, 2003; Bernstein et al., 1977; wwPDB Consortium, 2019). The technique is based on the interaction between X-rays and the electrons of the atoms in a crystalline sample. Major peaks in Fourier maps equate to atomic positions that can be used to determine the structure. As a hydrogen atom (H) possesses only one valence electron, it does not produce enough signal to be observed in electron density maps, unless the data resolution is very high.

By contrast, neutron diffraction relies on the interaction of neutrons with atomic nuclei. As hydrogen and deuterium atoms (D) have scattering cross sections of similar magnitude to those of heavier atoms (O, N, C), neutron diffraction can be used to locate H and D atoms even at medium data resolution. This opens up the possibility of determining the protonation states of amino acid side chains and the orientation of some water molecules, which is important for understanding the reaction pathways of proteins (Ankner, Heller, Herwig, Meilleur, & Myles, 2013). However, there are practical reasons that impede routine application of the technique; in particular, refinement is challenging due to low data completeness, low signal-to-noise ratio and the increased number of parameters (H atoms can introduce more variable parameters). The data completeness is typically low for several reasons, including the relatively low flux of neutron beams, reduced signal to noise due to incoherent scattering when hydrogen atoms are present in the sample, and the limited data-collection time available on neutron crystallography instruments (due to oversubscribed facilities). As a consequence, the completeness of neutron diffraction data reaches only about 80% on average across all neutron depositions. By contrast, an X-ray dataset is considered to be of good quality if the completeness is at least 95% (Dauter & Dauter, 2017). Neutron data are typically collected in Laue mode yielding harmonic and spatially overlapped reflections, particularly affecting the completeness of low resolution reflections (Cruickshank, Helliwell, & Moffat, 1987). Furthermore, a 95% complete dataset may lack 20% of low-resolution reflections (Dauter, 1999), so the completeness expressed as an overall value is not particularly informative (Rupp, 2018). Because all reflections contribute to each point of an electron density map, the map quality decreases if the data are not complete and if reflections are missing (Urzhumtsev, 1991; Urzhumtsev, Afonine, Lunin, Terwilliger, & Adams, 2014; Wlodawer, Minor, Dauter, & Jaskolski, 2008). The deterioration can be mild or severe: random or weak absent reflections have little effect on maps (Evans, 2011) while missing strong reflections and missing wedges (Wennmacher et al., 2019) can distort maps considerably. The completeness also affects coordinate errors as estimated by the Cruickshank DPI (diffraction-component precision index) (Cruickshank, 1999). In this work we quantify the influence of missing reflections on nuclear density maps by comparing model-calculated Fourier syntheses ($F_{\text{model}}e^{i\varphi_{\text{model}}}$) and histograms of $F_{\text{model}}e^{i\varphi_{\text{model}}}$ density at atomic centers.

The method of joint X-ray/neutron refinement (joint XN refinement) addresses the low data-to-parameter ratio by refining a single model simultaneously against X-ray and neutron data (Adams, Mustyakimov, Afonine, & Langan, 2009; Afonine et al., 2010; Coppens, 1967; Orpen, Pippard, Sheldrick, & Rouse, 1978; Wlodawer, 1980; Wlodawer & Hendrickson, 1982). Ideally, the X-ray and neutron datasets should be collected at the same temperature and from the same or a highly isomorphous crystal, although this cannot always be accomplished.

The number of neutron models deposited in the PDB has nearly doubled in the past 5 years, reaching 161 as of October 2019.

Therefore, an assessment of these models is timely as the number of structures derived from neutron diffraction is likely to increase further in the coming years. Special focus is put on protonation states and properties of H (D) atoms and water molecules, because this is the distinctive information that can be derived from neutron diffraction experiments. In particular, we investigate clashes and hydrogen bonds involving H (D) atoms, distinguishing in our analyses between hydrogen atoms that have a degree of freedom ("rotatable" H atoms) and those that are fully determined by their nonhydrogen parent atoms ("fixed" H atoms).

A prerequisite of applying the joint XN refinement method is that the models derived from the X-ray and neutron data should be isomorphous. It has been shown that water molecules are the least reproducible part of structural models (Blakeley, Kalb, Helliwell, & Myles, 2004; Fields et al., 1994; Fujinaga, Delbaere, Brayer, & James, 1985; Ohlendorf, 1994). One fundamental reason why water is difficult to model is that it has a very simple shape—just one peak—that often can be similar in appearance to noise (Weichenberger, Afonine, Kantardjieff, & Rupp, 2015). Ohlendorf (1994) reported that water is the least reliable species when re-refining four models of human interleukin 1β. Fujinaga et al. (1985) found that 15% of water molecules did not reappear if they were deleted and redetermined. Fields et al. (1994) showed that 85% of water molecule positions occurred within 1 Å in two independently refined models of platocyanin, while the root mean square deviation (*rmsd*) of protein Cα atoms was only 0.08 Å. The comparison of a 1.65 Å X-ray model with a 2.5 Å neutron model of concavanalin A revealed that only 19% of water molecules had matching positions (Blakeley et al., 2004). While these analyses show a trend that about 80% of water positions are conserved, a systematic study has not yet been performed. To address this issue, we compared the water structures in neutron models to (a) the water structure in high-resolution X-ray homologue models and (b) to joint XN models re-refined against X-ray data alone.

2. Materials and methods

2.1 Retrieving models and data from the PDB

Computations were carried out with *Phenix* tools (Liebschner et al., 2019) and scripts based on the *CCTBX* library (Grosse-Kunstleve, Sauter, Moriarty, & Adams, 2002). Model and data files were retrieved from the Protein Data Bank (PDB). To identify models determined by neutron diffraction, the PDB file header ("EXPERIMENTAL METHOD" in PDB format) was parsed. Other information retrieved from the file header includes resolution limits, *sigma* cutoffs, the twin law (if present), crystallographic *R*-factors and deposition year. Some models had to be curated before they could be used in *Phenix* tools; for example, atoms were renamed to obey conventions in *CCTBX* tools. Details of model and data curation are described in Liebschner, Afonine, Moriarty, Langan, and Adams (2018). For some analyses, such as refinement, map calculation and computation of completeness, the high-resolution limit is required. If the resolution limit in the PDB file header was inconsistent with that deduced from the reflections in the data file, the maximum (numerically larger value) of the two resolution limits was used. This prevents underestimating data completeness if the deposited dataset contains more reflections than indicated by the user-specified limit. If the limit value was missing or ambiguous, the high-resolution limit was determined from the data file.

2.2 Elongation to neutron X-H distances

Density peaks for H atoms appear at different distances in electron and nuclear density maps. In an electron density map, the density of the H atom is shifted away from the nucleus along the X-H bond towards atom X. To account for this difference, it is standard practice to apply different ideal X-H bond lengths for X-ray and neutron data, with neutron models using distances that are around 0.1 Å longer. However, the X-H distances in the deposited neutron models are not consistent: a nonnegligible number have H atoms at X-ray distances (Gruene, Hahn, Luebben, Meilleur, & Sheldrick, 2014; Liebschner et al., 2018). To compare models, it is important to compute clashes and hydrogen bonds from models with consistent X-H distances. Therefore, the X-H distances were elongated to neutron dictionary target values (Allen, 1986; Allen & Bruno, 2010) in all neutron models using *CCTBX* functions.

2.3 Fixed and rotatable H atoms

The position of fixed H atoms can be inferred from the coordinates of neighboring, non-H atoms (see also chapter "Implementation of the riding hydrogen model in CCTBX to support the next generation of X-ray and neutron joint refinement in *Phenix*" by Liebschner et al.). An example is the amide H atom in the peptide group; it is located in a plane formed by the atoms C, O and N. These fixed H atoms may deviate somewhat from

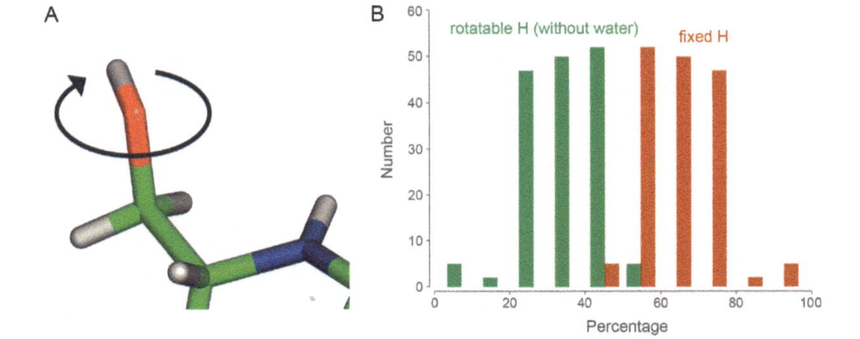

Fig. 1 (A) The Hγ atom in Ser has a rotational degree of freedom while the other H atoms have fixed geometry. (B) Percentage of fixed and rotatable H atom labels in neutron models. About two-thirds are fixed and one-third is rotatable. Water molecules are excluded as it depends on individual choice how many H atoms were added.

their ideal position, but the geometrical configuration will be preserved: an H atom in a tetragonal geometry will not switch into planar geometry. Rotatable H atoms have a degree of freedom to rotate, for example the Hγ atom in a serine residue, which can rotate around the axis formed by the Oγ and Cβ atoms (Fig. 1A), or "propeller" groups such as atoms Hγ21, Hγ22 and Hγ23 in an isoleucine residue. While the Hγ atom can rotate 360 degrees, the propeller group superposes after 120 degrees, so its effective degree of freedom is diminished. Water molecules are a special case because the molecule can orient to optimize local interactions, such as hydrogen bonds. Therefore, some analyses focus on the different H atom types: fixed, rotatable and water H atoms.

2.4 Calculating clashscores for subsets of H atoms

The Molprobity clashscore is defined as the number of steric overlaps $\geq= 0.4\,\text{Å}$ per 1000 atoms (Word et al., 1999). It is typically calculated for all atoms, but it can also be calculated for a subset of atoms. For this study, the clashscore was calculated for a subset of hydrogen atoms (rotatable, fixed and water H atoms):

$$\text{clashscore}_{\text{subset}} = n_{\text{overlaps in subset}} \cdot \frac{1000}{n_{\text{atoms in subset}}}$$

We note that interactions that qualify as hydrogen bond (see Section 2.5) are not counted as a clash.

2.5 Hydrogen bond analysis

The general chemical definition describes a hydrogen bond X-H...A-Y as a local, noncovalent bond where the X-H group acts as proton donor to the acceptor atom A (Steiner, 2002). A specialized definition in the context of crystal structures involves interaction geometries (distances and angles). Different cut-offs for bond lengths and angles are used in the literature (Fabiola, Bertram, Korostelev, & Chapman, 2002; Koch, Bocola & Klebe, 2005; Thomas, Benhabiles, Meurisse, Ngwabije, & Brasseur, 2001; Torshin, Weber, & Harrison, 2002); here we interpreted a X-H... A-Y interaction as a H-bond if (a) the H...A distance is between 1.2 and 2.2 Å, (b) the X...A distance is between 2.2 and 3.2 Å, (c) the X-H...A angle is larger than 110° and (d) the H...A-Y angle is larger than 90°, inline with recommendations given by Steiner (2002). The number of hydrogen bonds and their geometry was determined for subsets of hydrogen atoms (rotatable, fixed and water molecules).

2.6 Obtaining homologous models

Similar protein sequences for each neutron model chain were found with the program BLAST (*blastp*, Altschul, Gish, Miller, Myers, & Lipman, 1990; Madden, 2003). The model with the highest sequence identity and highest resolution was assigned as homologous structure. The following cut-offs were applied: minimum sequence identity of 90% and minimum data resolution of 2 Å.

2.7 Refinement of joint XN models against X-ray data alone

To investigate if the positions of water molecules change, joint XN models were re-refined against the X-ray data alone with *phenix.refine* (Afonine et al., 2012). The following parameters are different from default options: number of macrocycles (set to 10), optimization of the stereochemistry/ X-ray weight and the weight of atomic displacement parameters (ADP). Water molecules were automatically updated during refinement, the defaults were changed so that water molecules with ADPs >60 Å^2 or with occupancy less than 0.5 were removed, ensuring that less reliable water molecules are excluded from the model.

2.8 Comparing water structures

The water clustering method as described in Moriarty et al. (2018) was used to determine if a water molecule in the neutron model matches a water in another structure (X-ray homology model or newly refined joint XN model against X-ray data). Prior to this analysis, the models were superposed using the *CCTBX* script *phenix.superpose_pdbs*. If no water molecule was found within 1 Å, the water was considered as nonpreserved. The comparison was restricted to proteins with one chain, as the current version of the clustering algorithm only compares water molecules within a particular chain.

2.9 Calculated maps for incomplete and complete neutron data

To investigate the influence of data completeness on nuclear density maps, we calculated $F_{\text{model}}e^{i\varphi_{\text{model}}}$ Fourier syntheses that we scaled by the standard deviation of the map values (for details, see Section 3.3.1 in Urzhumtsev et al., 2014). For each neutron dataset, we computed Fourier syntheses from model-calculated data using a complete dataset up to highest resolution cutoff or only the reflections present in the experimental dataset. The peak correlation coefficient $CC_{<\text{peaks}>}$ at the 90 percentile rank between the two maps for each neutron model was calculated as described in Urzhumtsev et al. (2014). Furthermore, histograms of density values interpolated at atomic centers as well as data completeness profiles were calculated using *CCTBX* functions. Completeness was calculated for the entire resolution range (minimum to maximum resolution) and for resolution bins. Two different binning schemes were used: one containing an approximately equal number of reflections (cubic binning, number of reflections: 500), the other applied a logarithmic scale. The logarithmic scaling allows more detailed binning at low resolution without disproportionally increasing the number of bins at high resolution (Afonine, Grosse-Kunstleve, Adams, & Urzhumtsev, 2013).

2.10 Percentage of observed D atoms

To find out how many D atoms in deposited neutron models can be reliably located, we evaluated how many D atoms have significant nuclear density peaks. The script used for this task loops over all D atoms in a neutron model, omitting one D atom at a time and calculating an $m\text{F}_{\text{obs}}$-$D\text{F}_{\text{model}}$ map. If a density peak above 3σ exists in the near vicinity of the D atom, the D atom is considered to be "observed." This test was restricted to D atoms with occupancy >0.95, which excludes all D atoms forming multiple conformations or that are partly exchanged. If the D atom is in an alternative conformation, applying a criterion of 3σ for the observation of a peak is most likely too strict.

If the D atom is only partly exchanged, it is superposed to an H atom, which has a negative nuclear density peak. The resulting superposed density of D and H is therefore diminished, so that the 3σ criterion is not appropriate. This analysis was performed using a *Phenix* tool that is currently under development (Xu et al., unpublished) and will be described elsewhere.

3. Results and discussion

3.1 Overview of deposited neutron models and data

As of October 2019, the PDB contains 161 models from originating from neutron diffraction experiments. Fig. 2 shows the cumulative number of neutron models. The oldest model in the PDB is from 1984 (5PTi, Wlodawer, Walter, Huber, & Sjölin, 1984; for the PDB code naming convention used in this article, please see Moriarty, 2015), but some structural reports were performed before the PDB was established (Schoenborn, 1969) or the models were not deposited (Teeter & Kossiakoff, 1984). In the early 1990s, no new models were deposited because neutron facilities such as the Institute Laue-Langevin in Grenoble and the High Flux Beam Reactor at Brookhaven were unavailable (Chen & Unkefer, 2017). In the 2000s, the

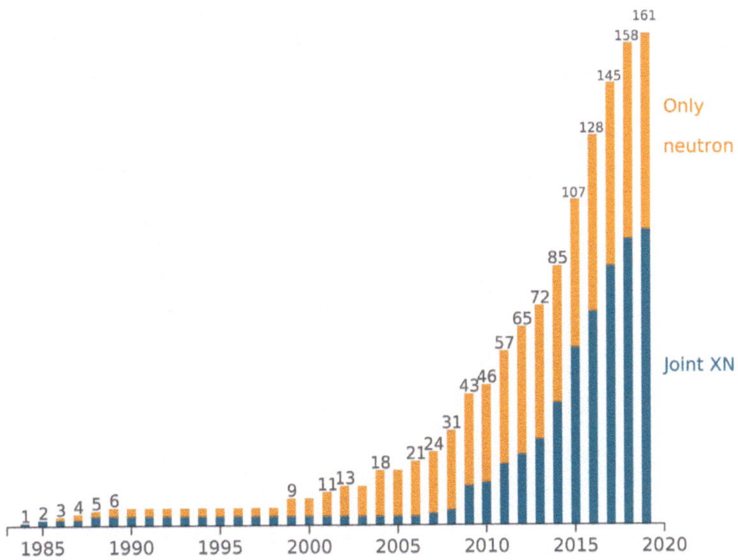

Fig. 2 Cumulative number of models determined by neutron diffraction in the PDB. Orange: models from refinement with neutron data alone. Blue: models from joint XN refinement. The number of cumulative models is written above the bar if it changed compared to the previous year.

rate of model deposition increased as a result of new advanced neutron sources (SNS in USA, FRM-2 in Germany, J-PARC in Japan) and developments in methods and technologies, such as the neutron image-plate detector (Niimura et al., 1994) and time-of-flight data collection techniques (Langan, Greene, & Schoenborn, 2004).

While the total number of structures has grown, the number of depositions per year is very low compared to X-ray crystallography (around 9000 in 2018). Among the 161 deposited models, 97 were determined from joint XN refinement (blue in Fig. 2) and 64 were obtained from neutron data alone (orange in Fig. 2). Most of the recent models were refined using the joint XN refinement method.

Some properties of deposited models and data are displayed in Fig. 3. Neutron diffraction requires relatively large (typically $>0.1\,\text{mm}^3$), well diffracting crystals. This is typically the case for relatively small proteins. As shown in Fig. 3A, most of the neutron models have less than 500 protein residues grouped in one or two chains (Fig. 3B). A large majority of structures (156) contains protein while only a few consist of nucleic acids (6). About half of neutron models (89) contain at least one ligand. Neutron diffraction is indeed attractive for the investigation of binding modes due to its ability to locate H atoms. The resolution of neutron diffraction data ranges from 1.05 Å to 2.75 Å, with an average of 1.97 Å. This reflects the fact that nuclear density maps at resolutions lower than 2.5 Å are often difficult to

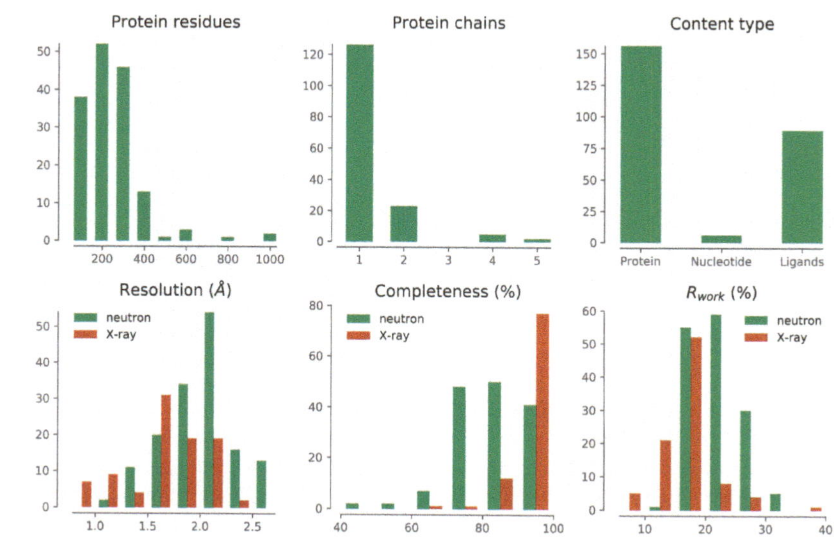

Fig. 3 Properties of neutron models and X-ray/neutron data.

interpret (see for example water molecule density in Fig. 2B of chapter "Interactive model building in neutron macromolecular crystallography" by Logan). For joint XN experiments, the X-ray data resolution is generally better than that of the neutron data; the average is 1.65 Å, extending from 0.93 Å to 2.30 Å. The data completeness differs strikingly between the experiments. The average completeness of neutron datasets is 82.3% compared to 94.1% for X-ray data. The crystallographic R-factor (R_{work}) is systematically smaller when calculated against X-ray data than against neutron data; the difference occurs because the data sets contain different sets of reflections and therefore cannot be compared meaningfully.

3.2 Clashscore

The clashscore represents sterically unlikely atom-atom overlaps (distance cut-off for overlaps: 0.4 Å) (Williams et al., 2018). It can reflect tight packing, but it is usually indicative of local fitting problems, such as flipped sidechains, misplaced residues, or misidentified groups. About two-thirds of H/D atoms are fixed and one-third are rotatable (Fig. 1B). We note that this number does not necessarily represent the number of sites, as there can be both H and D atoms at some sites when the crystal was subject to H/D exchange. We wanted to test if one H atom type (rotatable or fixed) has a higher propensity to form clashes than the other. As rotatable H atoms have a degree of freedom, it may be anticipated that they have fewer clashes.

Fig. 4A shows the clashscore of rotatable H atoms in each neutron model (including H from water) versus the clashscore of fixed H. The value of the

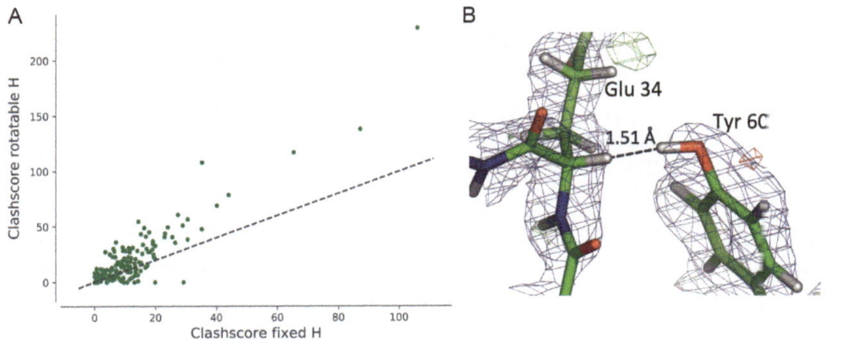

Fig. 4 (A) Comparing clashscores for fixed H and rotatable H (water included). (differences > zero: 117, difference < zero: 41, difference equal zero: 3). (B) Example for a clash (dashed line) in model 4C3Q. The $2mF_{obs}$-DF_{model} nuclear density is represented at 1σ in blue, mF_{obs}-DF_{model} nuclear density at ±3σ is represented in green (positive) and red (negative).

clashscore is typically around four for mid-resolution structures (Richardson et al., 2018). The values for H atom subsets are higher, because the number of clashes in the subset is divided by the number of the atoms in the subset instead of the total number of atoms. The numerical value, therefore, should not be interpreted like the usual clashscore. The distribution of points around the bisector shows that the majority of models have a larger clashscore for rotatable H and water (117 points are above the bisector, 41 are below). This is unexpected as rotatable hydrogen atoms and water molecules can theoretically adjust their position to fit into local geometry. Four models have clashscores larger than 100: in 1iO5, all 251 water molecules have two deuterium atoms, which form many clashes. PDB entries 1LZN and 3iNS have some modeling errors, such as water oxygen atoms in close proximity to protein residues and erroneous alternative conformations. Structure 2iNQ has several clashing H atoms in propeller groups. An example for a clashing rotatable D atom in model 4C3Q is shown in Fig. 4B. The Dη atom of Tyr_{60} clashes with the Dα atom of Glu_{34}. The distance between the two deuterium atoms is 1.51 Å, which corresponds to an overlap of 0.66 Å. There are two putative H-bond acceptors in the vicinity of the Dα atom: Oε1 of Glu_{37} (2.4 Å distance) and the O atom of water DOD_{1125} (3.2 Å distance). Also, the Dα atom of Glu_{34} is not covered by nuclear density. Altogether, this suggests that the orientation of the Dα atom could be improved to better fit the chemical environment. It is conceivable that some of the clashes involving rotatable H atoms in other neutron models can be also corrected by adjusting the H atom position.

3.3 Hydrogen bonds

The properties of H-bonds for different H atom types are summarized in Fig. 5. Less than 25% of fixed H atoms are involved in H-bonds in any neutron model. This means that at least 75% of fixed H atoms do not form H-bonds. Fixed H atoms include those attached to carbon (tetragonal Hα, H atoms in aromatic rings, CH_2 groups, etc.) that have weaker polarity and therefore lower propensity to form an H-bond. Similarly, less than 20% of rotatable atoms are involved in H-bonds in most neutron models (Fig. 5A); however, there are models with a higher percentage, reaching up to 65%. This may reflect the possibility of rotatable H atoms using their degrees of freedom to optimize local interactions. It is not clear why the majority of models have fewer rotatable H atoms forming an H-bond, but as there are no prior reports available, it is uncertain if this represents genuine differences between structures or modeling errors of rotatable H atoms.

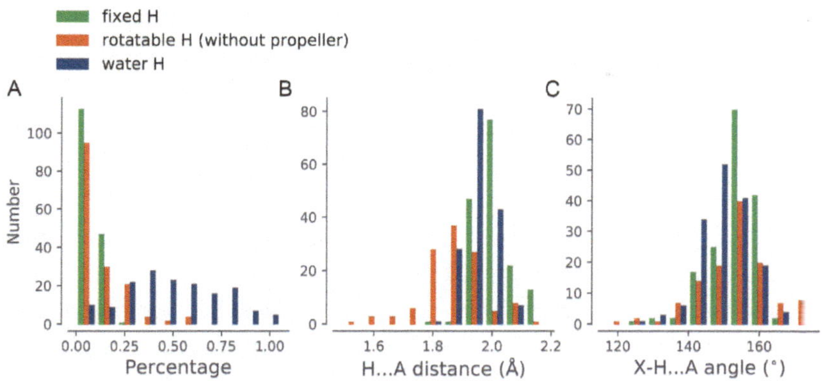

Fig. 5 Properties of hydrogen bonds for different H atom types. (A) Number of H-bonds. (B) Distance between H(D) and acceptor atom. (C) X-H…A angle.

Water molecules are a special case because they may act as both H-bond donor and acceptor, with up to four possible interactions per molecule. For water molecules modeled as oxygen or as oxygen with a single D atom, the number of possible interactions is two and three, respectively, as we did not take into account putative H-bonds. The shape of the distribution is very different compared to that of fixed or rotatable H atoms; it spans the entire range, from 0% to 5% of water molecules forming an H-bond up to 100–105% (as water can have several interactions, the percentage may be larger than 100%). There is no single peak, but in most models 25–80% of water molecules are involved in an H-bond. While the number of interactions are generally larger for water molecules than for fixed or rotatable H atoms, it still means that many models have water molecules without any H-bond interaction at all, as the percentage is mostly less than 100% at which, on average, each water molecule forms one H-bond. This can be partly due to the fact that some water molecules are modeled as oxygen atom without any D atoms, decreasing the number of possible interactions that can be determined with our algorithm. However, it also indicates that a nonnegligible number of water molecules are without any H-bond making it questionable if these waters should be placed.

The average donor-acceptor distance for each model is shown in Fig. 5B. The shorter the H…A distance, the stronger is the H-bond. Rotatable H atoms tend to form shorter H…A bonds, with an average of 1.8(1)Å, while that for fixed H atoms and water molecules is about 0.1 Å longer (2.01(6)Å and 1.95(5)Å for fixed H and water, respectively). However, at lower resolutions, this is within the experimental coordinate error and cannot be deemed statistically significant. The bond lengths are within typical values for H-bonds (Steiner, 2002). It can be noted that bond lengths <1.6 Å fall into the regime of low barrier hydrogen bonds between two oxygen atoms, where the energy barrier drops so that the H atom can move freely between the two O atoms (Cleland, 2000; Cleland, Frey, & Gerlt, 1998). For rotatable H, only one model has an average H…A distance <1.5 Å (6FJJ) because it has exactly one very short H-bond involving a rotatable H atom (Hγ of Ser_{50} with the O atom of DOD_{405}). The distributions and the average values for X-H…A angles (Fig. 5C) are similar for all three H atom types (153(7)°, 153(10)° and 149(7)° for fixed, rotatable H and water, respectively). These values agree with the expectation that H-bonds approach linear angles.

3.4 Homologue models

Neutron diffraction is not typically used to determine a structure de novo; in most cases, the model has been previously determined in an X-ray diffraction experiment. We parsed the PDB to find homologue structures for all neutron models. The BLAST search yielded homologues for 156 of all chains in the 161 neutron models. A complete list of homologues for each neutron model chain is available for download at https://doi.org/10.7941/D18907. Only five neutron models lack a high sequence identity X-ray equivalent: 6D54, 3QBA, 1V9G, 1WQZ, 6D4L, all of which are RNA/DNA structures. Fig. 6A shows the data resolution at which the

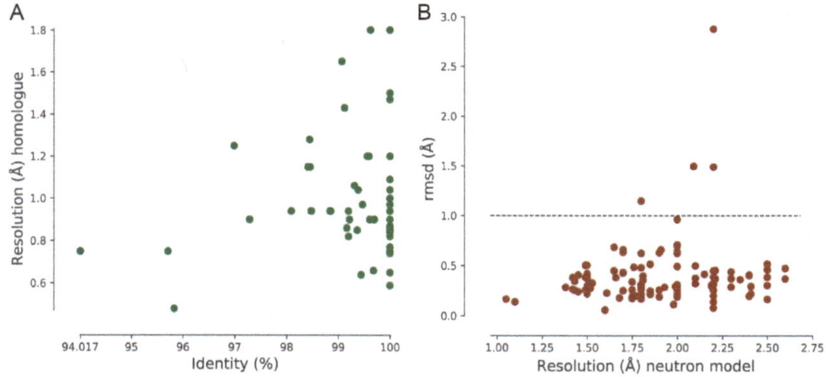

Fig. 6 (A) For each neutron model, the homologue X-ray model has high sequence identity and the concomitant X-ray data have high resolution. (B) *rmsd* between corresponding chains in neutron and homologue models. The following models have a *rmsd* larger than 1 Å (shown as dashed horizontal line): 5XPE, 5VNQ, 5PTi, 4QXK.

homologue structure was determined against the sequence identity with the neutron model. Most homologues have more than 98% sequence identity and have resolutions better than 1.2 Å. We also tested how well the homologues superposed to the neutron models (Fig. 6B). The *rmsd* between the homologue and neutron models is less than 1 Å in most cases; in particular, this appears to be independent of neutron data resolution, meaning that the protein structures are reproducible. However, the *rmsd* is calculated between protein residues only, ignoring water molecules. Section 3.6 focuses on the water structure comparison in neutron and homology models.

3.5 Refinement of joint XN models against X-ray data alone

Of the 97 joint XN models, 91 have X-ray data available and could be re-refined. The R_{work} and R_{free} factors are shown before and after re-refinement in Fig. 7. For the large majority of models, both R-factors decrease after refinement against X-ray data alone. With one exception, R_{work} and R_{free} are less than 25% for the re-refined models. This is not unexpected as the joint XN model is refined to simultaneously fit both the neutron and X-ray data, which can lead to a model that doesn't quite reach

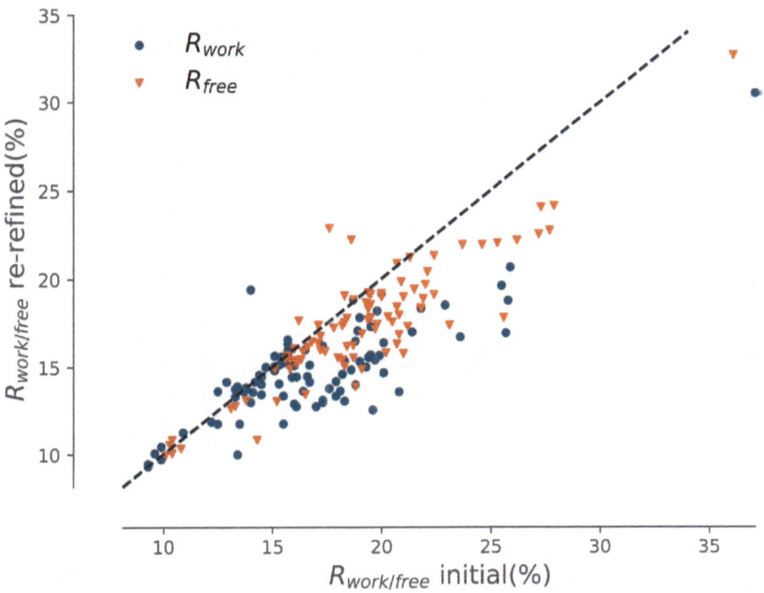

Fig. 7 R_{work} and R_{free} before and after re-refinement of joint XN models against X-ray data alone. Most of points are below the bisector (dashed line), meaning that the R-factors improved after re-refinement of the joint XN model against the X-ray data.

the optimal fit to either of the two datasets. However, the coordinates did not change significantly after re-refinement, as evidenced by an average *rmsd* of 0.07 Å for main chain atoms between deposited and re-refined models. The largest *rmsd* is 0.16 Å for model 3QBA. In particular, we were interested how the water network changes after re-refinement, which is described in Section 3.6.

3.6 Water structures

3.6.1 Water structure in homologues

Fig. 8 shows the results of the water cluster analysis that was performed for 121 of the 161 neutron models. The 40 nonevaluated models have either more than one protein chain (a requirement to have one chain is the limitation of our clustering analysis tool), have no water molecules modeled or their *rmsd* to the homologue is larger than 1 Å. Of the 121 compared water structures, 58 have at least 80% conserved water molecule positions, i.e., located closer than 1 Å to each other, in the superposed structures. However, more than half of the tested models have a lower percentage of matching waters; moreover, 20 models have more than 50% of waters without an equivalent in X-ray homologues. The model with the least matching water molecules is 1iO5 (Niimura et al., 1997). It contains 251 DOD molecules, while the 0.65 Å resolution X-ray homologue (2VB1) has only 165.

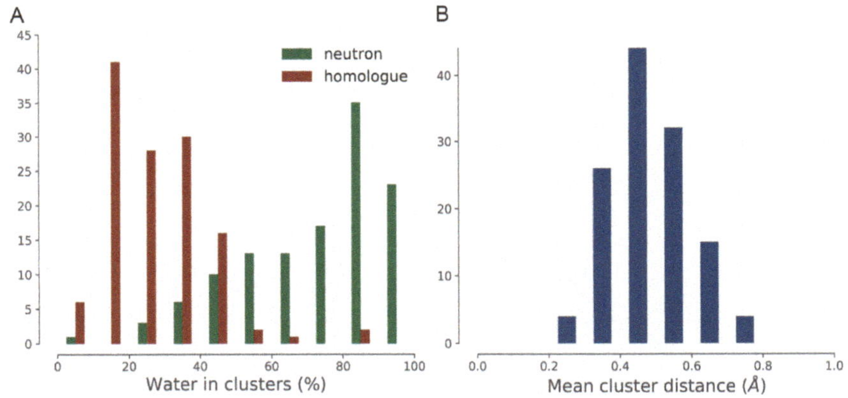

Fig. 8 (A) Comparing the water structure in neutron models and homologues. Water positions in neutron models are reproducible, but the spread is quite large. The water positions in the homologue are less conserved, most likely due to the fact that the high-resolution homologue models contain more water molecules than the neutron counterpart. (B) Histogram of the mean distance between matching waters. Most matching waters are 0.3–0.6 Å apart. In both figures, the Y-axis represents counts.

We note that 1iO5 has space group P 43 21 2 while the 2VB1 has P1. The packing is likely to affect the arrangement of water molecules. As no diffraction data are available for 1iO5, it is not possible to calculate maps to verify water positions in the neutron model. For all models, the mean distance between equivalent water molecule positions is 0.3–0.7 Å (Fig. 8B). Therefore, if the water molecules are conserved, they are located close to each other in the two structures.

There may be several possible reasons why some water molecules do not have equivalents: if sidechain rotamers are different, water molecules can occupy the space in one structure, while the same space in the other structure is occluded. Similarly, ligands that are only present in one of the structures can occupy available space for water molecules. Furthermore, even if the overall *rmsd* of the models is <1 Å, it may be that the "core" of the protein superposes well, but the outer regions, in particular surface sidechains, may deviate, forcing the water molecules to shift accordingly. Then, even if the protein superposes well everywhere, water molecules in the core of the protein might be more conserved than those at the surface, as they have fewer degrees of freedom being surrounded by protein residues. Furthermore, modeling of water molecules is quite subjective. Although guidelines exist (Levitt & Park, 1993), there are no community-wide accepted standards to validate water molecules. It is customary to reject water molecules that have low occupancy, are located too far from the protein, have large *ADPs* or weak density peaks but the cut-off values for these criteria vary from person to person and are typically adapted to data resolution: for example, a water with 0.5 occupancy can be acceptable at 1 Å while it may reflect over-optimistic modeling at 2.5 Å. This is also obvious from the proportion of unmatched waters in the homologue models (Fig. 8A, red bars); because they were mostly determined at high X-ray resolution where typically more water density peaks are visible, they have more lone water molecules than their neutron counterpart. Finally, some water molecules might be without map peaks due to data quality or they could be wrongly placed in noise peaks. A combination of all above reasons may explain why the proportions of conserved waters in neutron models vary so much.

3.6.2 Water structure in re-refined joint XN models

Of the 91 re-refined joint XN models, 73 have one chain so were used to compare their water structures. The results of the water structure comparison are shown in Fig. 9. While the majority of models (43) have more than 80% reproducible water molecules, a nonnegligible number of models (30)

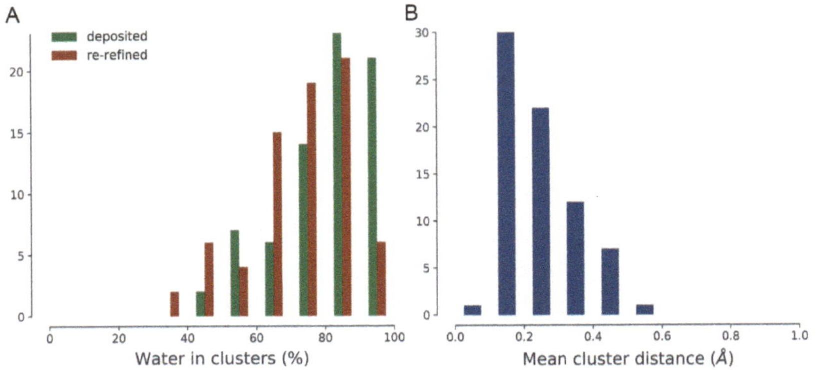

Fig. 9 (A) Comparing the water structure in deposited joint XN models and re-refined models against X-ray data alone. The majority of water positions are reproducible, but a significant number have a large proportion of unmatched waters. (B) Histogram of the mean distance between matching waters. Most waters in clusters are 0.1–0.3 Å apart. In both figures, the Y-axis represents counts.

have a lower proportion. This means that some joint XN models contain water molecules for which there is no or not enough signal in the X-ray data. If there are equivalent water molecules in both models, the distance between them is between 0.1 and 0.3 Å, which is closer than for the homologues. The position of waters that match is therefore well preserved. An example for matching and nonmatching water molecules in model 5EBJ (Langan et al., 2016) is shown in Fig. 10. DOD_{333} is present in the deposited and re-refined model and has clear density in both the electron and nuclear density maps. DOD_{360} has nuclear density but no electron density peak is present. The water molecule that was added with the *Phenix* water picking procedure during refinement against X-ray data, has clear electron density but no nuclear density peak. DOD_{373} has no peak at all at the given contour level in either map. Therefore, when using two datasets for the refinement of a single model, it is important to carefully model water molecules.

The analysis showed that the water structure in neutron models is more conserved for joint models re-refined against X-ray data alone than for homologue structures. The purpose of the joint refinement procedure is to produce a model that fits both datasets, so it is expected that the water positions should agree to a certain extent. Additionally, the crystals used for the neutron and X-ray experiments are usually produced with the same protocol while the crystals yielding the homologue models were most likely prepared under slightly different conditions (different reagents, procedure and handling). It is therefore possible that the reproducibility of water

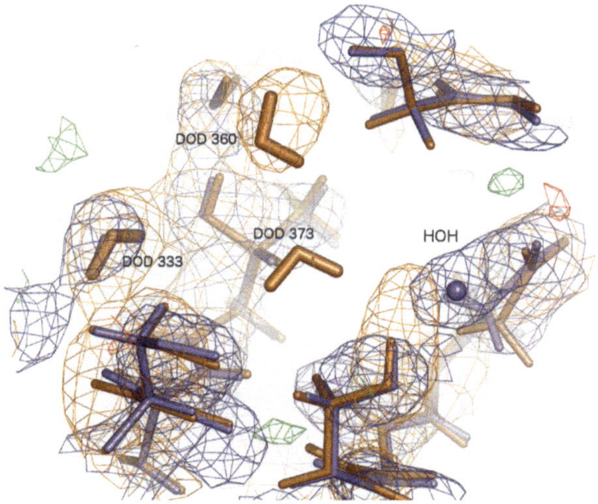

Fig. 10 Example for matching and nonmatching water molecules in 5EBJ. The deposited joint XN model (orange) and re-refined model (blue) are superposed. The $2mF_{obs}$-DF_{model} X-ray map calculated with the re-refined model (blue, 1.6σ) and $2mF_{obs}$-DF_{model} nuclear density map obtained from the deposited model (orange, 1.6σ) are displayed along with the mF_{obs}-DF_{model} nuclear density map represented in green (positive, +3σ) and red (negative, −3σ).

positions also depends on the crystallization conditions although it might be also due to the modeling or refinement protocol.

3.7 Density histograms

Unless completely deuterated, neutron models contain both hydrogen and deuterium atoms. Hydrogen has a negative scattering length while that of other elements typically found in macromolecules (C, N, O, S) and deuterium is positive. As a consequence, nuclear density maps can exhibit negative peaks near atomic centers, or the density can cancel out if H atom peaks overlap with positive peaks, such as when H is partially replaced by D or in groups of atoms with both positive and negative scattering lengths (for example CH_2 groups). We calculated histograms of model-calculated density map values interpolated at atomic centers for each neutron model that has deposited and usable diffraction data. We first gauged the impact of some model and data properties (H/D content, resolution, B-factors) for one particular example and then we investigated the impact of data completeness for all models.

Fig. 11 shows model based $F_{model}e^{i\varphi_{model}}$ Fourier map histograms at atomic centers for model 6BCC (Kovalevsky et al., 2018) (the model was

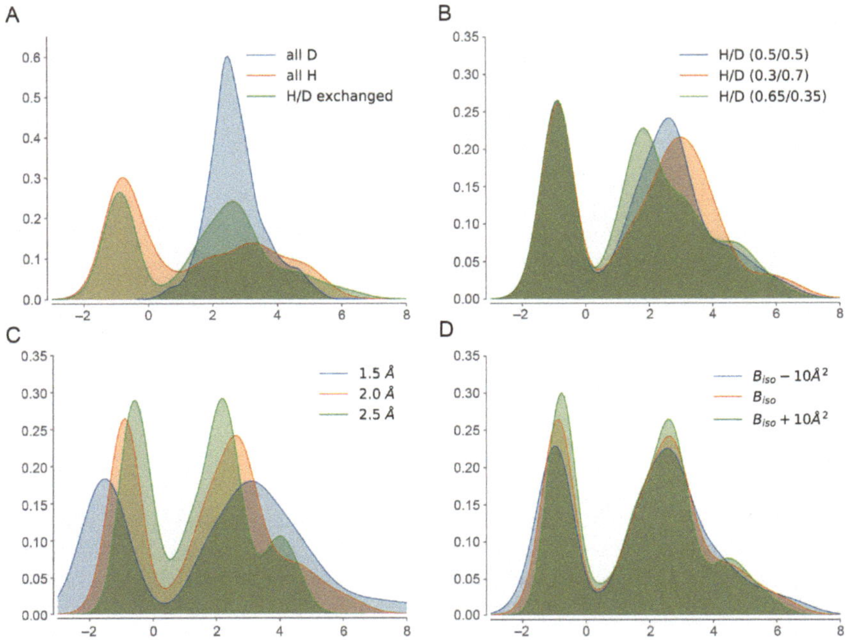

Fig. 11 Histogram of density values at atomic centers (the y-axes represent the probability density). The histograms are represented as smooth curve using a kernel density estimation, as implemented in the plot package Seaborn (www.seaborn.pydata.org). (A) Hydrogenation state. (B) H/D occupancy ratios if H is partially exchanged with D. (C) Data resolution. (D) *B*-factors.

chosen arbitrarily as an example). Fig. 11A shows the density histogram for different scenarios of H/D content; the histogram has a single peak for positive density points if the model contains only deuterium (blue). If the model contains only hydrogen atoms, there are two peaks (orange): one peak for the negative H atoms, and a broader peak in the positive region for all other protein atoms. The number of points with density equal to zero is about the same as the number of points at the positive peak. This means that the map has many regions of zero density, which can be challenging for model building and real-space refinement algorithms. If exchangeable sites are modeled as a superposition of H and D (with an occupancy ratio of 0.5/0.5, green), then there are two almost equally high peaks, located at similar positions to those observed when only D or only H is present. As the large majority of samples used for neutron diffraction experiments are deuterated (either completely or partially), the case of a sample with only hydrogen atoms is rare. Accordingly, there is currently only one model with only H atoms deposited in the PDB (Chatake & Fujiwara, 2016). For completely deuterated models density values at all atom centers are positive. Therefore, it is worth considering the case of models with partially exchanged H atoms in more detail because the histogram has an additional peak for negative densities. Fig. 11B shows the histogram for a partially exchanged model with different occupancy ratios at the exchanged sites. The shape and position of the negative peak is not affected, while the location of the positive peak shifts closer to zero if the H atom occupancy increases. We note that the number of points equal to zero does not increase if the H/D ratio is such that the sum of scattering lengths is zero (H/D occupancy ratio of 0.65/0.35), possibly because the density of the covalently bound heavy atom contributes at the H/D position. Fig. 11C shows the impact of resolution. The lower the resolution, the closer the peaks move together, and the narrower they become. Furthermore, the number of density points close to zero increases. The effect of isotropic *B*-factors is illustrated in Fig. 11D. With increasing *B*-factor, the histograms get narrower, but the peak position remains similar. These considerations illustrate that data resolution and H atom content are the main factors responsible for shifting the density histogram peaks towards zero.

3.8 Effect of incomplete neutron data on maps

Fig. 12 shows the peak correlation coefficient between complete and incomplete model-based maps as a function of data completeness. The data points are colored according to the hydrogenation state, which governs the shape of the density histogram (as discussed in Section 3.7). The distribution of the points suggests that the lower the completeness, the lower the correlation between the maps. Examples for different hydrogenation states are discussed below.

One model has CC values less than 0.5: 1WQZ ($CC_{<peaks>} = 0.39$) (Arai, 2005) contains both H and D atoms with a majority of H atoms. The calculated density for two thymine residues is shown in Fig. 13A and B. The density for the complete dataset covers most atoms, except CH_2 and CH_3 groups. For the incomplete dataset, the density covers fewer atoms, for example the sugar group of DT4 has very weak density. At other places, the density is smeared, such as between the two thymine bases. The histogram of density at atomic centers (Fig. 13C) has a drastically different shape for the incomplete dataset. Instead of two peaks, there is one peak with considerably more points of zero density. The neutron data completeness is

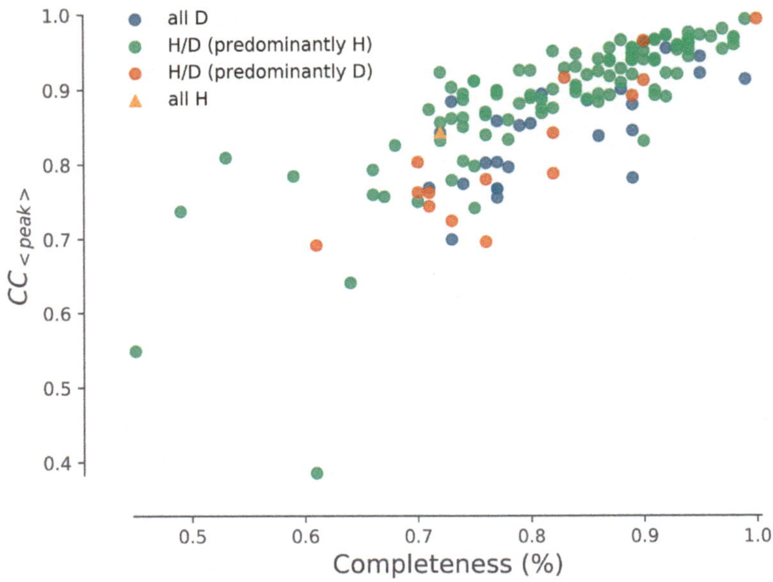

Fig. 12 Correlation coefficient between complete and incomplete calculated Fourier syntheses for all neutron models.

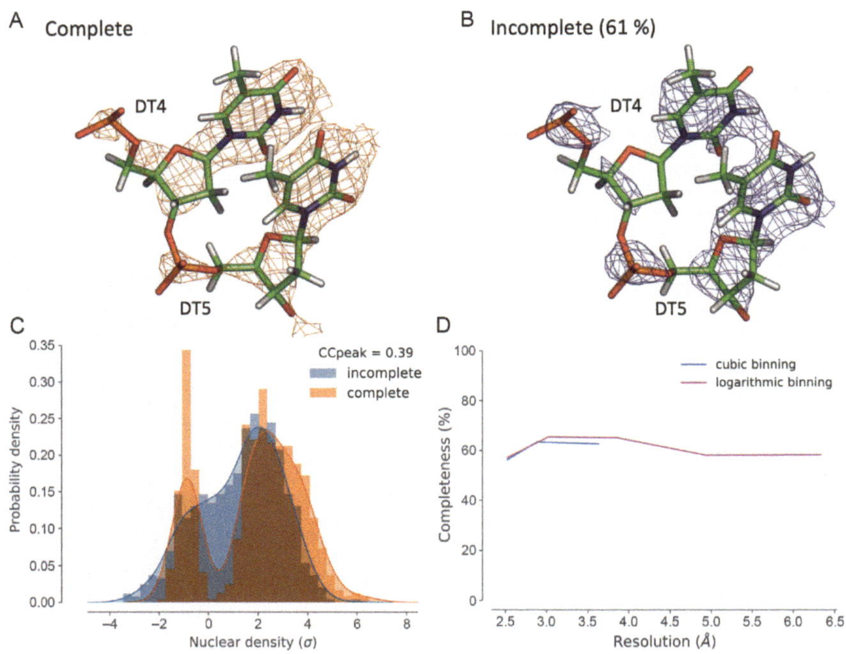

Fig. 13 Effect of data completeness, example of model 1WQZ. (A) Model-calculated map from complete data, contoured at 2σ. (B) Model-calculated map using the completeness of the deposited neutron data, contoured at 1.68σ. (C) Histogram of density at atomic centers. (D) Completeness vs. resolution for deposited neutron data. Contouring levels of maps were chosen with *phenix.map_comparison*.

61% overall and also across resolution ranges (Fig. 13D). As the data resolution is rather low (2.5 Å) and the space group symmetry is relatively high (tetragonal, $P3_2\ 2\ 1$), there are only a few resolution bins. This example shows that if data completeness is low across the entire resolution range, nuclear density maps can be severely distorted.

The nuclear density map for model 6E21 (Gerlits et al., 2019), which contains both H and D atoms with a majority of D, is shown in Fig. 14. In the complete map, the nuclear density covers most atoms, with the exception of most fixed H (D) atoms. At a comparable σ level, the density in the incomplete map covers much fewer atoms. We note that in both maps, the Oη group of Thr_{183} does not have a density peak impeding the location of both the O and the H (D) atoms, which are both rotatable. For the incomplete dataset, the density histogram is shifted to the left, as a result of fewer terms contributing to the Fourier transform. Consequently, more density points have a value of zero. The completeness is around 80% from the lowest limit to about 4 Å, then decreases slowly down to about 40%, leading to an

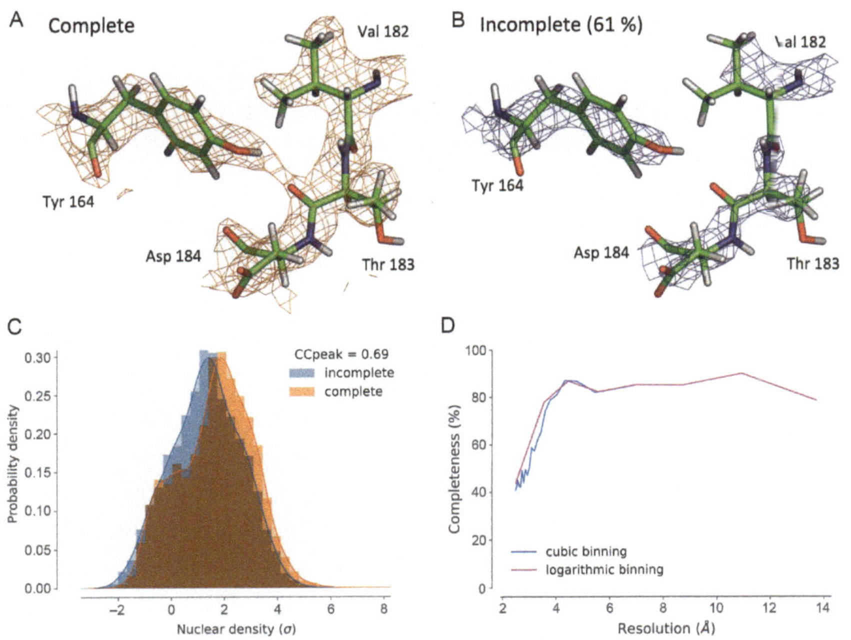

Fig. 14 Effect of data completeness, example of model 6E21. (A) Model-calculated map from complete data, contoured at 2σ. (B) Model-calculated map using the completeness of the deposited neutron data, contoured at 1.8σ. (C) Histogram of density at atomic centers. (D) Completeness vs. resolution for deposited neutron data. Contouring levels of maps were chosen with *phenix.map_comparison*.

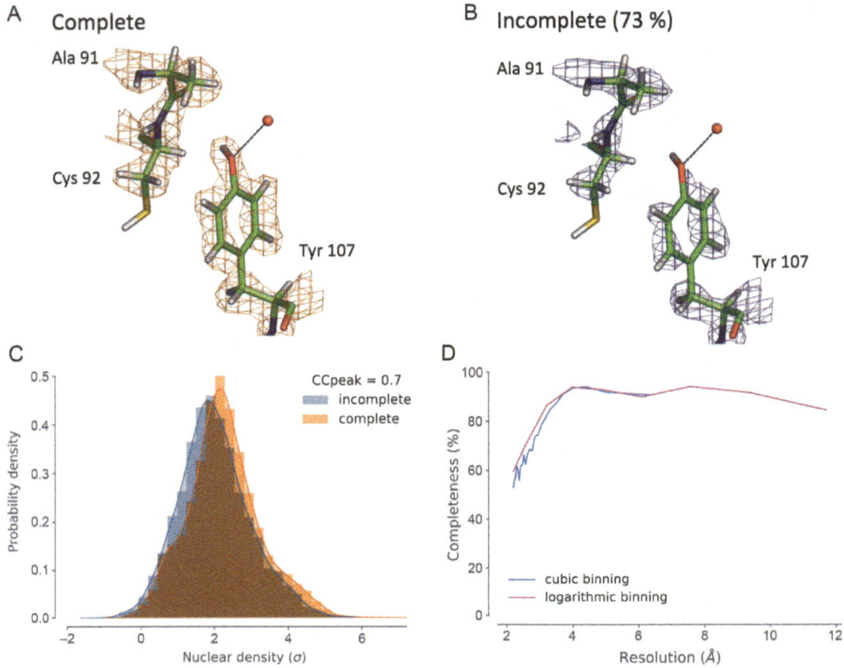

Fig. 15 Effect of data completeness, example of model 2R24. (A) Model-calculated map from complete data, contoured at 2σ. (B) Model-calculated map using the completeness of the deposited neutron data, contoured at 1.93σ. (C) Histogram of density at atomic centers. (D) Completeness vs. resolution for deposited neutron data. Contouring levels of maps were chosen with *phenix.map_comparison*.

overall completeness of 61%. As in the previous example (1WQZ), the completeness is therefore low over a wide resolution range, leading to a distorted nuclear density map.

Fig. 15 shows the nuclear density map for model 2R24 (Blakeley et al., 2008), which contains a majority of D atoms. The nuclear density calculated from the complete data covers almost all atoms at the 2σ contour. The density in the incomplete map is less clear as it has many gaps between atomic centers. In particular, the DH atom of Tyr_{107}, which is rotatable, does not have a density peak. We note further that the atom does not form any interaction with neighboring atoms while the oxygen atom of DOD_{376} could be an H-bond acceptor. The density histogram of the incomplete dataset is shifted to the left similar to the previous example. The overall completeness of the incomplete set is about 73%, dropping slowly from 90% at 4 Å and lower to 50% at the high-resolution limit of 2.19 Å.

The three examples show that nuclear density maps are typically distorted when the diffraction data are not complete. In particular, some H (D) atom density disappears or becomes weaker than that of neighboring atoms. This is the case for model-calculated maps, i.e. the structure factors are not subject to experimental noise or incorrectly measured intensities. It is very likely that nuclear maps calculated from measured structure factor amplitudes are even more distorted. Weak density makes it difficult to locate the H atoms; zero density will render it impossible to accurately determine the position based on the nuclear density itself. This suggests that nuclear density maps from incomplete data cannot be used to reliably locate all H (D) atom positions based on the experimental data alone.

3.9 Percentage of observed D atoms

The percentage of D atoms with nuclear density peaks above 3σ level is shown in Fig. 16. For nonwater D atoms, the percentage of reliably determined D atoms is typically less than 80%. Moreover, in the majority of models, less than 5% of D atom density has significant density peaks above 3σ. For water D atoms, the distribution is different: While the majority of models has less than 80% reliably located water D atoms, there are much fewer models with only 0–5% D atoms with peaks above the 3σ level. This result may appear counterintuitive: water D atoms are typically more mobile than those in the macromolecule. However, most neutron models

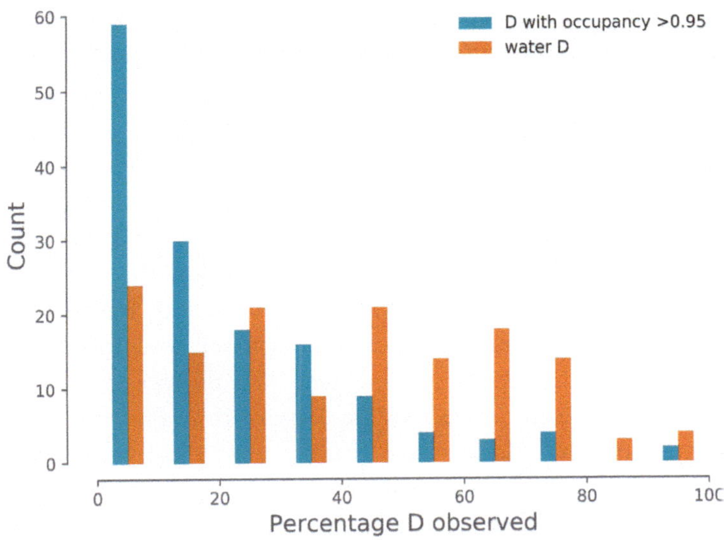

Fig. 16 Percentage of D atoms with nuclear density peaks >3σ in neutron models.

typically contain all possible D (or H) atoms, while only those water D atoms that have peaks in density maps are manually placed. It follows that most neutron models have a significant number of D atoms that do not have nuclear density peaks above the 3σ level. This means that these D atom positions cannot be accurately located on the basis of the experimental data.

4. Summary

This study evaluated currently available neutron models in the PDB. The clashscore analysis for subsets of H atoms showed that a majority of models have a larger clashscore for rotatable and water H atoms than for fixed H atoms. Inspection of deposited neutron structures suggests that some of the clashes involving rotatable H atoms may be corrected by adjusting the H atom position. This suggests that it may be possible to improve some of these neutron models, although whether this would lead to any change in the biological interpretation is unclear. As anticipated we observe that in any neutron model less than 25% of fixed H atoms form an H-bond. More surprisingly, for rotatable H atoms, the proportion of H atoms forming an H-bond is typically less than 20% (but can occasionally reach up to 65%), and many models have water molecules without any H-bond interaction. It is reassuring to note, that when present, the H-bond lengths and angles are within typical values found in the literature. These H-bond results again suggest that many neutron models could be improved in terms of local stereochemistry to better reflect the expected chemical interactions. It is possible that the use of refinement procedures that incorporate more realistic potential functions might automatically generate improved models (Moriarty et al., 2020).

Analysis of X-ray homologue structures for the deposited neutron models showed that the *rmsd* between the two models is less than 1 Å in most cases. We also saw that approximately 80% of water molecules have conserved positions, with mean distances between 0.3 and 0.7 Å. These results demonstrate that neutron models are well determined. Re-refinement of joint XN models against X-ray data alone showed that, in general, both R_{work} and R_{free} decreased after refinement. We also observed that, while more than 80% of water molecules are conserved in most models, a non-negligible number of models have fewer reproducible water molecules. These results highlight some limitations with current joint XN refinement procedures: (a) the final model is a compromise fit to the two datasets and (b) the model contains a single water structure that might not accurately reflect the water structures in the crystals used to collect the two datasets. These are particularly important issues when different data collection temperatures are used for the two datasets. Therefore, new joint refinement protocols are needed that combine the information available in neutron and X-ray datasets while simultaneously generating models that model the details of each dataset.

The analysis of nuclear density histograms showed that data resolution and H atom content can shift the density peaks towards zero, complicating their interpretation. Furthermore, nuclear density maps can be distorted when the diffraction data are not complete, as is typically the case for neutron datasets as a result of various experimental constraints. Analysis of the D atom nuclear density peaks, revealed that most neutron models have a significant number of D atoms with density peaks below the 3σ level. These results suggest that nuclear density maps from incomplete datasets cannot be used to accurately locate all D (H) atom positions on the basis of the experimental data. Experimental data collection protocols and data processing algorithms that can maximize data completeness would greatly improve this situation, as would the inclusion of evermore sophisticated descriptions of macromolecular stereochemistry in refinement algorithms.

Acknowledgments

This work was supported by the US National Institutes of Health (NIH) (grant P01GM063210), the Phenix Industrial Consortium and the NIH-funded (grant R01GM071939) Macromolecular Neutron Consortium between Oak Ridge National Laboratory and Lawrence Berkeley National Laboratory. This work was supported in part by the US Department of Energy under Contract No. DE-AC02-05CH1123.

References

Adams, P. D., Mustyakimov, M., Afonine, P. V., & Langan, P. (2009). Generalized X-ray and neutron crystallographic analysis: More accurate and complete structures for biological macromolecules. *Acta Crystallographica Section D: Biological Crystallography*, *65*, 567–573.

Afonine, P. V., Grosse-Kunstleve, R. W., Adams, P. D., & Urzhumtsev, A. (2013). Bulk-solvent and overall scaling revisited: Faster calculations, improved results. *Acta Crystallographica Section D: Biological Crystallography*, *69*, 625–634.

Afonine, P. V., Grosse-Kunstleve, R. W., Echols, N., Headd, J. J., Moriarty, N. W., Mustyakimov, M., et al. (2012). Towards automated crystallographic structure refinement with *phenix.refine*. *Acta Crystallographica Section D: Biological Crystallography*, *68*, 352–367.

Afonine, P. V., Mustyakimov, M., Grosse-Kunstleve, R. W., Moriarty, N. W., Langan, P., & Adams, P. D. (2010). Joint X-ray and neutron refinement with *phenix.refine*. *Acta Crystallographica Section D: Biological Crystallography*, *66*, 1153–1163.

Allen, F. H. (1986). A systematic pairwise comparison of geometric parameters obtained by X-ray and neutron diffraction. *Acta Crystallographica Section B: Structural Science, Crystal Engineering and Materials*, *42*, 515–522.

Allen, F. H., & Bruno, I. J. (2010). Bond lengths in organic and metal-organic compounds revisited: *X* —H bond lengths from neutron diffraction data. *Acta Crystallographica Section B: Structural Science, Crystal Engineering and Materials*, *66*, 380–386.

Altschul, S. F., Gish, W., Miller, W., Myers, E. W., & Lipman, D. J. (1990). Basic local alignment search tool. *Journal of Molecular Biology*, *215*, 403–410.

Ankner, J. F., Heller, W. T., Herwig, K. W., Meilleur, F., & Myles, D. A. A. (2013). Neutron scattering techniques and applications in structural biology. *Current Protocols in Protein Science*, *72*, 17.16.1–17.16.34.

Arai, S. (2005). Complicated water orientations in the minor groove of the B-DNA decamer d(CCATTAATGG)2 observed by neutron diffraction measurements. *Nucleic Acids Research*, *33*, 3017–3024.

Berman, H., Henrick, K., & Nakamura, H. (2003). Announcing the worldwide protein data bank. *Nature Structural & Molecular Biology*, *10*, 980.

Bernstein, F. C., Koetzle, T. F., Williams, G. J. B., Meyer, E. F., Brice, M. D., Rodgers, J. R., et al. (1977). The protein data bank: A computer-based archival file for macromolecular structures. *Journal of Molecular Biology*, *112*, 535–542.

Blakeley, M. P., Kalb, A. J., Helliwell, J. R., & Myles, D. A. (2004). The 15-K neutron structure of saccharide-free concanavalin A. *Proceedings of the National Academy of Sciences of the United States of America*, *101*, 16405–16410.

Blakeley, M. P., Ruiz, F., Cachau, R., Hazemann, I., Meilleur, F., Mitschler, A., et al. (2008). Quantum model of catalysis based on a mobile proton revealed by subatomic x-ray and neutron diffraction studies of h-aldose reductase. *Proceedings of the National Academy of Sciences of the United States of America*, *105*, 1844–1848.

Chatake, T., & Fujiwara, S. (2016). A technique for determining the deuterium/hydrogen contrast map in neutron macromolecular crystallography. *Acta Crystallographica Section D: Structural Biology*, *72*, 71–82.

Chen, J. C.-H., & Unkefer, C. J. (2017). Fifteen years of the protein crystallography station: The coming of age of macromolecular neutron crystallography. *International Union of Crystallography*, *4*, 72–86.

Cleland, W. W. (2000). Low-barrier hydrogen bonds and enzymatic catalysis. *Archives of Biochemistry and Biophysics*, *382*, 1–5.

Cleland, W. W., Frey, P. A., & Gerlt, J. A. (1998). The low barrier hydrogen bond in enzymatic catalysis. *Journal of Biological Chemistry*, *273*, 25529–25532.

Coppens, P. (1967). Comparative x-ray and neutron diffraction study of bonding effects in s-triazine. *Science*, *158*, 1577–1579.

Cruickshank, D. W. J. (1999). Remarks about protein structure precision. *Acta Crystallographica Section D: Biological Crystallography*, *55*, 583–601.

Cruickshank, D. W. J., Helliwell, J. R., & Moffat, K. (1987). Multiplicity distribution of reflections in Laue diffraction. *Acta Crystallographica Section A: Foundations of Crystallography*, *43*, 656–674.

Dauter, Z. (1999). Data-collection strategies. *Acta Crystallographica Section D: Biological Crystallography*, *55*, 1703–1717.

Dauter, M., & Dauter, Z. (2017). Many ways to derivatize macromolecules and their crystals for phasing. *Methods in Molecular Biology (Clifton, NJ)*, *1607*, 349–356.

Evans, P. R. (2011). An introduction to data reduction: Space-group determination, scaling and intensity statistics. *Acta Crystallographica Section D: Biological Crystallography*, *67*, 282–292.

Fabiola, F., Bertram, R., Korostelev, A., & Chapman, M. S. (2002). An improved hydrogen bond potential: Impact on medium resolution protein structures. *Protein Science*, *11*, 1415–1423.

Fields, B. A., Bartsch, H. H., Bartunik, H. D., Cordes, F., Guss, J. M., & Freeman, H. C. (1994). Accuracy and precision in protein crystal structure analysis: Two independent refinements of the structure of poplar plastocyanin at 173 K. *Acta Crystallographica Section D: Structural Biology, 50*, 709–730.

Fujinaga, M., Delbaere, L. T., Brayer, G. D., & James, M. N. (1985). Refined structure of alpha-lytic protease at 1.7 A resolution. Analysis of hydrogen bonding and solvent structure. *Journal of Molecular Biology, 184*, 479–502.

Gerlits, O., Weiss, K. L., Blakeley, M. P., Veglia, G., Taylor, S. S., & Kovalevsky, A. (2019). Zooming in on protons: Neutron structure of protein kinase a trapped in a product complex. *Science Advances, 5*, eaav0482.

Grosse-Kunstleve, R. W., Sauter, N. K., Moriarty, N. W., & Adams, P. D. (2002). The computational crystallography toolbox: Crystallographic algorithms in a reusable software framework. *Journal of Applied Crystallography, 35*, 126–136.

Gruene, T., Hahn, H. W., Luebben, A. V., Meilleur, F., & Sheldrick, G. M. (2014). Refinement of macromolecular structures against neutron data with *SHELXL2013*. *Journal of Applied Crystallography, 47*, 462–466.

Koch, O., Bocola, M., & Klebe, G. (2005). Cooperative effects in hydrogen-bonding of protein secondary structure elements: A systematic analysis of crystal data using Secbase. *Proteins: Structure, Function, and Bioinformatics, 61*, 310–317.

Kovalevsky, A., Aggarwal, M., Velazquez, H., Cuneo, M. J., Blakeley, M. P., Weiss, K. L., et al. (2018). "To be or not to be" protonated: Atomic details of human carbonic anhydrase-clinical drug complexes by neutron crystallography and simulation. *Structure, 26*, 383–390.e3.

Langan, P. S., Close, D. W., Coates, L., Rocha, R. C., Ghosh, K., Kiss, C., et al. (2016). Evolution and characterization of a new reversibly photoswitching chromogenic protein, Dathail. *Journal of Molecular Biology, 428*, 1776–1789.

Langan, P., Greene, G., & Schoenborn, B. P. (2004). Protein crystallography with spallation neutrons: The user facility at Los Alamos neutron science center. *Journal of Applied Crystallography, 37*, 24–31.

Levitt, M., & Park, B. H. (1993). Water: Now you see it, now you don't. *Structure, 1*, 223–226.

Liebschner, D., Afonine, P. V., Baker, M. L., Bunkóczi, G., Chen, V. B., Croll, T. I., et al. (2019). Macromolecular structure determination using X-rays, neutrons and electrons: Recent developments in Phenix. *Acta Crystallographica Section D: Structural Biology, 75*, 861–877.

Liebschner, D., Afonine, P. V., Moriarty, N. W., Langan, P., & Adams, P. D. (2018). Evaluation of models determined by neutron diffraction and proposed improvements to their validation and deposition. *Acta Crystallographica Section D: Structural Biology, 74*, 800–813.

Madden, T. (2003). *The BLAST sequence analysis tool.* USA: National Center for Biotechnology Information.

Moriarty, N. W. (2015). Human readable PDB codes: An editorial. *Computational Crystallography Newsletter, 6*, 26.

Moriarty, N. W., Janowski, P. A., Swails, J. M., Nguyen, H., Richardson, J. S., Case, D. A., et al. (2020). Improved chemistry restraints for crystallographic refinement by integrating the Amber force field into Phenix. *Acta Crystallographica Section D: Structural Biology, 76*, 51–62.

Moriarty, N. W., Liebschner, D., Klei, H. E., Echols, N., Afonine, P. V., Headd, J. J., et al. (2018). Interactive comparison and remediation of collections of macromolecular structures. *Protein Science, 27*, 182–194.

Niimura, N., Karasawa, Y., Tanaka, I., Miyahara, J., Takahashi, K., Saito, H., et al. (1994). An imaging plate neutron detector. *Nuclear Instruments and Methods in Physics Research A, 349*, 521–525.

Niimura, N., Yoshiaki, M., Nonaka, T., Castagna, J.-C., Cipriani, F., Hoghoj, P., et al. (1997). Neutron Laue diffractometry with an imaging plate provides an effective data collection regime for neutron protein crystallography. *Nature Structural Biology, 4*, 909–914.

Ohlendorf, D. H. (1994). Acuracy of refined protein structures. II. Comparison of four independently refined models of human interleukin 1β. *Acta Crystallographica Section D: Biological Crystallography, 50*, 808–812.

Orpen, A. G., Pippard, D., Sheldrick, G. M., & Rouse, K. D. (1978). Decacarbonyl-μ-hydrido-μ-vinyl-triangulo-triosmium: A combined X-ray and neutron diffraction study. *Acta Crystallographica Section B: Structural Crystallography and Crystal Chemistry, 34*, 2466–2472.

Richardson, J. S., Williams, C. J., Hintze, B. J., Chen, V. B., Prisant, M. G., Videau, L. L., et al. (2018). Model validation: Local diagnosis, correction and when to quit. *Acta Crystallographica Section D: Structural Biology, 74*, 132–142.

Rupp, B. (2018). Against method: Table 1—Cui bono? *Structure, 26*, 919–923.

Schoenborn, B. P. (1969). Neutron diffraction analysis of myoglobin. *Nature, 224*, 143–146.

Steiner, T. (2002). The hydrogen bond in the solid state. *Angewandte Chemie International Edition, 41*, 48–76.

Teeter, M. M., & Kossiakoff, A. A. (1984). The neutron structure of the hydrophobic plant protein Crambin. In B. P. Schoenborn (Ed.), *Neutrons in biology* (pp. 335–348). Boston, MA: Springer US.

Thomas, A., Benhabiles, N., Meurisse, R., Ngwabije, R., & Brasseur, R. (2001). Pex, analytical tools for PDB files. II. H-Pex: Noncanonical H-bonds in?-helices. *Proteins: Structure, Function, and Genetics, 43*, 37–44.

Torshin, I. Y., Weber, I. T., & Harrison, R. W. (2002). Geometric criteria of hydrogen bonds in proteins and identification of 'bifurcated' hydrogen bonds. *Protein Engineering, Design and Selection, 15*, 359–363.

Urzhumtsev, A. G. (1991). Low-resolution phases: Influence on SIR syntheses and retrieval with double-step filtration. *Acta Crystallographica Section A Foundations of Crystallography, 47*, 794–801.

Urzhumtsev, A., Afonine, P. V., Lunin, V. Y., Terwilliger, T. C., & Adams, P. D. (2014). Metrics for comparison of crystallographic maps. *Acta Crystallographica Section D: Biological Crystallography, 70*, 2593–2606.

Weichenberger, C. X., Afonine, P. V., Kantardjieff, K., & Rupp, B. (2015). The solvent component of macromolecular crystals. *Acta Crystallographica Section D: Biological Crystallography, 71*, 1023–1038.

Wennmacher, J. T. C., Zaubitzer, C., Li, T., Bahk, Y. K., Wang, J., van Bokhoven, J. A., et al. (2019). 3D-structured supports create complete data sets for electron crystallography. *Nature Communications, 10*, 1–6.

Williams, C. J., Headd, J. J., Moriarty, N. W., Prisant, M. G., Videau, L. L., Deis, L. N., et al. (2018). MolProbity: More and better reference data for improved all-atom structure validation: PROTEIN SCIENCE.ORG. *Protein Science, 27*, 293–315.

Wlodawer, A. (1980). Studies of ribonuclease-A by X-ray and neutron diffraction. *Acta Crystallographica Section B: Structural Science, Crystal Engineering and Materials, 36*, 1826–1831.

Wlodawer, A., & Hendrickson, W. A. (1982). A procedure for joint refinement of macromolecular structures with X-ray and neutron diffraction data from single crystals. *Acta Crystallographica Section A: Foundations and Advances, 38*, 239–247.

Wlodawer, A., Minor, W., Dauter, Z., & Jaskolski, M. (2008). Protein crystallography for non-crystallographers, or how to get the best (but not more) from published macromolecular structures. *The FEBS Journal, 275*, 1–21.

Wlodawer, A., Walter, J., Huber, R., & Sjölin, L. (1984). Structure of bovine pancreatic trypsin inhibitor. *Journal of Molecular Biology*, *180*, 301–329.
Word, J. M., Lovell, S. C., LaBean, T. H., Taylor, H. C., Zalis, M. E., Presley, B. K., et al. (1999). Visualizing and quantifying molecular goodness-of-fit: Small-probe contact dots with explicit hydrogen atoms. *Journal of Molecular Biology*, *285*, 1711–1733.
wwPDB Consortium. (2019). Protein Data Bank: The single global archive for 3D macromolecular structure data. *Nucleic Acids Research*, *47*, D520–D528.

CHAPTER TWELVE

Proton transfer and drug binding details revealed in neutron diffraction studies of wild-type and drug resistant HIV-1 protease

Andrey Kovalevsky[a,*], Oksana Gerlits[b], Kaira Beltran[b], Kevin L. Weiss[a], David A. Keen[c], Matthew P. Blakeley[d], John M. Louis[e], Irene T. Weber[f,g]

[a]Neutron Scattering Division, Oak Ridge National Laboratory, Oak Ridge, TN, United States
[b]Department of Natural Sciences, Tennessee Wesleyan University, Athens, TN, United States
[c]ISIS Facility, Rutherford Appleton Laboratory, Harwell Campus, Didcot, United Kingdom
[d]Large Scale Structures Group, Institut Laue–Langevin, Grenoble, France
[e]Laboratory of Chemical Physics, National Institute of Diabetes and Digestive and Kidney Diseases, National Institutes of Health, DHHS, Bethesda, MD, United States
[f]Department of Biology, Georgia State University, Atlanta, GA, United States
[g]Department of Chemistry, Georgia State University, Atlanta, GA, United States
*Corresponding author: e-mail address: kovalevskyay@ornl.gov

Contents

Abstract

HIV-1 protease is an essential therapeutic target for the design and development of antiviral inhibitors to treat AIDS. We used room temperature neutron crystallography to accurately determine hydrogen atom positions in several protease complexes with clinical drugs, amprenavir and darunavir. Hydrogen bonding interactions were carefully mapped to provide an unprecedented picture of drug binding to the protease target.

Methods in Enzymology, Volume 634
ISSN 0076-6879
https://doi.org/10.1016/bs.mie.2019.12.002

We demonstrate that hydrogen atom positions within the enzyme catalytic site can be altered by introducing drug resistant mutations and by protonating surface residues that trigger proton transfer reactions between the catalytic Asp residues and the hydroxyl group of darunavir. When protein perdeuteration is not feasible, we validate the use of initial H/D exchange with unfolded protein and partial deuteration in pure D_2O with hydrogenous glycerol to maximize deuterium incorporation into the protein, with no detrimental effects on the growth of quality crystals suitable for neutron diffraction experiments.

1. Introduction

Acquired immunodeficiency syndrome (AIDS) caused by the human immunodeficiency virus type 1 (HIV-1) is a deadly disease unless treated properly. AIDS has remained a global pandemic for over three decades, although the therapeutic intervention options available to the infected individuals have continuously improved. Currently, the combination antiretroviral therapy, involving simultaneous treatment with several drugs that target different stages of the viral replication cycle, constitutes the gold standard for treating HIV-1-infected patients (Laskey & Siliciano, 2014; Mehellou & De Clercq, 2010). HIV-1 protease (PR) enzyme is one of the crucial targets for the design and development of clinical drugs, with nine FDA-approved small-molecule inhibitors currently on the market. PR inhibitors have been designed using structure-assisted (or rational) drug design methodology; however, their continued success has been limited by the emergence of drug resistant PR variants due to the virus mutating and multiplying very rapidly (Ghosh, Osswald, & Prato, 2016; Weber & Harrison, 2016).

Low-temperature X-ray crystallography has played an indispensable role in guiding the design of PR inhibitors by providing accurate high-resolution structures of the enzyme-ligand complexes (Ghosh, Anderson, Weber, & Mitsuya, 2012; Mitsuya, Maeda, Das, & Ghosh, 2008). It has also proven important in studying the structural basis of drug resistance by identifying intermolecular interactions between inhibitors and specific residues of the enzyme that are altered by introducing drug resistant mutations (Weber, Kneller, & Wong-Sam, 2015). For single resistance mutations, the diminished inhibitor's affinity for a resistant PR mutant has been attributed to significant decreases in hydrogen bonding and hydrophobic interactions between the inhibitor and both mutated and native PR residues (Agniswami et al., 2012; Shen, Wang, Kovalevsky, Harrison, & Weber, 2010; Weber & Agniswami, 2009). Multiply mutated PRs with high levels of resistance, however, often show little apparent change in hydrogen bonding interactions with clinical inhibitors despite drastically decreased affinity of several orders of magnitude (Agniswamy, Louis, Roche, Harrison, & Weber, 2016). Such structures, however, lack information on the locations of hydrogen atoms, although half of all atoms in a protein and in small-molecule inhibitors are hydrogen (H). H atoms are integral players in most intermolecular interactions, especially in highly directional H bonds. Interestingly, we were not able to detect H atoms even in the ultra-high-resolution (0.84 Å) X-ray structure of the darunavir complex with HIV-1 PR bearing the drug resistant V32I mutation (Kovalevsky et al., 2006). Moreover, the X-ray structures are normally obtained in cryo-conditions at −173 °C, which is over 200 °C below the physiological temperature and require the use of cryo-protecting chemicals that can also bind to a protein. Performing structural studies at room temperature, which is only ~17 °C below the physiological temperature, utilizing a technique capable of directly determining H atom positions and visualizing intermolecular interactions would be greatly beneficial for rational drug design and advance our understanding of the molecular mechanisms for drug resistance. Therefore, we chose macromolecular neutron crystallography for our structural studies of clinical drugs binding to the wild-type and drug resistant PR.

Macromolecular neutron crystallography is the technique of choice to directly determine positions of H atoms and to visualize H bonds in proteins, nucleic acids and protein-ligand complexes (Blakeley, Hasnain, & Antonyuk, 2015; Fisher, Jackson, Kovalevsky, Oksanen, & Wacklin, 2017; Niimura & Podjarny, 2011). Neutrons are scattered by the atomic nuclei and their scattering power is independent of the atomic number. Hence, H atoms scatter neutrons almost as strongly as do carbon, oxygen and nitrogen that are the primary non-H constituents of proteins, nucleic acids and other biomacromolecules. Positions of H atoms can be accurately determined from neutron diffraction data and this permits direct visualization of intermolecular interactions, including H bonds and other conventional and unconventional contacts (e.g., O–H...π, C–H...O, C–H...π, etc.). Interestingly, in many instances, the functional H atoms that are most critical to detect remain invisible even in ultra-high-resolution X-ray structures (Oksanen, Chen, & Fisher, 2017), because they are highly polarized or participate in possible low-barrier H bonds (LBHB) with little electron density around them. In addition, macromolecular neutron crystallography affords radiation-damage-free macromolecular structures because neutrons (λ ≈ 2–5 Å) used to collect neutron diffraction data do not ionize atoms, unlike X-rays, and consequently do not cause direct radiation damage to macromolecular crystals. Macromolecular neutron crystallographic experiments can therefore be carried out at room temperature (~20 °C). It is worth noting that H has a negative neutron scattering length (−3.739 fm) that manifests by H atoms appearing as troughs, instead of peaks, in the neutron scattering length density maps. Hence, at the average resolution of a neutron diffraction experiment of 2.0–2.2 Å, aliphatic groups (CH_3, CH_2, CH) are not visible in the maps due to the density cancelation effects. Also, H has a very high incoherent neutron scattering cross section, which dramatically increases the background and decreases its signal in the neutron diffraction images. These drawbacks are overcome by replacing titratable H atoms with deuterium (D) atoms by H/D exchange in heavy water (D_2O) and by protein perdeuteration to replace H in the C–H bonds that is achieved through expression in the deuterated media. The neutron scattering length of D (6.671 fm) is positive and its incoherent scattering cross section is negligible. D atoms are observed as peaks in the neutron scattering length density maps at similar levels as carbon, nitrogen and oxygen atoms, and the density cancelation effects are avoided.

We have used macromolecular neutron crystallography to obtain four neutron structures of HIV-1 PR in complex with clinical drugs amprenavir (APV) and darunavir (DRV) (Gerlits et al., 2016, 2017; Weber et al., 2013) (Fig. 1). We investigated the atomic details of drug binding to the wild-type

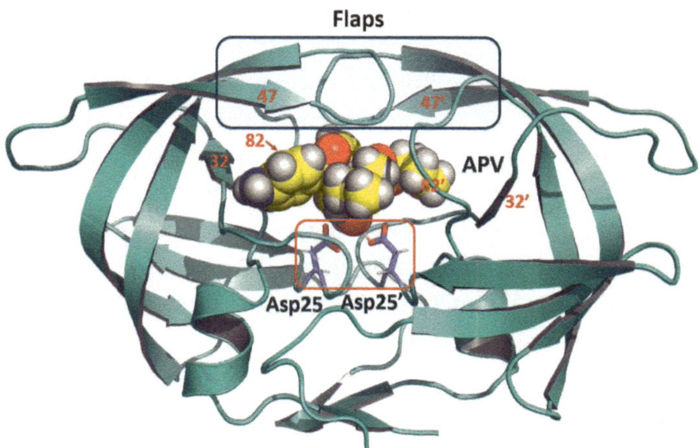

Fig. 1 Overall structure of PR_{WT}/APV complex in cartoon representation. The clinical drug APV is shown in space-fill representation and is colored by atom type (carbon is yellow). Catalytic aspartates are shown as sticks.

PR (PR_{WT}) and to the triple mutant PR containing drug resistant substitutions of V32I, I47V and V82I (PR_{TM}). PR_{WT} used for the current study harbors five mutations Q7K, L33I, L63I, C67A and C95A to restrict autoproteolysis and prevent cysteine-thiol oxidation (Louis, Clore, & Gronenborn, 1999). We have located all D atoms in the structures and examined their movements due to pH changes in the crystal. All neutron diffraction measurements were carried out at room temperature and compared to our previous 100 K X-ray structures to reveal how intricate details of the structure are altered by temperature.

Placing H atoms correctly, and therefore deducing protonation states of amino acid residues, chemical groups of ligands and water species, is critical information to fully understand the biological function of a macromolecule being studied. Neutron crystallography offers a unique solution to answer long-standing questions in biochemistry (Langan & Chen, 2013; O'Dell, Bodenheimer, & Meilleur, 2016).

2. Preparation of PR samples

Preparation of a sample for a successful neutron diffraction experiment is not a trivial task. Compared to X-ray crystallography, where producing a small crystal may very well result in a crystal structure, it requires much more significant effort in optimizing protein expression, possible deuteration and crystal growth to get a crystal suitable for neutron diffraction. Although neutron sources and instruments have improved over the past two decades, crystals on the order of 0.1–1 mm^3 are still required for macromolecular neutron crystallography. Such crystals are at least two orders of magnitude larger than those necessary for X-ray diffraction. For example, HIV PR crystals of 0.006 mm^3 are suitable for collecting X-ray diffraction data on high intensity synchrotron sources. Growing sufficiently large protein crystals for neutron crystallography always involves careful (re)optimization of the crystallization conditions. The two most critical issues are deuteration of the protein sample and growing crystals of sufficient size for neutron crystallography.

2.1 Protease deuteration

The simplest way to deuterate a protein sample for a neutron diffraction experiment is to exchange labile H atoms bound to OH, NH, SH functional chemical groups with D using D_2O. Because protein crystals contain at least 50% water and about 15–20% of H atoms in a protein are exchangeable, this approach takes care of most of the H content in the sample. If suitably large crystals of a hydrogenous protein can be grown, performing H/D exchange may be sufficient to get a neutron structure. When large crystals of a hydrogenous protein cannot be produced, further protein deuteration, that is introduction of D into the non-exchangeable C–H groups, becomes necessary. Protein expression in deuterated media permits incorporation of D atoms into C–H groups. Proteins can be perdeuterated by expression in minimal media made with pure D_2O (99.8%) and deuterated glycerol (or deuterated glucose) as a sole carbon source. Preparing a protein sample in this manner guarantees virtually complete conversion of C–H into C–D groups, significantly improving the signal-to-noise ratio of the neutron diffraction. Perdeuterated crystals also diffract neutrons much better than hydrogenous ones and therefore can be several-fold smaller for a successful data collection. The caveat of perdeuteration is that it changes the hydrophobicity of proteins. For many proteins this results in their diminished ability to crystallize. We performed perdeuteration of PR_{WT}. The perdeuterated enzyme expressed well; however, it did not produce crystals suitable for neutron diffraction. Furthermore, hydrogenous protein crystallized as plates that could only grow to volumes of ~0.2 mm^3 that were insufficient to get decent neutron diffraction.

To solve this problem, we elected to make partially deuterated PR. We expressed PR_{WT} in *Enfors* minimal medium made with pure D_2O and hydrogenous glucose for *E. coli* bacterial growth. The cells grew at about half the rate in D_2O compared to that in H_2O, and after induction with IPTG (1 mM) the expression was allowed to proceed for ~8 h (about twice as long as in H_2O) at 37 °C. The partially deuterated PR_{WT} was purified using the same protocol described for the hydrogenous enzyme in buffer solutions made with H_2O (Mahalingam, Louis, Hung, Harrison, & Weber, 2001). We then assessed the level of the enzyme deuteration using mass spectrometry measurements. We determined that PR_{WT} was ~85% deuterated, indicative of the *E. coli* metabolic pathways capable of incorporating a substantial amount of D into the enzyme's C–H groups from D_2O alone. Our partially deuterated protein grew crystals of about the same volume as did the hydrogenous PR_{WT}, and these crystals were found to be suitable for neutron diffraction experiments. Based on the success with PR_{WT}, we prepared the deuterated drug resistant PR_{TM} variant in the same way, except for using hydrogenous glycerol instead of glucose as the sole carbon source, and the expression was performed in a bioreactor instead of shake flasks thus significantly enhancing the yield of deuterated PR_{TM}. We also demonstrated that the variant could produce larger crystals, up to ~0.5 mm^3 (Fig. 2), than did PR_{WT}.

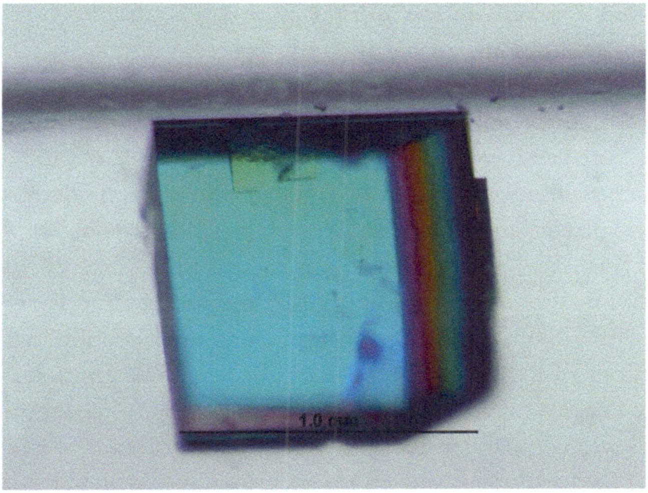

Fig. 2 Crystal of deuterated PR_{TM}/DRV complex of ~0.5 mm^3 volume mounted in a 2 mm inner diameter quartz capillary and used for neutron diffraction data collection. Notice stunning light interference pattern inside the crystal.

2.2 Crystal growth

Preparation of HIV-1 PR is unusual because of its accumulation in inclusion bodies upon expression in *E. coli*. Hence, its purification is carried out under denaturing conditions upon dissolving the inclusion bodies in guanidine hydrochloride. This scheme has the added benefit of restricting PR self-degradation, termed autoproteolysis. The purified PR is stored unfolded in the H_2O/acetonitrile/0.05% trifluoroacetic acid mixture following its final high performance liquid chromatography step. In an unfolded state, all exchangeable sites are fully accessible by D_2O, which is not the case for a folded protein where buried amino acids may have extremely slow rates of H/D exchange. We took advantage of this to maximize H/D exchange in our unfolded PR_{WT} and PR_{TM} samples. The first step in the PR refolding procedure is the removal of acetonitrile from the sample by dialysis in 25 mM formic acid that keeps the protein unfolded at low pH. We performed this dialysis in pure D_2O, essentially exchanging all labile H with D. Then the enzymes were refolded in 50 mM NaOAc at pH 5.0 in D_2O to preserve H/D exchange. Subsequently, PR_{WT} and PR_{TM} were concentrated and stored frozen, taking precautions to avoid D-to-H back exchange. Complexes of PR_{WT} and PR_{TM} with APV and DRV inhibitors dissolved in dimethylsulfoxide were made immediately before setting up crystallization plates, also paying attention to prevent possible back exchange. By preparing the PR/drug complexes using this methodology, we preserved D atoms in the catalytic site of PR that becomes solvent-inaccessible after binding of high-affinity protease inhibitors such as APV and DRV and cannot be H/D exchanged again after crystals grow in H_2O-containing crystallization solutions. We optimized crystal growth conditions very carefully (0.1 M MES, 0.9–1.0 M NaCl, pH 6.0), increased the size of crystallization drops to 300–400 μL and utilized siliconized 9-well glass plates placed in Hampton Research sandwich boxes (Hampton Research, Aliso Viejo, CA) to grow crystals of PR_{WT}/APV, PR_{TM}/APV and PR_{TM}/DRV complexes at pH 6.0 by the sitting-drop vapor diffusion method. The large crystallization drops contained protein with starting concentrations of 2–3 mg/mL to make sure there was enough material to grow large crystals suitable for neutron crystallography.

Our additional goal was to obtain a neutron structure of PR_{TM}/DRV complex at acidic pH. All our efforts to grow crystals at pH values below 5.0 failed to produce crystals suitable for neutron diffraction. Importantly, soaking large crystals in solutions of different pH relative to the initial crystal growth conditions always destroyed PR crystals. We therefore adopted an approach that we developed previously to obtain a neutron structure of the xylanase enzyme at acidic pH (Wan et al., 2015). The pH in the crystallization drops was changed by removing the initial well solution and replacing it three times with 0.1 M NaOAc, 1.0 M NaCl at pH 5.5, 5.0 and 4.3 made with H_2O. The pH in the crystallization drops was measured after the last buffer change with a microelectrode to be 4.3. A crystal was harvested and mounted in capillaries containing the same acetate buffer at pH 4.3 made with D_2O.

All neutron diffraction data were collected from the PR complexes at room temperature using the quasi-Laue neutron diffractometers: IMAGINE beamline (Coates et al., 2018; Meilleur et al., 2013) at the High Flux Isotope Reactor (Oak Ridge National Laboratory, USA) and the LADI-III beamline (Blakeley et al., 2010) at the Institut Laue Langevin (France).

3. Analysis of joint XN structures

Neutron diffraction data are normally weaker (i.e., overall smaller $<I/\sigma(I)>$) and extend to lower resolution than the corresponding X-ray data. With twice as many atoms to include, the neutron structure refinement suffers from a very low data-to-parameter ratio. To overcome this limitation the joint X-ray/neutron (XN) refinement procedure has been developed (Adams, Mustyalimov, Afonine, & Langan, 2009). In joint XN refinement,

X-ray and neutron diffraction data collected preferably from the same crystal are combined to greatly improve the data-to-parameter ratio, resulting in joint XN structures that contain accurate information on the positions of all atoms in a protein molecule. The neutron diffraction data for the studied PR complexes extended to 2.0–2.3 Å resolution, whereas the corresponding X-ray diffraction data were collected to 1.75–2.0 Å resolution. In this section we will illustrate how neutron crystallography improved our understanding of drug binding to the wild-type and drug resistant mutant PR, thus providing insights for future drug design that were hidden in low temperature X-ray crystallographic structures. Specifically, proton positions and their movement in the catalytic site were mapped in different complexes and at different pH values, and hydrogen bonding between the drug molecules, PR residues and water molecules were directly visualized uncovering limitations of X-ray crystallography.

3.1 Observing protons in the catalytic site of PR

HIV-1 PR is a homodimeric aspartic protease that is comprised of two identical 99-amino acid subunits with the residues numbered 1–99 and 1′-99′ (Fig. 1). Each PR monomer contributes a conserved catalytic triad, Asp25-Thr26-Gly27, to form the catalytic site in the dimer. The side chains of the two catalytic aspartates, Asp25 and Asp25′, face each other, which is a common feature for all aspartic proteases. The active site cavity is shaped like a flattened cylinder with two flexible flaps encompassing residues 45–55 and 45′–55′ covering the catalytic site (Weber, Kovalevsky, & Harrison, 2007). HIV-1 PR catalyzes hydrolysis of peptidic bonds in the viral Gag and Gag-Pol polyproteins during the maturation stage of the HIV-1 replication cycle when newly formed immature viral particles transform into infectious virions with well-defined internal structure (Beck, Morris, & Elder, 2002). It is generally accepted that the catalytic Asp dyad is monoprotonated in the apo-state prior to substrate binding and in the inhibitor-bound form (Huang, Sun, Yao, & Liu, 2017; Smith, Brereton, Chai, & Kent, 1996; Torbeev & Kent, 2012), but the exact proton location and the distribution of H bonds have always been debated. Until recently, direct observation of the protons in the PR catalytic site, in the active-site cavity and throughout the enzyme has not been possible. Neutron crystallography, on the other hand, has been able to reveal locations of the protons, uncovering unusual hydrogen bonding patterns, proton transfer and revealing fewer hydrogen bonds in PR complexes with clinical inhibitors than suggested by the

X-ray structures. In addition, deuteration of PR_{WT} and PR_{TM} has alleviated a common problem found in most PR/ligand X-ray structures, i.e., the disorder of the ligand over two alternate orientations related by a 180° rotation which complicates structural analysis. Such disorder of PR inhibitors is not observed in the room-temperature joint XN structures of deuterated PR complexes, permitting accurate determination of the proton positions.

Examination of the neutron scattering length density maps for the joint XN structure of PR_{WT}/APV complex confirmed that the Asp dyad is monoprotonated, with the D atom covalently bound to Oδ1 (inner) oxygen of Asp25 (Weber et al., 2013). Asp25 donates the D atom in a short 2.5 Å hydrogen bond with the hydroxyl oxygen of APV (Fig. 3A). Asp25′ is not protonated, accepting the D atom of APV hydroxyl in a short 2.5 Å hydrogen bond with its Oδ2 oxygen. Despite such short O...O distances, the D atom positions clearly show that the two interactions are regular, albeit possibly very strong, hydrogen bonds rather than LBHBs. Remarkably, the 2.8 Å contact between the two inner oxygen atoms of Asp25 and Asp25′ is probably not significant, because the O–D...O angle is 95°, although the interatomic distance would indicate a hydrogen bond (Fig. 3B). Therefore, although the juxtaposition of the oxygen atoms in the PR catalytic site with the bound inhibitor suggests possibility of five hydrogen bonds, only two are in fact formed. In contrast to our observation, in an earlier neutron structure of wild-type PR in complex with a nonclinical inhibitor KNI-272 (Adachi et al., 2009) the Oδ2 (outer) oxygen of one catalytic Asp residue is protonated, possibly because it donates its D in a hydrogen bond with the carbonyl group of KNI-272 instead of the hydroxyl group present in APV. The hydroxyl group of KNI-272 instead hydrogen bonds with the inner oxygen of the other Asp.

In the joint XN structure of PR_{TM}/APV complex at pH 6.0 (Gerlits et al., 2017), one D atom was found belonging to the catalytic Asp dyad, as expected. To our surprise, however, the D was observed located halfway between the inner oxygen atoms of Asp25 and Asp25′ being possibly involved in an LBHB. In an LBHB, the proton has very little electron density and in our joint XN structure the D atom can be characterized as D^+ (deuterium ion). The D^+ is positioned equidistantly to the aspartic acid residues, with the D...O distances of 1.5–1.6 Å. As in PR_{WT}/APV complex, the hydroxyl of the inhibitor in PR_{TM}/APV is rotated toward Asp25′ and makes a strong hydrogen bond of 2.6 Å. D^+ is farther away from the APV hydroxyl (D...O distance of 2.0 Å), indicating that this interaction is probably very weak. As a result, the analysis of the H bond pattern in the

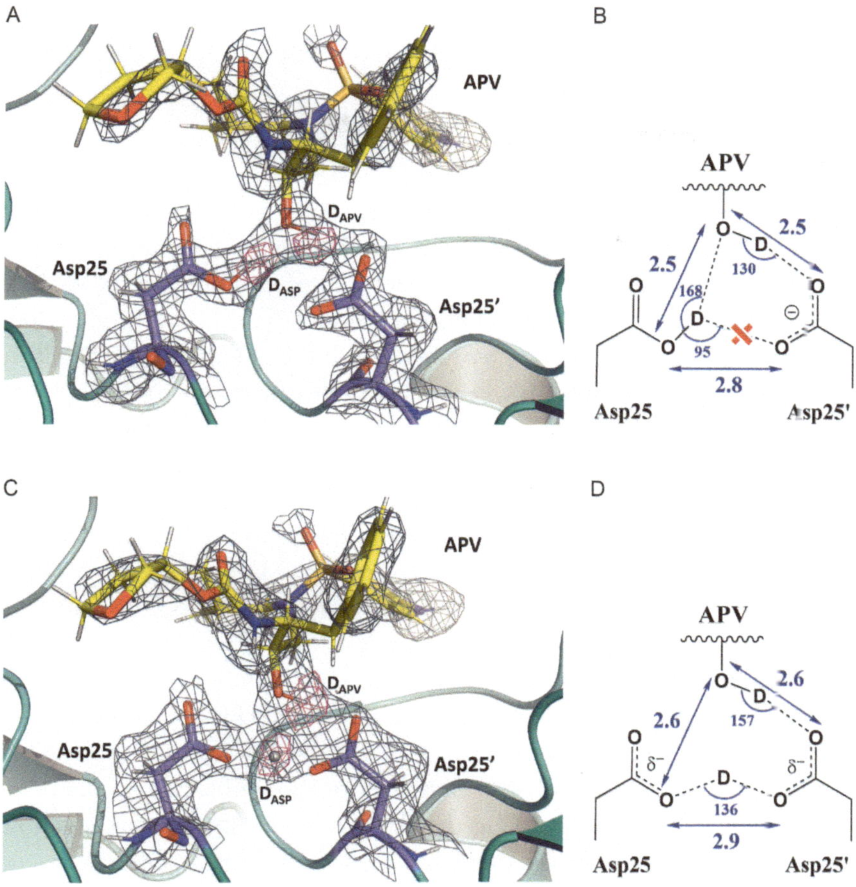

Fig. 3 (A and B) Neutron scattering length density maps and H bond interactions between APV and the catalytic Asp dyad in PR_{WT}/APV complex (PDB ID 4JEC). The $2F_O - F_C$ map is contoured at 1.5 σ level, and the omit difference $F_O - F_C$ map (magenta) directly showing protonation of Asp25 and orientation of APV hydroxyl group is contoured at 3 σ level. (C and D) Neutron scattering length density maps and H bond interactions between APV and the catalytic Asp dyad in PR_{TM}/APV complex (PDB ID 5T8H). The $2F_O - F_C$ map is contoured at 1.0 σ level, and the omit difference $F_O - F_C$ map (magenta) showing possible LBHB between inner oxygen atoms of Asp25 and Asp25′ and orientation of APV hydroxyl group is contoured at 3 σ level. Notice density cancelation effects in the $2F_O - F_C$ maps of APV. Distances are in Å, O–D…O angles are in degrees.

catalytic site of PR_{TM}/APV demonstrates that the drug makes one less H bond with the Asp dyad compared to PR_{WT}/APV, which possibly contributes to the 15-fold diminished affinity of APV for PR_{TM} relative to PR_{WT} (Tie et al., 2012). The formation of the LBHB in the PR catalytic site was previously proposed according to QM/MM calculations (Porter & Molina, 2006), and considered an essential feature on the PR catalytic mechanism (Northrop, 2001).

3.2 Electrostatics-driven proton transfer

Proton transfer is arguably the most important and common chemical reaction governing biochemical processes, including ion channel function, enzymatic reactions, etc. Direct observation of proton transfer in biomacromolecules has not been possible previously. The reasons include inability to detect H atoms and/or to capture the reactants or products in a proton transfer system mainly because H can easily escape to the bulk solvent. HIV-1 PR is unique in that the catalytic site becomes inaccessible to solvent upon binding to a tight inhibitor, such as DRV with an extremely high affinity (low picomolar). To visualize proton transfer two joint XN structures of PR_{TM}/DRV complex were obtained at pH 6.0 and pH 4.3 (Gerlits et al., 2016). In the pH 6.0 structure the D atom of the catalytic Asp dyad is shared by the Asp25′ carboxylate and the DRV hydroxyl. It is therefore a D^+ and participates in an unusual LBHB not previously seen in any neutron structures (Fig. 4). The DRV hydroxyl is rotated to face the carboxylate of Asp25, making a moderate bifurcated H bond with the aspartate's inner and outer oxygen atoms of 2.7 and 2.9 Å, respectively. The H bonding pattern is therefore different from those observed in the complexes with APV, demonstrating that DRV has the strongest interactions with the Asp dyad among the three discussed complexes, in agreement with

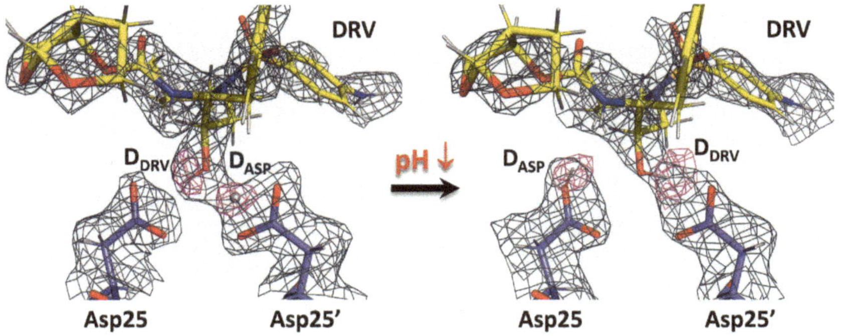

Fig. 4 Electrostatics-driven two-proton transfer in PR_{TM}/DRV complex. The catalytic site at pH 6 is on the left (PDB ID 5E5J), whereas the catalytic site at pH 4.3 is on the right (PDB ID 5E5K). $2F_O - F_C$ neutron scattering length density maps are contoured at 1.0 σ level, and the omit difference F_O-F_C maps (magenta) showing the movement of protons due to pH changes is contoured at 3 σ level.

its tighter binding affinity than that of APV. After the pH in the crystal was altered to 4.3, the D atom arrangement in the catalytic site clearly changed. In the pH 4.3 structure, Asp25 is now protonated on the outer oxygen, whereas DRV hydroxyl faces Asp25′ carboxyl (Fig. 4). Moreover, the inhibitor's OD makes a good H bond of 2.6 Å with the outer carboxylic oxygen of Asp25′, but the protonated Asp25 forms a very weak H bond with DRV, with the distance of 3.1 Å and the D atom rotated away from the inhibitor. On the enzyme surface, four residues—Asp30, Asp30′, Glu34, and Glu34′—became protonated in the acidic conditions, decreasing the negative charge and thereby altering surface electrostatics of the protein. Comparison of the D atom positions in the joint XN structures at pH 6.0 and 4.3 clearly suggests that a net two-proton transfer has occurred between the Asp residues and the hydroxyl of the drug. The shared D atom near Asp25′ is fully transferred to the drug's oxygen, while the latter donates a D atom to the Asp25 carboxylic group. Because bulk water cannot enter the catalytic site, it can be concluded that the proton transfer is triggered by the changes in the electrostatics of the PR surface. Notably, the residues that changed their protonation states due to pH are 11–14 Å away from the site of proton transfer. Therefore, long-range electrostatic effects of distant surface residues can influence the PR catalytic site and can facilitate movement of protons.

These neutron structures unequivocally demonstrate that, in many instances, proton positions and H bonding patterns cannot be deduced solely based on analysis of donor-acceptor distances in X-ray crystal structures.

3.3 Analysis of hydrogen bonding in PR/drug complexes

H bonds are directional interactions. It is important to consider several geometric parameters when analyzing H bonding contacts, including distance between the donor and acceptor heavy atoms (O...O, N...O, etc.), distance from the H to the acceptor heavy atom (H...O, H...N, etc.), the H bond angle (O–H...O, N–H...O, etc.), and to take into account the orientation of the lone electron pairs (e.g., lone pairs on the carbonyl oxygens are in the plane of the carbonyl group). Careful examination of the possible H bonds in PR_{WT}/APV joint XN structure revealed that APV makes only four direct H bonds (two of which are with catalytic aspartates) and two water-mediated interactions (Weber et al., 2013) (Fig. 5A). Of all the H bonds, only those formed with Asp25 and Asp25′ can be designated as strong (D...O distances <2.0 Å), others are weak with long D...O and D...N

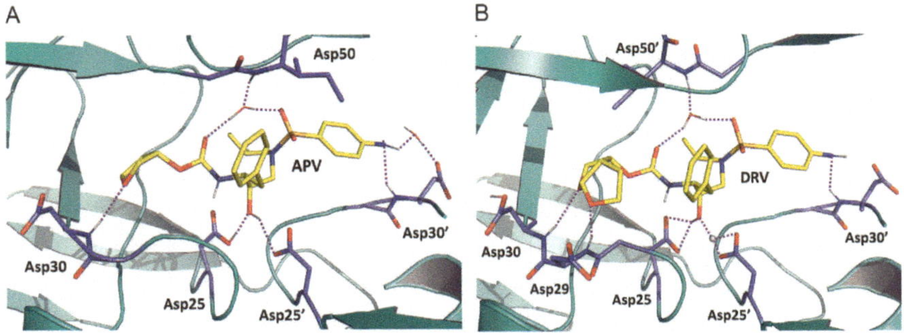

Fig. 5 H bonding interactions (violet dashed lines) in PR_{WT}/APV and PR_{TM}/DRV (pH 6.0) complexes. APV forms four direct H bonds with PR, whereas DRV makes five.

distances of 2.4–2.5 Å. Note that these distances are expressed between a D atom and the O or N atoms, whereas in X-ray structures H bonding distances are conventionally given between the two non-H atoms in the range of 2.5–3.5 Å due to the absence of density for the H atoms. Yet, in the atomic-resolution low-temperature X-ray structure of this complex we previously suggested eight direct H bonds and three water-mediated contacts between PR and inhibitor (Shen et al., 2010). Two fewer H bonds are formed in the catalytic site of PR_{WT}/APV. In addition, the APV carbamate's N–D makes a $<90°$ angle with the main-chain carbonyl of Gly27, suggesting the presence of a weak N–D…π interaction rather than an H bond. The aniline substituent of APV also appears to be quite poorly anchored to the Asp30′ site, making a weak N-D…N H bond with its main chain amide and a water-mediated contact with its side chain carboxylate. We proposed that replacing this water-mediated contact with a direct H bond to either Asp30′ backbone or side chain might improve inhibitor binding affinity to the PR.

The flap water connecting a ligand to Ile50 and Ile50′ is present in almost all structures of PR/ligand complexes, except for tipranavir. It is thought to be essential for the inhibitor and substrate binding in the PR active-site cavity. For the majority of X-ray structures, it has been inferred that this H_2O utilizes all four of its H bonding capabilities, making two H bonds with ligands and two H bonds with the main chain amides of Ile50 and Ile50′. Neutron crystallography makes it possible to determine the correct orientation of water molecules and to clarify what H bonds are formed. In the joint XN structure of PR_{WT}/APV complex the flap water is oriented in such a way that neither of the lone pairs is directed toward the main-chain amide of Ile50′, resulting in only three H bonds made by this D_2O. Thus, APV is connected to the flap through one, instead of two, interactions.

H bonding interactions are virtually identical in the PR_{TM}/APV complex (Gerlits et al., 2017). This indicates that no H bonds, other than those with the catalytic aspartates discussed above, are altered due to the introduction of the three drug resistant mutations. Similarly, in the PR_{TM}/DRV complex the interactions made by the drug are largely conserved. The difference comes from the replacement of the tetrahydrofuran moiety of APV with the bis-tetrahydrofuran group in DRV, allowing it to make an extra H bond with the main-chain amide of Asp29 that appears to be stronger than the one formed with Asp30 backbone amide (Fig. 5B). Comparison of the H bonds in the joint XN structures of PR_{TM}/APV and PR_{TM}/DRV complexes clearly demonstrates why DRV is more effective against drug resistant mutant variants of HIV-1 PR.

3.4 H/D exchange of backbone amides in PR complexes

Protein dynamics can also be assessed by determining the H/D exchange levels of the main-chain amides in a joint XN structure. This allows identification of the regions in a protein that are more flexible where H/D exchange is higher, or less flexible where exchange is lower and/or where access by solvent molecules is restricted. H/D exchange analysis of our PR/drug complexes should produce both interesting and unusual results, because the denatured enzymes were first subjected to dialysis and subsequently refolded in D_2O; second, crystals were grown in H_2O solutions and the other round of H/D exchange was performed by vapor diffusion after the crystals were mounted in quartz capillaries for the neutron crystallographic experiments. Thus, one can expect that the overall level of H/D exchange in our PR/drug complexes will be higher than in other reported neutron structures, because some buried D atoms will not back-exchange in H_2O after the protein is folded. The H/D exchange patterns in the PR_{WT}/APV and PR_{TM}/APV complexes can be directly compared because they contain the same inhibitor as depicted in Fig. 6. We consider the backbone amide sites to be fully exchanged when the refined D atom occupancies are in the range of 0.51–1.0, partially exchanged D atoms have occupancies in the range of 0.0–0.50, whereas when the D occupancy is refined to be −0.56–0.0 the site is considered non-exchanged. Because the neutron scattering length of H is −0.56 times the scattering length of D, the amount of D (%D) at each site can be computed using the following formula: $\%D = [(D_{occup.} + 0.56)/1.56]*100$. Thus, the amount of D is >2/3 for fully exchanged, between 1/3 and 2/3 for partially

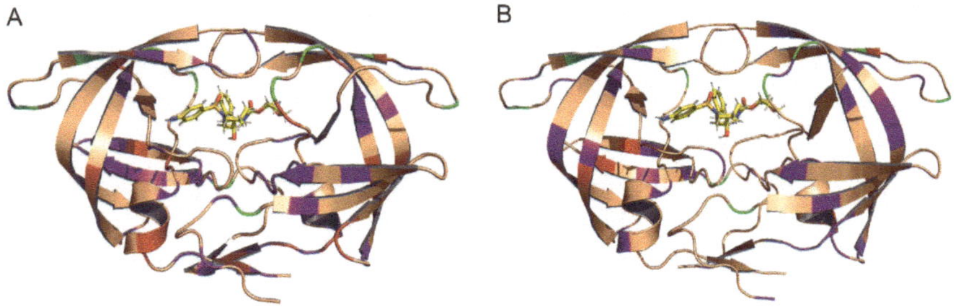

Fig. 6 H/D exchange patterns for the main-chain amide protons in the deuterated PR_{WT}/APV (A) and PR_{TM}/APV (B) complexes at pH 6.0. Fully exchanged sites are colored wheat, partially exchanged amides are purple and non-exchanged amides are dark red. APV is shown in stick representation and colored by atom type.

exchanged and <1/3 for non-exchanged amide sites. As expected, there are very few non-exchanged sites in these structures—11% in PR_{WT}/APV and 5% in PR_{TM}/APV. 24% and 19% of the amide protons are partially exchanged in the wild-type and triple mutant complexes, respectively. Predictably, PR_{WT} and PR_{TM} have many backbone regions fully H/D exchanged, at 65% and 76%, respectively (excluding proline residues that lack exchangeable protons); thus, two-thirds in PR_{WT} and three-quarters in PR_{TM} of all main-chain amides are completely substituted with D. Compared to other neutron structures of proteins crystallized from H_2O solutions and H/D exchanged by vapor (Dajnowicz et al., 2016; Gerlits et al., 2019), PR in our structures has 10–20% more fully exchanged and 6–12% fewer non-exchanged backbone amides. We can conclude that the procedure used for PR involving the unfolded protein treatment with D_2O solutions and then refolding it also in D_2O can generate a highly H/D exchanged crystal even after it is grown in H_2O buffers and subjected to another round of H/D exchange. Such a crystal, especially if the protein is perdeuterated, would create minimal incoherent background in the neutron diffraction images, which is highly desired. However, this procedure of using unfolded protein for deuteration may not be feasible for all proteins.

3.5 Analysis of D incorporation into C–H bonds

A more detailed analysis of the neutron structures can be done by calculating how much D is incorporated into the non-exchangeable C–H bonds of PR_{WT} and PR_{TM} during expression. This is interesting because the enzymes were expressed in pure D_2O, but the sole carbon sources used were different—fully hydrogenous glucose for PR_{WT} and fully hydrogenous glycerol for PR_{TM} expression. Glucose contains seven C–H bonds and glycerol has five C–H bonds; it would be informative to compare differences, if any, in the D incorporation into the C–H bonds for the two PR variants created by feeding *E. coli* bacteria with different compounds. Tables 1 and 2 present this analysis for PR_{WT} and PR_{TM}, respectively. To simplify the analysis, we considered the whole residue as a single entity by averaging occupancies of all D atoms in the C–D bonds for each amino acid. Thus, each residue in the two enzymes was identified as having no D incorporation, partial D incorporation or full D incorporation based on the D occupancy distribution discussed above. Remarkably, the level of D incorporation is very high, with only two residues in PR_{WT} and one residue in PR_{TM} having no D incorporation. In PR_{WT} one Met and one Trp residues belong to this class, whereas in PR_{TM} it is a single Trp. On average, D incorporation is split evenly (~50/50) between residues with partial or full incorporation of D into their C–H bonds for PR_{WT} expressed using glucose. On the other hand, almost 80% of residues contain full D incorporation and just ~20% have partial D incorporation in case of PR_{TM} expressed using glycerol. Interestingly, in PR_{WT} about half of 186 hydrophobic and hydrophilic residues have partial D incorporation, whereas the other half has full D incorporation. But, compared to PR_{WT}, there appears to be much more significant D incorporation in PR_{TM}—86% of hydrophobic and 78% of hydrophilic residues have full D incorporation into their C–H bonds. Arg is the only amino acid in PR_{WT} that has full D incorporation for all eight (i.e., 100%) Arg residues. In PR_{TM}, 87% of Arg show full D incorporation. No other amino acids were observed to have all their residues with full D incorporation. On the other hand, none of the 12 (i.e., 0%) aromatic residues in each PR variant have full D incorporation, demonstrating the poorest levels of D incorporation and indicating that not much D is inserted into the aromatic C–H bonds from D_2O. This analysis thus clearly shows that using glycerol as a sole carbon source for *E. coli* growth and protein expression is beneficial to obtain substantially more deuterated protein than would be afforded by using glucose. It is nonetheless noteworthy to mention that by expressing a protein simply in pure D_2O without using perdeuterated carbon source, such as d_8-glycerol, can afford an almost 80% deuterated protein. Such partial protein deuteration may both alleviate crystallization challenges arising from perdeuteration and be a considerably cheaper option.

Table 1 Incorporation of D into the C–H bonds in PR_{WT}.

Amino acid	% No incorporation	% Partial incorporation	% Full incorporation	Number of residues
Hydrophobic				
Gly	0	31	69	26
Ala	0	30	70	10
Val	0	75	25	12
Ile	0	73	27	30
Leu	0	35	65	20
Pro	0	25	75	12
Met	25	50	25	4
All hydrophobic	*0.9*	*47.4*	*51.7*	*114*
Hydrophilic[a]				
Ser	0	100	0	2
Thr	0	63	27	16
Asp	0	25	75	8
Asn	0	83	17	6
Glu	0	25	75	8
Gln	0	30	70	10
Arg	0	0	100	8
Lys	0	86	14	14
All hydrophilic	*0*	*50*	*50*	*72*
Aromatic				
His	0	100	0	2
Tyr	0	100	0	2
Phe	0	100	0	4
Trp	25	75	0	4
All aromatic	*8*	*92*	*0*	*12*
All residues	*1*	*51*	*48*	*198*

[a]There are no Cys residues in this PR sequence.

Percentage values indicate how many residues of the same type relative to their total number have a particular level of D incorporation. The level of D incorporation for each residue is calculated by averaging the occupancies of D atoms bound to C atoms.

Table 2 Incorporation of D into the C–H bonds in PR_{TM}.

Amino acid	% No incorporation	% Partial incorporation	% Full incorporation	Number of residues
Hydrophobic				
Gly	0	8	92	26
Ala	0	10	90	10
Val	0	10	90	10
Ile	0	16	84	32
Leu	0	10	90	20
Pro	0	25	75	12
Met	0	25	75	4
All hydrophobic	*0*	*14*	*86*	*114*
Hydrophilic[a]				
Ser	0	100	0	2
Thr	0	13	87	16
Asp	0	13	87	8
Asn	0	17	83	6
Glu	0	38	62	8
Gln	0	20	80	10
Arg	0	13	87	8
Lys	0	29	71	14
All hydrophilic	*0*	*22*	*78*	*72*
Aromatic				
His	0	100	0	2
Tyr	0	100	0	2
Phe	0	100	0	4
Trp	25	75	0	4
All aromatic	*8*	*92*	*0*	*12*
All residues	*0.5*	*21*	*78.5*	*198*

[a]There are no Cys residues in the sequence.
The numbers are calculated in the same way as in Table 1.

4. Summary

Macromolecular neutron crystallography is an unrivaled structural biology technique capable of mapping catalytic mechanisms of enzymes, ligand (drug) binding and protein-protein interactions in exquisite detail by visualization and accurate determination of H atom positions in proteins, ligands and water molecules. It is, by no means, an easy technique to apply to important biochemical questions and structure-assisted drug design. One needs to consider growing larger crystals than required for standard X-ray crystallography and making deuterated protein, but the answers that await if a neutron crystallographic experiment is successful are well worth the often-herculean effort. We obtained several HIV-1 PR neutron structures in complex with clinical drugs amprenavir and darunavir and explored changes in H atom positions and protonation states by varying pH inside the crystals. Such changes generally cannot be detected by X-rays. We directly determined positions of H atoms and visualized hydrogen bonds responsible for the high affinity of amprenavir and darunavir, demonstrating that these drugs make fewer interactions with PR than inferred from X-ray structures. We showed that protonation of surface residues at acidic pH alters the electrostatics enough to trigger a two-proton transfer in the catalytic site >10 Å away between the catalytic Asp residues and the hydroxyl of darunavir. Analysis of the main-chain amide H/D exchange levels demonstrated the utility of using unfolded protein to do initial H/D exchange, whereas analysis of D incorporation into C–H bonds provided the evidence that partial deuteration in pure D_2O with hydrogenous glycerol is an attractive method to introduce as much of D as possible into a protein without compromising its crystal growth. Hence, with careful optimization of protein expression, deuteration and crystal growth, neutron crystallography offers unparalleled advantages in understanding biochemical mechanisms and binding of ligands.

Acknowledgments

The neutron diffraction experiments were performed on IMAGINE beamline at Oak Ridge National Laboratory's High Flux Isotope Reactor sponsored by the Scientific User Facilities Division, Office of Basic Energy Sciences (BES), U.S. Department of Energy (DOE), and on beamline LADI-III at the Institut Laue-Langevin. Protein deuteration, purification and crystallization were carried out at ORNL's Center for Structural Molecular Biology (CSMB) supported by the DOE Office of Biological and Environmental Research (BER), using facilities supported by the Scientific User Facilities Division (DOE BES). This work was supported by the NIH and by the DOE BES.

References

Adachi, M., Ohhara, T., Kurihara, K., Tamada, T., Honjo, E., Okazaki, N., et al. (2009). Structure of HIV-1 protease in complex with potent inhibitor KNI-272 determined by high-resolution X-ray and neutron crystallography. *Proceedings of the National Academy of Sciences of the United States of America*, *106*, 4641–4646.

Adams, P. D., Mustyalimov, M., Afonine, P. V., & Langan, P. (2009). Generalized X-ray and neutron crystallographic analysis: More accurate and complete structures for biological macromolecules. *Acta Crystallographica. Section D, Biological Crystallography*, *65*, 567–573.

Agniswami, J., Shen, C.-H., Aniana, A., Sayer, J. M., Louis, J. M., & Weber, I. T. (2012). HIV-1 protease with 20 mutations exhibits extreme resistance to clinical inhibitors through coordinated structural rearrangements. *Biochemistry*, *51*, 2819–2828.

Agniswamy, J., Louis, J. M., Roche, J., Harrison, R. W., & Weber, I. T. (2016). Structural studies of a rationally selected multi-drug resistant HIV-1 protease reveal synergistic effect of distal mutations on flap dynamics. *PLoS One*, *11*, e0168616.

Beck, Z. Q., Morris, G. M., & Elder, J. H. (2002). Defining HIV-1 protease substrate selectivity. *Current Drug Targets*, *2*, 37–50.

Blakeley, M. P., Hasnain, S. S., & Antonyuk, S. V. (2015). Sub-atomic resolution X-ray crystallography and neutron crystallography: Promise, challenges and potential. *IUCrJ*, *2*, 464–474.

Blakeley, M. P., Teixeira, S. C., Petit-Haertlein, I., Hazemann, I., Mitschler, A., Haertlein, M., et al. (2010). Neutron macromolecular crystallography with LADI-III. *Acta Crystallographica. Section D, Biological Crystallography*, *66*(Pt. 11), 1198–1205.

Coates, L., Cao, H. B., Chakoumakos, B. C., Frontzek, M. D., Hoffmann, C., Kovalevsky, A. Y., et al. (2018). A suite-level review of the neutron single-crystal diffraction instruments at Oak Ridge National Laboratory. *The Review of Scientific Instruments*, *89*, 092802.

Dajnowicz, S., Seaver, S., Hanson, B. L., Fisher, S. Z., Langan, P., Kovalevsky, A. Y., et al. (2016). Visualizing the Bohr effect in hemoglobin: Neutron structure of equine cyanomethemoglobin in the R state and comparison with human deoxyhemoglobin in the T state. *Acta Crystallographica Section D, Structural Biology*, *72*, 892–903.

Fisher, Z., Jackson, A., Kovalevsky, A., Oksanen, E., & Wacklin, H. (2017). Chapter 1. Biological structures. In Vol. 49. *Experimental methods in the physical sciences* (pp. 1–75): Elsevier.

Gerlits, O., Keen, D. A., Blakeley, M. P., Louis, J. M., Weber, I. T., & Kovalevsky, A. (2017). Room temperature neutron crystallography of drug resistant HIV-1 protease uncovers limitations of X-ray structural analysis at 100K. *Journal of Medicinal Chemistry*, *60*, 2018–2025.

Gerlits, O., Weiss, K. L., Blakeley, M. P., Veglia, G., Taylor, S. S., & Kovalevsky A. (2019). Zooming in on protons: Neutron structure of protein kinase A trapped in a product complex. *Science Advances*, *5*, eaav0482.

Gerlits, O., Wymore, T., Das, A., Shen, C.-H., Parks, J. M., Smith, J. C., et al. (2016). Long-range electrostatics-induced two-proton transfer captured by neutron crystallography in an enzyme catalytic site. *Angewandte Chemie, International Edition*, *55*, 4924–4927.

Ghosh, A. K., Anderson, D. D., Weber, I. T., & Mitsuya, H. (2012). Enhancing protein baclbone binding—A fruitful concept for combating drug-resistant HIV *Angewandte Chemie, International Edition*, *51*, 1778–1802.

Ghosh, A. K., Osswald, H. L., & Prato, G. (2016). Recent progress in the development of HIV-1 protease inhibitors for treatment of HIV/AIDS. *Journal of Medicinal Chemistry*, *59*, 5172–5208.

Huang, J., Sun, B., Yao, Y., & Liu, J. (2017). Fast and reliable thermodynamic approach for determining the protonation state of the Asp dyad. *Journal of Chemical Information and Modeling*, *57*, 2273–2280.

Kovalevsky, A. Y., Liu, F., Leshchenko, S., Ghosh, A. K., Louis, J. M., Harrison, R. W., et al. (2006). Ultra-high resolution crystal structure of HIV-1 protease mutant reveals two binding sites for clinical inhibitor TMC114. *Journal of Molecular Biology*, *363*, 161–173.

Langan, P., & Chen, J. C.-H. (2013). Seeing the chemistry in biology with neutron crystallography. *Physical Chemistry Chemical Physics*, *15*, 13705–13712.

Laskey, S. B., & Siliciano, R. F. (2014). A mechanistic theory to explain the efficacy of antiretroviral therapy. *Nature Reviews. Microbiology*, *12*, 772–780.

Louis, J. M., Clore, G. M., & Gronenborn, A. M. (1999). Autoprocessing of HIV-1 protease is tightly coupled to protein folding. *Nature Structural Biology*, *6*, 868–875.

Mahalingam, B., Louis, J. M., Hung, J., Harrison, R. W., & Weber, I. T. (2001). Structural implications of drug-resistant mutants of HIV-1 protease: High-resolution crystal structures of the mutant protease/substrate analogue complexes. *Proteins*, *43*, 455–464.

Mehellou, Y., & De Clercq, E. (2010). Twenty-six years of anti-HIV drug discovery: Where do we stand and where do we go? *Journal of Medicinal Chemistry*, *53*, 521–538.

Meilleur, F., Munshi, P., Robertson, L., Stoica, A. D., Crow, L., Kovalevsky, A., et al. (2013). The IMAGINE instrument: First neutron protein structure and new capabilities for neutron macromolecular crystallography. *Acta Crystallographica. Section D, Structural Biology*, *69*, 2157–2160.

Mitsuya, H., Maeda, K., Das, D., & Ghosh, A. K. (2008). Development of protease inhibitors and the fight with drug-resistant HIV-1 variants. *Advances in Pharmacology*, *56*, 169–197.

Niimura, N., & Podjarny, A. (2011). *Neutron protein crystallography*. Oxford University Press. 250 pp.

Northrop, D. B. (2001). Follow the protons: A low-barrier hydrogen bond unifies the mechanisms of the aspartic proteases. *Accounts of Chemical Research*, *34*, 790–797.

O'Dell, W. B., Bodenheimer, A. M., & Meilleur, F. (2016). Neutron protein crystallography: A complementary tool for locating hydrogens in proteins. *Archives of Biochemistry and Biophysics*, *602*, 48–60.

Oksanen, E., Chen, J. C.-H., & Fisher, Z. (2017). Neutron crystallography for the study of hydrogen bonds in macromolecules. *Molecules*, *22*, 596.

Porter, M. A., & Molina, P. A. (2006). The low-barrier double well potential of the $O^{\delta1}$-H-$O^{\delta1}$ hydrogen bond in unbound HIV protease: A QM/MM characterization. *Journal of Chemical Theory and Computation*, *2*, 1675–1684.

Shen, C.-H., Wang, Y.-F., Kovalevsky, A. Y., Harrison, R. W., & Weber, I. T. (2010). Amprenavir complexes with HIV-1 protease and its drug-resistant mutants altering hydrophobic clusters. *The FEBS Journal*, *277*, 3699–3714.

Smith, R., Brereton, I. M., Chai, R. Y., & Kent, S. B. (1996). Ionization states of the catalytic residues in HIV-1 protease. *Nature Structural Biology*, *3*, 946–950.

Tie, Y., Wang, Y.-F., Boross, P. I., Chiu, T.-Y., Ghosh, A. K., Tozser, J., et al. (2012). Critical differences in HIV-1 and HIV-2 protease specificity for clinical inhibitors. *Protein Science*, *21*, 339–350.

Torbeev, V. Y., & Kent, S. B. H. (2012). Ionization state of the catalytic dyad Asp25/25' in the HIV-1 protease: NMR studies of site-specifically ^{13}C labelled HIV-1 protease prepared by total chemical synthesis. *Organic & Biomolecular Chemistry*, *10*, 5887–5891.

Wan, Q., Parks, J. M., Hanson, B. L., Fisher, Z., Osterman, A., Schrader, T., et al. (2015). Direct determination of protonation states and visualization of hydrogen bonding in a glycoside hydrolase with neutron crystallography. *Proceedings of the National Academy of Sciences of the United States of America*, *112*, 12384–12389.

Weber, I. T., & Agniswami, J. (2009). HIV-1 protease: Structural perspectives on drug resistance. *Viruses*, *1*, 1110–1136.

Weber, I. T., & Harrison, R. W. (2016). Tackling the problem of HIV drug resistance. *Postepy Biochemii*, *62*, 273–279.

Weber, I. T., Kneller, D. W., & Wong-Sam, A. (2015). Highly resistant HIV-1 proteases and strategies for their inhibition. *Future Medicinal Chemistry*, 7, 1023–1038.

Weber, I. T., Kovalevsky, A. Y., & Harrison, R. W. (2007). Structures of HIV protease guide inhibitor design to overcome drug resistance. In *Vol. 3. Frontiers in drug design & discovery* (pp. 45–62). Oak Park, IL: Bentham Science Publishers.

Weber, I. T., Waltman, M. J., Mustyakimov, M., Blakeley, M. P., Keen, D. A., Ghosh, A. K., et al. (2013). Joint X-ray/neutron crystallographic study of HIV-1 protease with clinical inhibitor amprenavir—Insights for drug design. *Journal of Medicinal Chemistry*, *56*, 5631–5635.

CHAPTER THIRTEEN

Neutron crystallographic studies of carbonic anhydrase

Jacob E. Combs, Jacob T. Andring, Robert McKenna*
Department of Biochemistry and Molecular Biology, College of Medicine, University of Florida, Gainesville, FL, United States
*Corresponding author: e-mail address: rmckenna@ufl.edu

Contents

Abstract

The carbonic anhydrases (CAs; EC 4.2.1.1) are a family of metalloenzymes that catalyze the reversible hydration of carbon dioxide (CO_2) and bicarbonate (HCO_3^-). Since their discovery in 1933, CAs have been at the forefront of scientific discovery: the understanding of enzymatic reactions, structural biology, molecular dynamics, drug discovery, and clinical medicine. These ubiquitous enzymes equilibrate the reaction between CO_2, HCO_3^-, and protons. Hence, CAs have important roles in ion transport, acid–base regulation, gas exchange, photosynthesis, and CO_2 fixation. In this chapter, we describe the protocols leading to, and the analysis of CA neutron crystal structures. This accumulation of structural knowledge adds to our understanding of the enzymatic mechanism and development of CA inhibitors.

Methods in Enzymology, Volume 634
ISSN 0076-6879
https://doi.org/10.1016/bs.mie.2020.01.003

1. Introduction

The carbonic anhydrase (CA) family of metalloenzymes catalyze the reversible conversion of carbon dioxide (CO_2) and water to bicarbonate (HCO_3^-) and a proton (H^+) via a coordinated metal ion. The CAs coordinate their metal ion through histidine, cysteine, and/or glutamine residues at the core of their active sites, forming a tetrahedral-like coordination to the metal. While zinc is the most common metal utilized, many CAs have been shown to be catalytically active with other di-covalent metals (M^{2+}) such as cobalt, nickel, cadmium, and copper. The CA family of enzymes are able to catalyze their reaction at very fast rates with the human CA isoform, CA II, exhibiting a near diffusion limited rate, with a k_{cat} of 1.1 μs^{-1} (Steiner, Jonsson, & Lindskog, 1975). The CAs achieve this near catalytic perfection via a ping-pong mechanism through a metal-bound hydroxyl/water. The metal ion acts as a Lewis acid, inducing nucleophilic character on the hydroxyl, priming it for catalysis. The reaction in the hydration direction starts with the nucleophilic attack via the metal-bound hydroxyl onto the carbon atom of the CO_2 substrate (Silverman & Lindskog, 1988). This forms a metal-bound HCO_3^- that is displaced via a readily available water in the active site. The second step of the reaction is the regeneration of the metal-bound hydroxyl via proton transfer through a coordinated water network (Silverman & Lindskog, 1988). A series of hydrogen-bonded waters allows for this rapid transfer of a proton from the zinc-bound water to bulk solvent. For most CAs, this second step is the rate limiting step.

CAs perform a variety of physiological functions in humans: CO_2 sensing and transport, HCO_3^- sensing and transport, gas exchange, respiration, blood homeostasis, and pH balance of the extracellular and intracellular environment (Frost, 2014). In plants, CAs are expressed in chloroplasts and are involved in photosynthetic CO_2 fixation as well as CO_2 capture mechanisms in marine diatoms (Rowlett, 2010). In prokaryotes, CAs are involved in respiration, photosynthesis, and CO_2 capture and sensing, as well as other physiological functions (Capasso & Supuran, 2015).

CAs are expressed in nearly all forms of life including mammals, protozoans, diatoms, and even viruses. Currently, there are eight genetically unique CA families identified: α, β, γ, δ, ζ, η, θ, and the recently reported ι (Del Prete, Vullo, De Luca, et al., 2014; Del Prete, Vullo, Fisher, et al., 2014; Iverson, Alber, Kisker, Ferry, & Rees, 2000; Jensen, Clement, Kosta, Maberly, & Gontero, 2019; Kikutani et al., 2016; Meldrum & Roughton, 1933;

Mitsuhashi et al., 2000; Xu, Feng, Jeffrey, Shi, & Morel, 2008). These CA families all catalyze the reversible hydration of CO_2 but are unique from each other. The families are classified by differences in structural fold, active site residues coordinating the metal, and the metal used in catalysis. Within these families, the tetrahedral coordination of a metal to three active site residues and a water/hydroxyl molecule is conserved, but the metal and the coordinating amino acids differ. The α CAs are the most well-studied class as these are expressed mainly in mammals and have been associated with human diseases and are therefore recognized as therapeutic targets.

The α CAs utilize a zinc ion coordinated by three histidine residues and a solvent molecule in the core of the active site, that is divided into a hydrophilic and hydrophobic side. In humans, there are 15 isoforms of α CAs, each with unique sequences, cellular locations, and catalytic efficiencies. CA I, CA II, CA III, CA VII, CA VIII, CA X, CA XI, and CA XIII are cytosolic; CA IV, CA IX, CA XII, and CA XIV are membrane-bound; CA Va and CA Vb are mitochondrial; and CA VI is secreted (Frost, 2014).

The β CAs were initially discovered in plant chloroplasts and have since been identified in eubacteria, algae, and archa. β CAs form homo-oligomers, with their catalytic metal at the interface between monomers. The β CAs are mainly dimeric and are the only CA class to exhibit allosteric regulation. β CAs are also zinc metalloenzymes; however, the zinc is coordinated by two cysteines and one histidine in the active site (Rowlett, 2010). The catalytic activity of β CAs has been shown to be essential to the survival of bacteria and this class is structurally different from the CAs expressed in humans, and therefore β CAs are targets in the design of new antibiotics (Supuran, 2011).

γ CAs are found in archa and bacteria, however, only the CA Mt-Cam from *Methanosarcina thermophila* has been extensively studied. They are catalytically active as trimers with the metal ion coordinated between the monomers by a histidine residue from each monomer (Zimmerman, Tomb, & Ferry, 2010). Previous studies have hypothesized iron as the likely metal, however, this class of enzyme is also highly active with zinc and cobalt (Ferry, 2010).

The δ and ζ classes of CA are both expressed in marine diatoms and have yet to be identified elsewhere. δ CAs were first identified in *Thalassiosira weissflogii* (Roberts, Lane, & Morel, 1997). Although studied biochemically, there is currently no determined structure of a δ CA. The active site is hypothesized to be similar to that of the α class with a catalytic zinc ion coordinated by three histidines (Beauchemin & Morse, 2015; Del Prete, Vullo, Scozzafava, et al., 2014). The ζ class is known to bind and retain activity with cadmium and/or zinc, although it is hypothesized to preferentially bind cadmium in seawater, where there are low concentrations of zinc. The widely studied ζ CA known as CDCA1, has an active site similar to β CAs, coordinating the metal with a histidine and two cysteine residues (Amata, Marino, Russo, & Toscano, 2011).

The η CAs are closely related to and previously thought to belong to the α class. However in 2014, sequence and phylogenetic analyses identified the η CAs as a unique genetic class (Del Prete, Vullo, Fisher, et al., 2014). η CAs are predicted to utilize zinc as their catalytic metal and are unique in that the catalytic zinc is hypothesized to be coordinated by two histidine and one glutamine (De Simone, Di Fiore, Capasso, & Supuran, 2015). Currently, η CA has only been identified in *Plasmodium falciparum*, the pathogen that causes malaria, and is therefore recognized as a potential target of antiparasitic agents. Activation and inhibition studies have been performed on this η CA (known as PfACA) but more work is needed before it can be targeted as a malaria treatment (Del Prete, Vullo, Fisher, et al., 2014).

θ CA was discovered in *Phaeodactylum tricornutum* and identified as a unique CA zinc metalloenzyme due to its CO_2 hydration and esterase activity (Kikutani et al., 2016). This enzyme is known to be involved in algal CO_2 concentrating mechanism (CCM) and CO_2 fixation. While few studies have been performed with this family of CA, in *Phaeodactylum tricornutum* this CA is critical for growth as it supplies CO_2 for the Calvin Cycle (Kikutani et al., 2016).

The ι family of CAs have recently been identified as the latest genetically unique family of CAs. The enzyme LCIP63 (Low CO_2-inducible protein of 63 kDa) was isolated from a marine diatom, *Thalassiosira pseudonana*, and shown to have robust CA activity. Preliminary studies indicate the likely metal catalyst is manganese, making this family of CAs unique. This enzyme also has multiple predicted metal binding sites, with the primary manganese binding site consisting of two histidines, glutamate, and aspartate, with one of the acidic residues likely occupying the metal-bound solvent site, rendering the enzyme acatalytic in that configuration. LCIP63 has been shown to be involved with CCM (Jensen et al., 2019).

2. X-ray crystallographic studies of carbonic anhydrase

Despite the extensive structural research on CAs, there are currently structures for only half of the families. The α, β, γ, and ζ classes all have at least one structure from their families determined, while δ, η, θ, and ι have yet to have structures determined. Despite each family being genetically unique with low sequence similarity between them, each family retains some common motifs. The active site utilizes three residues, usually histidine and cysteine residues, to conjugate the M^{2+}. They also partition the active site into a hydrophobic and hydrophilic face to bind the substrate and aid in proton transfer, respectively. The most studied CAs are the α CAs with ~950 structures currently deposited in the Protein Data Bank (PDB) (Berman et al., 2000). The monomeric CA II is the most extensively studied of the α CAs, therefore its residue numbering and type will be used throughout this chapter, unless mentioned otherwise.

The catalytic domain of the α CAs are ~30 kDa with a solvent accessible active site cavity. The overall shape of the domain is an ellipsoid with dimensions ~40 × 40 × 50 Å (Eriksson, Jones, & Liljas, 1988). Seven right-handed α-helices decorate the surface of the enzyme, surrounding a 10 stranded mixed β-sheet core. The conically shaped active site is ~15 Å deep with a volume of ~900 $Å^3$ (Krishnamurthy et al., 2008). The catalytic zinc is situated at the base of the active site cavity, tetrahedrally coordinated to three histidine residues and a solvent molecule (Figs. 1 and 2). The histidine

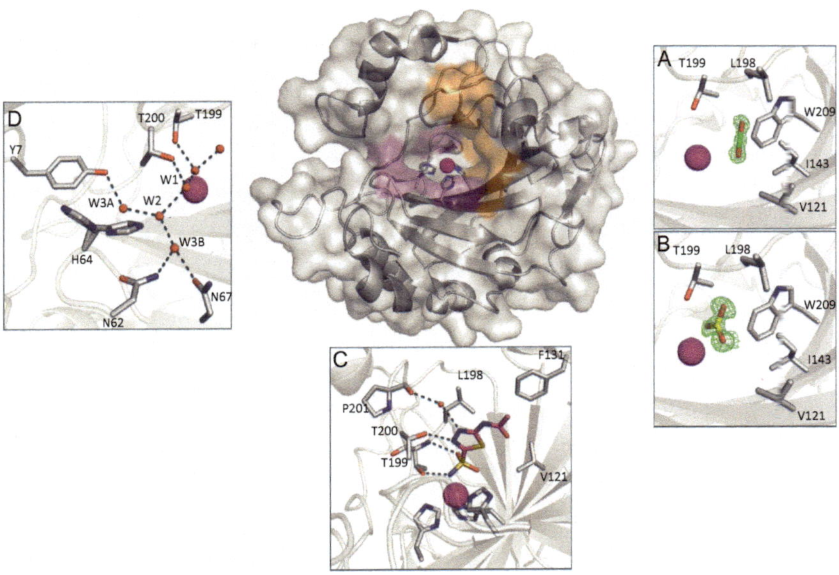

Fig. 1 Structure of CA II. Center, surface rendition showing zinc coordination (orange shading indicates hydrophobic face and purple shading indicates the hydrophilic face). Inset (A) CO_2 binding site (Domsic et al., 2008; Domsic & McKenna, 2010), (B) HCO_3^- binding site (Domsic et al., 2008; Domsic & McKenna, 2010), (C) inhibitor (Acetazolamide) binding site (Sippel et al., 2009), and (D) proton wire (Boone, Pinard, McKenna, & Silverman, 2014). A magenta sphere represents the zinc ion and specific residues and water molecules discussed in the text are as labeled.

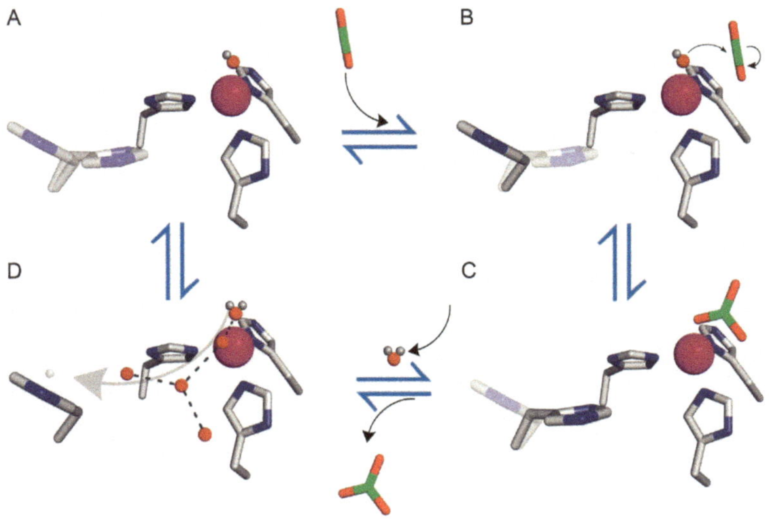

Fig. 2 The structures of the steps in the catalytic mechanism of CA II. (A) Empty active site primed for CO_2 binding (Avvaru et al., 2010). (B) Nucleophilic attack of Zn-bound hydroxyl onto CO_2 (Domsic et al., 2008). (C) Bicarbonate bound to Zn (Xue et al., 1993). (D) Regeneration of active site through proton transfer (Fisher et al., 2011).

residues coordinate the zinc via their imidazole rings, aligning the lone pairs of their nitrogens to interact with the zinc and form multiple hydrogen bonds with nearby polar residues (Avvaru et al., 2010). The catalytic zinc-bound hydroxyl/water is stabilized through a hydrogen-bond with residue T199 and further bridged to E106. The active site cavity consists of two faces, termed the hydrophobic and hydrophilic sides because of the residues on their respective surfaces. The hydrophobic side consists of residues I91, V121, F131, V135, L141, V143, L198, P202, L204, V207, and W209 (Fig. 1) (Avvaru et al., 2010). The hydrophobic side is associated with a substrate CO_2/HCO_3^- binding pocket, at the base of the active site; formed by residues V121, V143, L198, and W209 (Fig. 1A and B) (Avvaru et al., 2010). The hydrophilic side consists of resides N62, H64, N67, Q92, T199, and T200 (Fig. 1D) (Avvaru et al., 2010). These residues stabilize several water molecules within the active site via hydrogen-bonding, forming a proton wire. These waters, termed W1, W2, W3A, and W3B, transfer H^+ to and from the catalytic zinc-bound hydroxyl/water and residue H64 (Fig. 1D) (Fisher et al., 2010; Silverman & Vincent, 1983). This histidine can occupy two conformations dictated by substrate binding or pH. At alkaline pHs, the H64 is in the "in" (pointing towards the zinc) conformation while at acidic pHs, H64 is in the "out" (pointing away from the zinc) conformation (Fisher et al., 2005). This flexibility of conformational change promotes H^+ transfer in/out of the active site and to/from the bulk solvent. A water molecule, termed "deep water" (Dw) is located 2.4 Å from the zinc-bound solvent, and primed to occupy the space created during substrate released (Kim et al., 2016).

Although less extensively studied, X-ray crystal structures of β, γ, and ζ CAs have been determined, providing mechanistic insight into these families of the enzyme. Approximately 90 β CA structures have been determined and deposited to the PDB from a variety of organisms including yeast, plants, fungus, and bacteria. All β CAs exist as dimers or a multimer of dimers, with the tetrameric oligomer being the most common (Rowlett, 2014). Although each monomer of β CA contains a zinc-coordinated active site, the monomers of these enzymes are acatalytic as they require the allosteric activation of the dimerized state (Capasso & Supuran, 2015; Rowlett, 2014). All β CA monomers have a four- or five-stranded parallel sheet core. Upon dimerization, the β-sheets of each monomer align to form extended sheets. Several α helices surround the core, two of which interact with the neighboring monomer to support dimerization. As seen in α CAs, the β CAs retain a hydrophobic face along the dimer interface that leads into the catalytic site that binds CO_2 (Aggarwal, Chua, Pinard, Szebenyi, & McKenna, 2015). There are two subclasses of β CA characterized by the coordination of the zinc ion. In Type I, zinc is coordinated by two cysteine, one histidine, and a water molecule/ligand (Ferraroni, Del Prete, Vullo, Capasso, & Supuran, 2015). In Type II the coordination sphere consists of two cysteines, one histidine, and one aspartate, which replaces the water. While Type I are active under most pHs, Type II are acatalytic below pH 8.3. This is due to the aspartic acid residue, when deprotonated, binding to the zinc and blocking the catalytic hydroxyl from coordinating to the zinc (Cronk et al., 2006).

There are currently 26 structures of the γ CA family that have been determined and deposited in the PDB, most of which are CAs from *Methanosarcina thermophila*. These enzymes form trimers in their crystal structures which is assumed to be maintained at physiological conditions Their active sites are formed at the interfaces between monomers, giving three active sites in the biological assembly. These enzymes can utilize zinc or cobalt as their catalytic metal and structures of both complexes have been determined (Iverson et al., 2000). Unlike the other families, the metal is coordinated by residues from the adjacent monomers. One monomer contributes two histidine while the other comes from a symmetry related monomer (Kisker, Schindelin, Alber, Ferry, & Rees, 1996). Due to the active site interface differing from other CA families, these enzymes are thought to be acatalytic as monomers (Capasso & Supuran, 2015; Iverson et al., 2000).

There are currently only two solved structures of ζ CA deposited in the PDB. Both of which are the cadmium CA from *Thalassiosira weissflogii* (CDCA1). While similar to β CA structures, this enzyme is active as a monomer and uses cadmium as the catalytic metal ion. CDCA1 exists as a cambialistic enzyme, meaning that it can interchangeably use zinc or cadmium for catalysis. The structures with either metal show the same coordination through one histidine and two cysteines, similar to the β CAs (Viparelli et al., 2010; Xu et al., 2008).

2.1 Mechanism of carbonic anhydrase

CAs are among the most catalytically efficient enzymes known, with CA II exhibiting a turnover rate of $1.1\,\mu s^{-1}$ and SazCA a rate of $4.4\,\mu s^{-1}$ (Steiner et al., 1975; Vullo et al., 2012). Additionally, the recently characterized ζ enzyme CDCA R1 also has an efficient rate of $1.5\,\mu s^{-1}$ (Iverson et al., 2000). CA II and presumably all α CAs, catalyze their reaction via the same two-step, ping-pong mechanism where the first step is a nucleophilic attack of CO_2, generating HCO_3^-. The second step is the regeneration of the zinc-bound hydroxyl through a proton transfer (Fig. 2) (Silverman & Lindskog, 1988).

In the hydration direction, the reaction starts with an empty active site presenting a zinc-bound hydroxyl primed for catalysis. A CO_2 molecule enters the active site, binding into the hydrophobic pocket ~2.8 Å away from the zinc-bound hydroxyl (Figs. 1A and 2A) (Merz, 1991). As CO_2 binds, H64 starts to undergo a conformational change from the "out" to the "in" position (Fig. 2B) (Maupin & Voth, 2007). This triggers a nucleophilic attack from the hydroxyl group to the positively charged carbon of the CO_2 (Boone et al., 2014). The result is the formation of a zinc-bound HCO_3^- (Figs. 1B and 2B). This nucleophilic attack is nearly instantaneous and therefore does not affect the overall rate of the reaction.

The second step of catalysis is the release of HCO_3^- and regeneration of the zinc-bound hydroxyl. HCO_3^- is bound to the zinc via the oxygen that performed the nucleophilic attack (Fig. 2B). The negatively charged oxygen also forms a hydrogen-bond with T199 (Silverman & Lindskog, 1988). At this point, the HCO_3^- is displaced by a water molecule in the active site and is released (Fig. 2C) (Kim et al., 2016). The final step of the mechanism is a proton transfer from the newly zinc-bound water to bulk solvent. With the release of HCO_3^-, H64 is primed for proton transfer and completes its conformational change to the out conformation (Fig. 2D) (West et al., 2012). The extensive hydrogen-bonding of the water network allows for rapid H^+ transfer through these waters and H64, away from the active site, and regenerating the catalytic hydroxyl (Fig. 1D and 2D) (Boone et al., 2014). Despite being very efficient, proton transfer is the rate limiting step of the mechanism. The other families of CA all follow a similar mechanism comprising the two main steps of nucleophilic attack and hydroxyl regeneration.

The α CA class are all efficient enzymes with k_{cat} rates $>0.1\ \mu s^{-1}$ with the exception of the acatalytic CARPS (CA VIII, X, and XI) and CA III. In contrast to the other active α CAs, CA III utilizes a lysine instead of a histidine for proton transfer. This difference in proton shuttle residue and several other active site residue differences leads to a less ordered water network and thus inefficient proton wire (Jewell et al., 1991). In the γ CA from *Methanosarcina thermophila*, we see a similar trend of proton shuttle deviations from optimum, leading to lower catalytic rates. This form of CA uses two glutamic acid residues as the proton shuttle instead of the histidine residues in most of the α family (Tripp & Ferry, 2000). Not only do residue modifications change the catalytic rates, but also different metals in the active site can reduce/increase catalytic rates, as seen in the ζ CA family. In CDCA1, the k_{cat} and k_{cat}/K_m vary depending on the binding of zinc or cadmium in the active site. While this enzyme normally utilizes cadmium under physiological conditions, it has increased catalytic efficiency when zinc is substituted in the active site (Iverson et al., 2000).

2.2 Inhibition of carbonic anhydrase

Many *α* CA isoforms are recognized as therapeutic targets in diseases such as glaucoma (CA II, CA IV, CA XII), edema (CA II, CA IV, CA XII, CA XIV), epilepsy (CA VII), and cancer (CA IX, CA XII). Therefore, a significant portion of CA research has focused on the design of CA inhibitors (CAIs). Studies on the inhibition of CAs began in tandem with its discovery in the 1930s. Initial research on the kinetic parameters of α CAs included inhibition studies with anions; sulfanilamide, cyanide, and azide, showing nearly complete inhibition at micromolar concentrations (Keilin & Mann, 1940). Through subsequent testing, it was quickly determined that sulfonamides (SO_2NH_2) were the most effective CAIs. Sulfonamide-based inhibitors bind to the catalytic zinc in a tetrahedral geometry, displacing the zinc-bound water/hydroxyl and also forming hydrogen bonds with T199 and E106, two active site residues conserved among the α CAs and thought of as "gate keeper" residues (Supuran, 2016).

Subsequent studies revealed that adding ring structures to sulfonamides greatly increased their affinity to CAs, making them sub micromolar inhibitors. Furthermore, adding heterocyclic structures to the sulfonamides exhibited >1000-fold increase in affinity (Miller, Dessert, & Roblin, 1950). These ring bound sulfonamide molecules became the first clinically used CAIs, and are still used in the clinic today. Examples of these CAIs include acetazolamide (AZM, Diamox) (Fig. 1C), methazolamide (Neptazane), ethoxzolamide, and dichlorphenamide (Keveyis). Further research led to the development of more complex, higher affinity inhibitors that treat a variety of the aforementioned diseases, including topiramate, sulpiride (Dogmatil), sulthiame, valdecoxib, zonisamide, irosustat (COUMATE), and esterone sulfamate (EMATE).

Besides sulfonamides, there are several other classes of compounds known to be potent CAIs. Examples of such compounds include phenol based and carboxylic based inhibitors (Lomelino, Supuran, & McKenna, 2016). Unlike sulfonamides, these classes of inhibitors do not displace the zinc-bound solvent, but instead anchor through their hydroxyl moiety. Phenol-based and carboxylic acid-based compounds exhibit binding affinities within the micromolar range and are hypothesized to inhibit activity via obstruction of the CO_2 binding site caused by interactions of the phenyl group with hydrophobic residues in the active site (Lomelino et al., 2016; Nair, Ludwig, & Christianson, 1994). Increasing the length or complexity of phenols, as seen in phenol-based natural products, has been shown to improve inhibitory properties and increase their affinities (Innocenti, Beyza Oztürk Sarikaya, Gülçin, & Supuran, 2010; Sechi et al., 2012).

3. Neutron crystallography

X-ray crystallography is one the most useful structural biology techniques we have, allowing discrete localization of protein atoms in a 3-dimensional structure. This method allows for the study of proteins with substrates and cofactors bound, leading to structure related mechanistic studies and structure related drug design. This technique, however, has a large limitation in the inability to determine hydrogen locations. X-ray crystallography requires the electrons in a sample to interact with the incoming X-ray beam. The resulting scattered X-rays interfere to produce a distinct pattern. However, the X-ray scattering power (scattering factor) of an atom to diffract is dependent on how electron rich it is; heavier more electron dense atoms will scatter more efficiently than smaller atoms (Groeneweger & Feil, 1969). Hydrogens are electron-poor and thus have an inherently low scattering power (Table 1). Therefore, hydrogens are not normally seen in X-ray crystallography electron density maps. However, neutrons can be utilized in leu of X-rays in the technique known as neutron crystallography. In this technique, the atomic nucleus scatters the neutrons and this alternative crystallography route allows the determination of hydrogen locations. For neutrons, the scattering lengths of hydrogen (H) and deuterium (D) are comparable in magnitude to the other atoms commonly found in proteins (Table 1) (Blakeley, Langan, Niimura, & Podjarny, 2008; Oksanen, Chen, & Fisher, 2017). Neutron crystallography can therefore be used to identify the position and accessibility of H or D atoms, providing insight into side chain protonation states and hydrogen-bonding networks that may improve the understanding of catalytic mechanisms (Lomelino, Andring, & McKenna, 2018a; Norvell, Nunes, & Schoenborn, 1975). CAs are an ideal protein that can benefit from neutron crystallographic studies, as the second half of their mechanism relies on a hydrogen-bonded water network for proton transfer.

While an advantage of neutron over X-ray crystallography is the ability to "see" H atoms, there are challenges that arise when using neutrons as a source of diffraction. Current challenges in neutron crystallography include;

Table 1 Neutron and X-ray scattering lengths for atom types (Blakeley, Hasnain, & Antonyuk, 2015; National Institute of Standards and Technology, 2013).

	Neutron scattering length (10^{-15} m)	Incoherent cross section σ_I (10^{-28} m^2)	X-ray scattering length (10^{-15} m), $\sin\theta = 0$	X-ray scattering length (10^{-15} m), $(\sin\theta)/\lambda = 0.5$ Å^{-1}
^{1}H	−3.74	80.27	2.80	0.20
^{2}H (D)	6.67	2.05	2.80	0.20
^{12}C	6.65	0.00	16.90	4.80
^{14}N	9.37	0.50	19.70	5.30
^{16}O	5.80	0.00	22.50	6.20
^{32}S	2.80	0.00	45.00	1.90

the need for large crystals for data collection because of the low beam flux; the production of perdeuterated protein; and the issue of residual hydrogens from H/D exchange that lead to background noise which complicate the interpretation of the neutron density maps (Table 1).

The major issue with neutron crystallography is the inherently low neutron beam flux (Meilleur et al., 2013). Compared to X-ray sources, neutron flux is several orders of magnitude less, requiring data collection that can take weeks to accumulate good signal to noise data (Shu, Ramakrishnan, & Schoenborn, 2000). For example, a CA neutron dataset can take ~200 h to collect at LADI-III, while a X-ray synchrotron dataset at MAXIV line I911-2 takes less than an hour to collect (Koruza et al., 2019). This problem is being resolved by several avenues, including growing larger crystals, developing better neutron sources, and more efficient neutron detectors.

In a typical protein crystal, the protein is repeated in a uniform lattice over 10^{12} times. These repeating units amplify the scattering from the individual protein thus the more molecules, the stronger the signal. Larger protein crystals will therefore reduce the time needed for data collection and increase the signal-to-noise ratio (Shu et al., 2000). This can be a challenge, however, as growing large crystals requires a combination of higher protein concentrations and further optimization of crystallization conditions. Even with optimized crystal growing conditions, it may take longer to grow larger crystals and may not even be possible. However, if technological advances improve neutron source flux or better detection methods are developed, then the need for larger crystals can be circumvented.

For example, the development of the IMAGINE beamline along with others such as MaNDi, have increased the flux density, improved detection methods, and improved sample handling, so that submillimeter-cubed protein crystals can now be used (Meilleur et al., 2013; Oksanen et al., 2017). Improvements include using an accelerator-driven spallation source to provide a high flux of neutrons in discrete pulses, and using instrument detector arrays with approximate cylindrical or spherical geometries to record both forward- and back-scattering reflections (O'Dell, Bodenheimer, & Meilleur, 2016). These improvements in flux and detection allow more data at higher resolution to be collected in less time. Ultimately, these advances should allow smaller protein crystals to be used similar to those for X-ray crystallography.

Besides the issue of low flux, the other factors limiting neutron crystallography are the incoherent scattering of H atoms and its negative coherent scattering factor (Table 1). The large incoherent neutron scattering factor results in isotropic background noise, leading to poor signal to noise ratios for the reflections and thus degrades the quality of the resultant density map (Shu et al., 2000). The coherent scattering length of H is negative while that of carbon, nitrogen, and oxygen, are positive. This can cause density cancelation in Fourier maps which at moderate resolution makes interpretation and analysis difficult (Oksanen et al., 2017). Significant portions of the protein (e.g., methylene units) are not seen in the density maps due to this phenomenon. As a protein is comprised of ~50% H atoms, these issues can lead to nuclear density map interpretation in the absence of X-ray information being difficult. These issues are somewhat negated by substituting the H atoms with D. Compared to H, D has a positive scattering length as well as a sixfold decrease in incoherent scattering cross section. Thus the phenomenon of density cancelation does not occur and the background noise is much reduced (Shu et al., 2000). Therefore, replacing H with D atoms results in better neutron diffraction data quality and interpretable neutron density maps.

The most common method of replacing H is to express the hydrogenated protein and then perform H/D exchange with D_2O buffers. This H/D exchange can be done via dialysis before crystallization or during/after crystallization with D reagents. A major limitation to this technique is only the solvent/chemically accessible H will be exchanged. On average, it takes 4–6 weeks of exchange to partially deuterate (~80%) a protein crystal (Bennett, Gardberg, Blair, & Dealwis, 2008). Non-exchangeable H will not be observed in the neutron map and also contribute to background noise, thus ideally should be eliminated. To fully deuterate a crystal, the entire process of protein expression and purification must be done with deuterated media/buffers. This process can take even longer to streamline as the subtly altered fully deuterated protein may require further optimization of the synthesis and purification steps. Cells may grow poorly, protein expression may be lessened, and therefore yields decreased.

A further issue arising from the use of deuterated protein for neutron crystallography is the requirement to reoptimize crystallization conditions from those for the hydrogenated protein. The crystal conditions for a specific protein may be well documented, but deuterium may affect a protein's dynamics, thus the original crystal conditions may not work. Mother liquor, crystalizing agents, concentrations, and ratios may all need to be reoptimized for the most efficient crystal growth. Lastly, this entire process is extremely costly (Koruza, Lafumat, Végvári, Knecht, & Fisher, 2018). These steps will take additional resources, money, and most importantly time.

While neutron crystallography does have its challenges, the ability to determine H or D atom locations is invaluable for elucidating the fine workings of a protein mechanism. A great example of this, is the enzyme presented in this chapter, CA II. While CA II has been extensively studied through X-ray crystallography, the solved structures are still limited based on the issues outlined above, the absence of the location of H atoms. Through the use of perdeuterated CA II crystals and technological advances in the neutron crystallography field, the neutron structure of CA II was solved and extensively studied, and this issue resolved.

4. Protocols

4.1 Expression and purification of hydrogenated carbonic anhydrase II

CA II is expressed in *Escherichia coli* BL21 pLysS (DE3) cells and purified according to standard protocol (Pinard, Boone, Rife, Supuran, & McKenna, 2013; Tanhauser, Jewell, Tu, Silverman, & Laipis, 1992). Briefly, cells are grown in Luria broth (LB) media and expression is induced via 1 mM isopropyl-β-d-thiogalactopyranoside (IPTG) and 1 mM $ZnSO_4$ for 3 h (Fisher et al., 2009). Cell pellets are lysed with a microfluidizer, sonicator, homogenizer, French press, etc. Cell lysate is loaded onto *p*-Aminomethylbenzenesulfonamide (pAMBS)-agarose resin and CA II is eluted with 0.4 M sodium azide after multiple wash steps (0.2 M sodium sulfate, 100 mM Tris pH 9.0 and 7.0) (Fisher et al., 2009). A Pharmacia desalting column is used to buffer exchange the eluted protein into 50 mM Tris 8.0 pH (Fisher et al., 2009). A Millipore Amicon Ultra centrifugal filter with a 10 kDa molecular weight cutoff is used to concentrate the purified protein to 15 mg/mL which is determined from the molar absorptivity at 280 nm (Fisher et al., 2009). Protein purity is checked by running an SDS-PAGE gel.

4.2 Expression and purification of perdeuterated carbonic anhydrase II

CA II is expressed in *Escherichia coli* BL21 pLysS (DE3) cells and purified following standard protocol with a few exceptions (Pinard et al., 2013; Tanhauser et al., 1992). Cells are grown in LB media and transferred into D_2O minimal media with deuterated glycerol. After cell adaption, culture is transferred to fully deuterated liquid media. Cell culture is induced with 200 μm IPTG at a final OD_{600} of 15 (Fisher et al., 2009). Cells are harvested after 16 h and cell lysate is loaded onto pAMBS-agarose resin to purify the sample with H_2O buffers (Fig. 3) (Khalifah, Strader, Bryant, & Gibson, 1977).

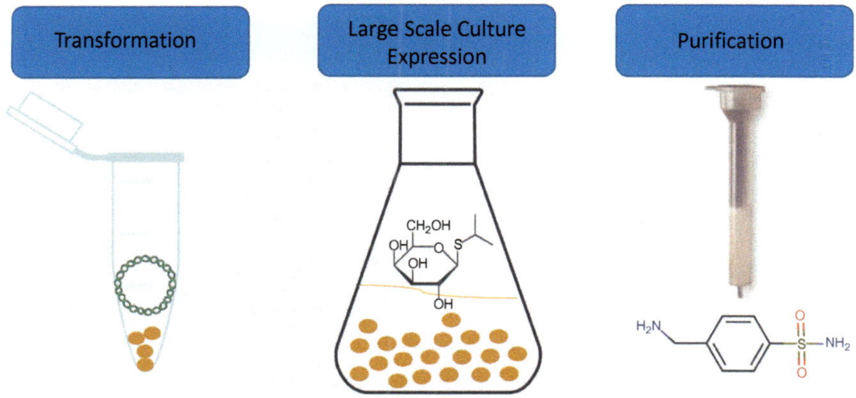

Fig. 3 CA transformation, expression, and purification. Cells are transformed via a plasmid (green circle) and IPTG induces expression. The protein is purified with pAMBS-agarose resin.

Deuterated crystallization buffer (100 m*M* Tris–DCl pH 7.8) is used to back-exchange exchangeable H to D atoms. Protein purity is checked by running a SDS-PAGE gel, concentration by taking the absorbance ratio, and deuteration levels of CA II via mass spectrometry with an expected yield of around 50 mg perdeuterated CA II from 2 L culture.

4.3 Deuterium exchanged hydrogenated carbonic anhydrase II crystallization

Using the protocol from Section 4.1, the sitting-drop method employing the Sandwich Box technique (Hampton Research) is used to grow crystals in the nine-well glass-plate setup, at 277 K (McPherson, 1982). A 1:1 protein (50 mM Tris 8.0 pH) to well solution (1.3 M sodium citrate, 0.1 M Tris pH 9.0) is set up with 0.4 mL of crystallization drops (Fig. 4 left) (Fisher et al., 2009). After a week, sodium citrate salt is added directly to the well solution to increase the concentration to ~1.6 M and after a couple weeks of vapor diffusion, crystals appear (Fisher et al., 2009). An individual ~1.0 mm^3 CA II crystal is mounted in a quartz capillary with D_2O mother liquor on one end. This mother liquor is exchanged every week for 2 months to allow H/D exchange of most H atoms (Fisher et al., 2009). Most labile H atoms are exchanged by D atoms fairly rapidly with up to 80% amide backbone exchange in a protein in one to 2 months (Bennett et al., 2008). Some highly accessible H atoms exchange to D atoms on a second to minute timescale (Yan & Maier, 2009). A crystal underwent a 24 h test exposure diffracting to ~2.4 Å resolution at the Protein Crystallography Station (PCS) at the Los Alamos Neutron Science Center (Fisher et al., 2009).

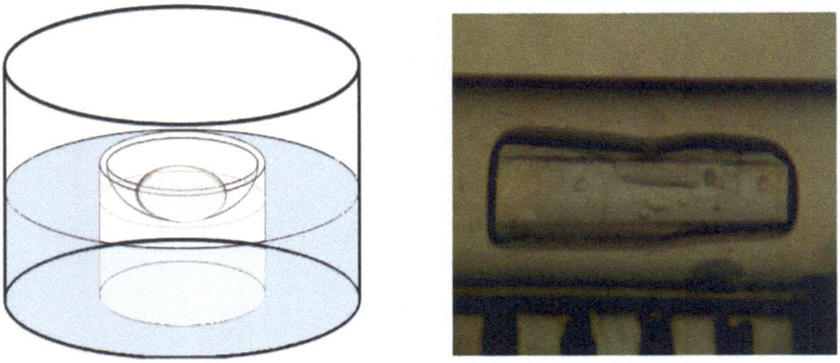

Fig. 4 Cartoon of sitting-drop vapor-diffusion and a CA II crystal. Method for crystallization of CA II prior to H/D exchange (left) and optical photograph of a ~1.3mm^3 CA II crystal H/D exchanged after 1 month (right). A crystal of this size is sufficient for neutron data collection. *Reproduced with permission of the International Union of Crystallography, https://journals.iucr.org/.*

4.4 Crystallization of perdeuterated carbonic anhydrase II

Using the methods from Section 4.2, the perdeuterated CA II crystals can be grown in a similar fashion as the hydrogenated crystals in Section 4.3. Perdeuterated protein (20–30 mg/mL) is mixed with crystallization solution (100 m*M* Tris–DCl pH 7.8, 1.15 M sodium citrate) in a 1:1 ratio (Fig. 4 left). A lower salt concentration of crystallization buffer is used compared to normal CA II crystals. Crystal quality and size is optimized with the EMBL, Grenoble temperature-control device (Budayova-Spano et al., 2005).

4.5 Data collection and structure determination of carbonic anhydrase II

Time-of-flight (TOF) Laue neutron diffraction data was collected at the Protein Crystallography Station in Los Alamos, New Mexico (an image is shown in Fig. 5) (Fisher et al., 2010). Presently available neutron diffraction facilities are described elsewhere in this volume. The method of data collection will differ with the experimental set-up, in this example 41 positional settings were used with 32 h data collection for each frame (Fisher et al., 2010).

Diffraction data is usually measured using derivatives of X-ray software customized for individual beam-lines. Phenix (Adams et al., 2010) is typically used to refine the structures. Neutron structures can be refined starting with the X-ray structures that already exists. Joint X-ray and neutron refinement are merged in Phenix refinements. The graphic interface software package Coot is used to manually fit the structure into neutron and electron

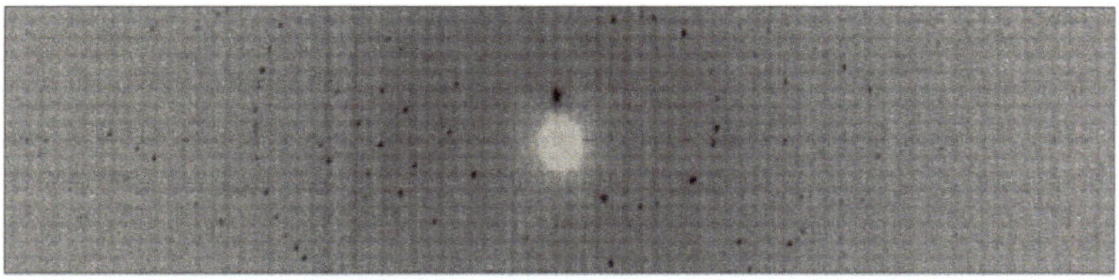

Fig. 5 A CA II crystal's TOF Laue neutron diffraction pattern (Fisher et al., 2009). The dataset was collected at PCS with 41 settings with each frame taking 32h to collect (Fisher et al., 2009). The typical 2-D Laue pattern was created from collapsing the 3-D TOF diffraction data for the dimension of time in each crystal setting (Fisher et al., 2009). *Reproduced with permission of the International Union of Crystallography, https://journals.iucr.org/.*

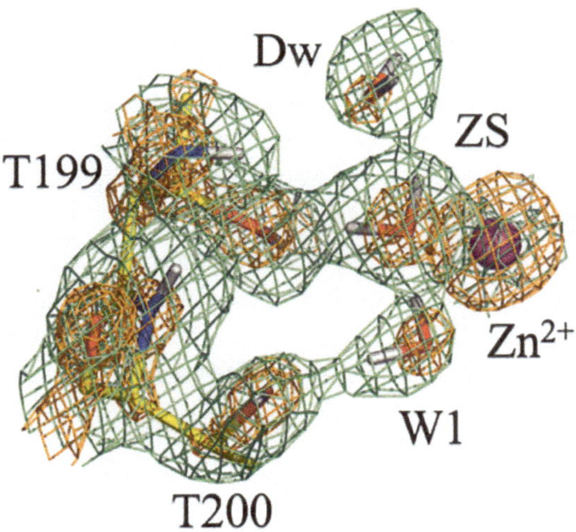

Fig. 6 Neutron and electron density maps of CA II active site. The orientations of D_2O (W1 and deep water (DW)) and the interactions between T199 and catalytic zinc-bound solvent can be seen clearly (Rowlett, 2010). Deuterium atoms are colored teal, oxygen is red, carbon yellow and nitrogen blue. The electron density map is shown in orange and neutron density map in green contoured to 2.0σ. *Adapted with permission from Fisher, S. Z., Kovalevsky, A. Y., Domsic, J. F., Mustyakimov, M., McKenna, R., Silverman, D. N., et al. (2010). Neutron structure of human carbonic anhydrase II: implications for proton transfer.* Biochemistry, *49, 415–421. Copyright 2010 American Chemical Society.*

density maps (Emsley & Cowtan, 2004). Positive and negative nuclear scattering length maps guide how to place H/D atoms within the protein and solvent. Final structures can be viewed in PyMOL or Chimera (Pettersen et al., 2004; PyMOL, 2006). The neutron density of the refined active site of CA is shown in Fig. 6.

The accessibility of the amide groups on the protein backbone can be assessed by the extent of H/D exchange after refining D atoms (Fisher et al., 2010). In the first determined CA II neutron structure, approximately a third of backbone amides were fully exchanged to D, a third were partially exchanged, and a third were not (Fisher et al., 2010). The more accessible residues are to solvent, the more likely they are to undergo H/D exchange. While the active site backbone residues have no H/D exchange, the side chains are in contrast fully exchanged (Fisher et al., 2010). This reflects the fixed nature of the backbone and the accessibility to solvent in the active site.

5. Analysis of neutron structures

Table 2 summarizes the neutron structural studies of CA II and IX. Expression, purification, and crystallization studies of deuterated CA were performed as these procedures were different than hydrogenated CAs (Table 2). Later publications showed the hydrogen-bonding network of the water proton wire, the protonation states of side-chain residues (including the proton shuttle residue H64), and the determination of the protonation state of different inhibitors bound to the zinc (Table 2).

5.1 Mechanism of carbonic anhydrase II

Neutron crystallography (Table 2) has been critical to observe the protonation states of active site polar residues and the orientation of the proton transfer water network in the catalytic mechanism of CA II (Fisher et al., 2009, 2010). The first neutron structure of CA II was reported in 2010 at pH 10 (Fisher et al., 2010). This structure provided the location of the hydrogens of the active site waters and detailed information of the inter-water hydrogen-bonded network (Fisher et al., 2010). This structure showed the active site solvent network divided into two clusters separated by W2, rather than the predicted complete hydrogen-bonding network. The first solvent cluster consists of DW, zinc-bound solvent, and W1 leading to W2. DW and zinc-bound solvent act as hydrogen-bond acceptors (Fig. 6) (Fisher et al., 2010). The second solvent cluster from W2, consists of W3a and W3b forming a hydrogen-bonding network with each other towards H64. Alas, this structure at pH 10 was disappointing as it did not show the existence of a complete proton wire.

This first neutron study was followed up with a structure of CA II at pH 7.8 (catalytically active pH) to see if a rearrangement of the

Table 2 CA II and IX neutron studies.

Publication	Findings	PDB	References
Purification, crystallization, and structure of CA			
Production and X-ray crystallographic analysis of fully deuterated human carbonic anhydrase II	Perdeuterated and hydrogenated CA II crystals are isomorphous	N/A	Budayova-Spano et al. (2005)
Preliminary joint neutron and X-ray crystallographic study of human carbonic anhydrase II	First determined neutron structure at 9.0 pH at 2.4 Å resolution	N/A	Fisher et al. (2009)
Enzymes for carbon sequestration: Neutron crystallographic studies of carbonic anhydrase	Neutron structure leads to new understanding of proton transfer mechanism	3kkx	Fisher et al. (2010)
Deuteration of human carbonic anhydrase for neutron crystallography: Cell culture media, protein thermostability, and crystallization behavior	Overall methods on deuterated CA II for optimal crystal growth	N/A	Koruza et al. (2018)
Mechanism of CA			
Neutron structure of human carbonic anhydrase II: implications for proton transfer	First CA II neutron structure submitted to PDB	3kkx	Fisher et al. (2010)
Neutron structure of human carbonic anhydrase II: A hydrogen-bonded water network "switch" is observed between pH 7.8 and 10.0	First time an entire hydrogen bond network has been observed in CA II	3tmj	Fisher et al. (2011)
Joint neutron crystallographic and NMR solution studies of Tyr residue ionization and hydrogen bonding: Implications for enzyme-mediated proton transfer	Provided structural understanding of CA II active site electrostatics	4q49 4y0j	Michalczyk et al. (2015)
Inhibition of CA			
Neutron diffraction of acetazolamide-bound human carbonic anhydrase II reveals atomic details of drug binding	First neutron structure of clinical drug complex, providing details of protonation	4g0c	Fisher, Aggarwal, Kovalevsky, Silverman, and McKenna (2012)

Continued

Table 2 CA II and IX neutron studies.—cont'd

Publication	Findings	PDB	References
Neutron structure of human carbonic anhydrase II in complex with methazolamide: mapping the solvent and hydrogen-bonding patterns of an effective clinical drug	Comparison of clinical dugs bound to CA II, providing details of solvent exclusion	5c8i	Aggarwal et al. (2016)
"To Be or Not to Be" Protonated: Atomic Details of Human Carbonic Anhydrase-Clinical Drug Complexes by Neutron Crystallography and Simulation	Neutron structures of clinical drugs interacting in the active site of CA II	6bbs 6bc9 6bcc	Kovalevsky et al. (2018)
Using neutron crystallography to elucidate the basis of selective inhibition of carbonic anhydrase by saccharin and a derivative	How selective inhibitors bind to CA IX over CA II	6fji 6fjj 6gcy	Koruza et al. (2019)

hydrogen-bonded water network occurred. This study was successful and exhibited a complete hydrogen-bonded network between the zinc-bound solvent to the proton transfer residue H64 at pH 7.8 (Fisher et al., 2011).

In brief, the information gained from this neutron structure allowed an understanding of the "gating" of the solvent network. W2 acts a switch connecting two groups of solvent, linking the zinc-bound solvent to H64. The two water clusters do not directly hydrogen-bond to each other, but instead rely on W2 as a proton wire "circuit" switch (Fisher et al., 2011). W2 has two conformations, rotating about its oxygen, which connects or disconnects the hydrogen-bonding network in the active site of CA II. The structure showed H64 partially protonated in an intermediate state (Fisher et al., 2011). It was hypothesized H64 rotates from the "in" to "out" conformer to deliver the excess protons to the bulk solvent. H64 and Y7 protonation changes likely cause the reordering of the water structure hydrogen-bonding (Fisher et al., 2011). In the high pH structure, H64 is unprotonated and only occupies the inward confirmation, but at pH 7.8, H64 has three confirmations (Fisher et al., 2011). DW and zinc-bound solvent are unchanged in the two pHs (Fisher et al., 2011). W1 flips to hydrogen-bond with W2 while also engaging with T200 as a hydrogen-bond acceptor (Fisher et al., 2011). W1, W2, and W3a remain hydrogen-bonded together at the high pH structure, but switch their orientations to achieve this completed

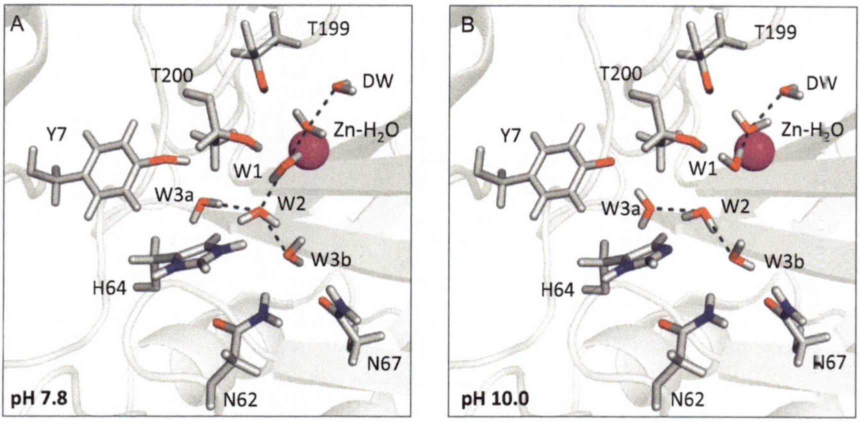

Fig. 7 Water (W2) circuit switch in CA II neutron structures. (A) pH 7.8 (PDB: 3TMJ), (B) pH 10.0 (PDB: 3KKX). Hydrogen bonds are depicted as dashes. Zinc (magenta), nitrogen (blue), and oxygen (red) atoms. Key active site amino acid residues are as labeled (Lomelino, Andring, & McKenna, 2018b).

network (Fisher et al., 2011). The new orientation of W2 allows a hydrogen-bond with H64 as a hydrogen-bond donor (Fisher et al., 2011). This completes the network from zinc-bound solvent to H64 which allows for efficient proton transfer in CA II (Table 2) (Fig. 7).

In CA catalysis for the hydration of CO_2, hydrophilic residues coordinate an ordered water network that mediates a proton transfer out of the active site. The role of one of these hydrophilic residues (Y7) was shown using neutron crystallography. Neutron diffraction data for apo CA II (without zinc in the active site) at pH 7.5 and holo CA II (with zinc in the active site) at pH 6 was collected. The maps showed the protonation state of Y7 varied depending on pH and whether the zinc was present in the active site or not. In holo CA II at pH 6.0 and 7.8, Y7 is protonated and at pH 10 Y7 is deprotonated. In apo CA II at pH 7.5, Y7 is protonated. However, Mass-spec titration studies revealed a lower pKa of Y7 in holo CA II compared to free tyrosine (Michalczyk et al., 2015).

It should be noted that caution should be taken when quoting pH in deuterated protein crystal structures. The measured pH is not equivalent to pD, and 0.4 should be added to the pH when assigning the pD value of the protein crystal structure.

5.2 Inhibition of carbonic anhydrase II

Acetazolamide (AZM) is a sulfonamide inhibitor with high affinity for CA II under physiological conditions. H/D exchanged CA II was

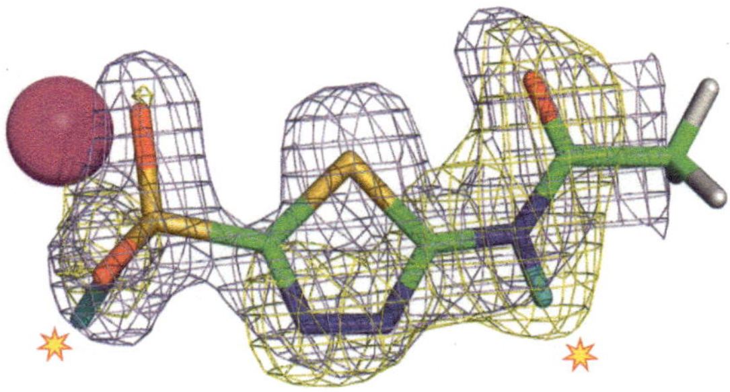

Fig. 8 Neutron and electron density map of AZM showing the anionic sulfamido form when complexed to CA II. Electron density map (blue) and neutron density map (yellow). Zinc (magenta), deuterium (cyan), sulfur (yellow), nitrogen (blue), carbon (green), and oxygen (red) atoms. Stars indicate ionizable groups. Most likely the methyl and sulfur are invisible in the neutron map because their hydrogen atoms had not undergone deuterium exchange. *Reprinted with permission from Fisher, S. Z., Aggarwal, M., Kovalevsky, A. Y., Silverman, D. N. & McKenna, R. (2012). Neutron diffraction of acetazolamide-bound human carbonic anhydrase II reveals atomic details of drug binding.* Journal of the American Chemical Society, 134, *14726–14729. Copyright 2012 American Chemical Society.*

co-crystalized with AZM. The CA II neutron structure showed how AZM bound within the active site; that four waters were displaced on AZM binding, what hydrogen-bonds were formed, and the drug's protonation state (Fig. 8). In brief, the lone pair of the sulfonamide amide coordinated with the zinc at a distance of ~2.4 Å. This binding configuration led to the observation of one D in the sulfonamide group acting as a hydrogen-bond donor to T199 and another D as a hydrogen-bond donor to E106. In addition, the T200 OD group formed a bifurcated hydrogen-bond donor with P201 and a solvent molecule. This solvent molecule acts as hydrogen-bond acceptor from the protonated acetoamido group and as a donor to the P201 backbone carbonyl, thus forming a hydrogen bonded bridging water between AZM and CA II. This study also highlighted the hydrophobic interactions that occur when AZM binds within CA II. This study was the first clinical drug observed bound to its target using neutron crystallography (Fisher et al., 2012).

Methazolamide (MZM) is another clinically used CAI. The neutron structure of CA II cocrystalized with MZM also showed the solvent displacement after MZM binding in the active site, water orientations, the hydrogen-bonding network, and protonation states, similar to the AZM structure. The sulfonamide MZM in the neutron structure also showed the sulfamido nitrogen deprotonated. The binding affinity (K_i) for MZM and AZM in CA II are both ~10 nM even though MZM displaces one more water than AZM and three additional waters in the MZM complex are disordered. MZM is a little more hydrophobic which may compensate for the loss of a hydrogen-bonding network compared to binding AZM. The methyl derivative has a substantial effect within the active site of CA II on the solvent organization. MZM has a loss of binding energy compared to AZM. These factors are mainly negated from MZM exhibiting favorable dispersion forces within the CA II active site (Aggarwal et al., 2016). Binding of MZM has a more favorable enthalpy of binding from dispersion forces compared to AZM. This results in the similar K_i observed between the two clinical drugs because of the overall similar Gibbs free energy of binding. The study allowed two clinical drugs bound to their target to have their hydrogen bonding compared (Aggarwal et al., 2016). Recently, additional clinical drugs such as BZM, DZM, and EZM have been co-crystalized to their target CA II and their neutron structure determined (Kovalevsky et al., 2018).

6. Summary

Increasing the flux for neutron crystallography is reducing the need to grow large crystals, but data collection times are still significantly longer than X-ray crystallography. In addition, obtaining useful crystals for neutron crystallography requires H/D exchange that can take several months and/or the optimization of perdeuterated protein expression (Fig. 4). All of which are costly processes, and therefore neutron crystallography should only be a method of choice, if deciphering the location of H atoms of solvent and knowing the protonation states of amino acids and ligands/drugs are essential to the structural study (Fig. 6).

Here we describe the methods that have been used towards obtaining neutron diffraction quality crystals of CA II (Fig. 4). Neutron studies have permitted assigning the orientations of the waters and protonation states of residues in the active site at various pHs (Fig. 6). This has allowed the complete mapping of the solvent hydrogen-bonding network of CA II. The data has provided fine details of the overall mechanism of proton transfer and how residue H64 delivers protons to and from the bulk solvent. In addition, these studies have demonstrated residue Y7 has a shifted pKa compared to free tyrosine and its involvement in orchestrating the water network. Also, a neutron study of CA II in complex with the drug AZM has provided details of the protonation state of AZM, information on water displacement, and hydrogen-bond interactions with residues. A similar study with the drug

MZM also has permitted a detailed comparison and provided information about the mode and binding energy of these clinically used inhibitors. With the recent crystallization of CA IX, this has now opened up the possibility of using the structural knowledge obtained from neutron crystals structures to advance the rational drug design of CA specific inhibitors (Aggarwal et al., 2016).

Acknowledgments

The authors would like to thank the following collaborators for their contributions to the studies described in this chapter: Mayank Aggarwal, S. Zoë Fisher, Katarina Koruza, Andrey Kovalevsky, Paul Langan, Brian Mahon, Ryszard Michalczyk, David Silverman, and Hector Velazquez.

References

Adams, P. D., Afonine, P. V., Bunkóczi, G., Chen, V. B., Davis, I. W., Echols, N., et al. (2010). PHENIX: A comprehensive Python-based system for macromolecular structure solution. *Acta Crystallographica Section D: Biological Crystallography*, *66*(2), 213–221.

Aggarwal, M., Chua, T. K., Pinard, M. A., Szebenyi, D. M., & McKenna, R. (2015). Carbon dioxide "trapped" in a β-carbonic anhydrase. *Biochemistry*, *54*(43), 6631–6638.

Aggarwal, M., Kovalevsky, A. Y., Velazquez, H., Fisher, S. Z., Smith, J. C., & McKenna, R. (2016). Neutron structure of human carbonic anhydrase II in complex with methazolamide: Mapping the solvent and hydrogen-bonding patterns of an effective clinical drug. *IUCrJ*, *3*(Pt. 5), 319–325.

Amata, O., Marino, T., Russo, N., & Toscano, M. (2011). Catalytic activity of a ζ-class zinc and cadmium containing carbonic anhydrase. Compared work mechanisms. *Physical Chemistry Chemical Physics*, *13*(8), 3468–3477.

Avvaru, B. S., Kim, C. U., Sippel, K. H., Gruner, S. M., Agbandje-McKenna, M., Silverman, D. N., et al. (2010). A short, strong hydrogen bond in the active site of human carbonic anhydrase II. *Biochemistry*, *49*(2), 249–251.

Beauchemin, M., & Morse, D. (2015). Chapter 19—δ-carbonic anhydrases: Structure, distribution, and potential roles. In C. T. Supuran & G. D. Simone (Eds.), *Carbonic anhydrases as biocatalysts* (pp. 337–349, Elsevier).

Bennett, B. C., Gardberg, A. S., Blair, M. D., & Dealwis, C. G. (2008). On the determinants of amide backbone exchange in proteins: A neutron crystallographic comparative study. *Acta Crystallographica Section D, Biological Crystallography*, *D64*(Pt. 7), 764–783.

Berman, H. M., Westbrook, J., Feng, Z., Gilliland, G., Bhat, T. N., Weissig, H., et al. (2000). The protein data bank. *Nucleic Acids Research*, *28*(1), 235–242.

Blakeley, M. P., Hasnain, S. S., & Antonyuk, S. V. (2015). Sub-atomic resolution X-ray crystallography and neutron crystallography: Promise, challenges and potential. *IUCrJ*, *2*(Pt. 4), 464–474.

Blakeley, M. P., Langan, P., Niimura, N., & Podjarny, A. (2008). Neutron crystallography. *Current Opinion in Structural Biology*, *18*(5), 593–600.

Boone, C. D., Pinard, M., McKenna, R., & Silverman, D. (2014). Catalytic mechanism of α-class carbonic anhydrases: CO_2 hydration and proton transfer. In *Subcellular biochemistry. carbonic anhydrase: Mechanism, Regulation, links to disease, and industrial applications* (pp. 31–52). Netherlands: Springer.

Budayova-Spano, M., Fisher, S. Z., Dauvergne, M.-T., Agbandje-McKenna, M., Silverman, D. N., Myles, D. A. A., et al. (2005). Production and X-ray crystallographic analysis of fully deuterated human carbonic anhydrase II. *Acta Crystallographica Section F: Structural Biology and Crystallization Communications*, *62*(Pt. 1), 6–9.

Capasso, C., & Supuran, C. T. (2015). An overview of the alpha-, beta- and gamma-carbonic anhydrases from Bacteria: Can bacterial carbonic anhydrases shed new light on evolution of bacteria? *Journal of Enzyme Inhibition and Medicinal Chemistry*, *30*(2), 325–332.

Cronk, J. D., Rowlett, R. S., Zhang, K. Y. J., Tu, C., Endrizzi, J. A., Lee, J., et al. (2006). Identification of a novel noncatalytic bicarbonate binding site in eubacterial beta-carbonic anhydrase. *Biochemistry*, *45*(14), 4351–4361.

De Simone, G., Di Fiore, A., Capasso, C., & Supuran, C. T. (2015). The zinc coordination pattern in the η-carbonic anhydrase from *Plasmodium falciparum* is different from all other carbonic anhydrase genetic families. *Bioorganic & Medicinal Chemistry Letters*, *25*(7), 1385–1389.

Del Prete, S., Vullo, D., De Luca, V., Supuran, C. T., & Capasso, C. (2014). Biochemical characterization of the δ-carbonic anhydrase from the marine diatom *Thalassiosira weissflogii*, TweCA. *Journal of Enzyme Inhibition and Medicinal Chemistry*, *29*(6), 906–911.

Del Prete, S., Vullo, D., Fisher, G. M., Andrews, K. T., Poulsen, S.-A., Capasso, C., et al. (2014). Discovery of a new family of carbonic anhydrases in the malaria pathogen *Plasmodium falciparum*—The η-carbonic anhydrases. *Bioorganic & Medicinal Chemistry Letters*, *24*(18), 4389–4396.

Del Prete, S., Vullo, D., Scozzafava, A., Capasso, C., & Supuran, C. T. (2014). Cloning, characterization and anion inhibition study of the δ-class carbonic anhydrase (TweCA) from the marine diatom *Thalassiosira weissflogii*. *Bioorganic & Medicinal Chemistry*, *22*(1), 531–537.

Domsic, J. F., Avvaru, B. S., Kim, C. U., Gruner, S. M., Agbandje-McKenna, M., Silverman, D. N., et al. (2008). Entrapment of carbon dioxide in the active site of carbonic anhydrase II. *The Journal of Biological Chemistry*, *283*(45), 30766–30771.

Domsic, J. F., & McKenna, R. (2010). Sequestration of carbon dioxide by the hydrophobic pocket of the carbonic anhydrases. *Biochimica et Biophysica Acta*, *1804*(2), 326–331.

Emsley, P., & Cowtan, K. (2004). Coot: Model-building tools for molecular graphics. *Acta Crystallographica Section D: Biological Crystallography*, *60*(12), 2126–2132.

Eriksson, A. E., Jones, T. A., & Liljas, A. (1988). Refined structure of human carbonic anhydrase II at 2.0 A resolution. *Proteins*, *4*(4), 274–282.

Ferraroni, M., Del Prete, S., Vullo, D., Capasso, C., & Supuran, C. T. (2015). Crystal structure and kinetic studies of a tetrameric type II β-carbonic anhydrase from the pathogenic bacterium Vibrio cholerae. *Acta Crystallographica. Section D, Biological Crystallography*, *71*(Pt. 12), 2449–2456.

Ferry, J. G. (2010). The γ class of carbonic anhydrases. *Biochimica et Biophysica Acta*, *1804*(2), 374.

Fisher, S. Z., Aggarwal, M., Kovalevsky, A. Y., Silverman, D. N., & McKenna, R. (2012). Neutron diffraction of acetazolamide-bound human carbonic anhydrase II reveals atomic details of drug binding. *Journal of the American Chemical Society*, *134*(36), 14726–14729.

Fisher, Z., Hernandez Prada, J. A., Tu, C., Duda, D., Yoshioka, C., An, H., et al. (2005). Structural and kinetic characterization of active-site histidine as a proton shuttle in catalysis by human carbonic anhydrase II. *Biochemistry*, *44*(4), 1097–1105.

Fisher, S. Z., Kovalevsky, A. Y., Domsic, J. F., Mustyakimov, M., McKenna, R., Silverman, D. N., et al. (2010). Neutron structure of human carbonic anhydrase II: Implications for proton transfer. *Biochemistry*, *49*(3), 415–421.

Fisher, S. Z., Kovalevsky, A. Y., Domsic, J. F., Mustyakimov, M., Silverman, D. N., McKenna, R., et al. (2009). Preliminary joint neutron and X-ray crystallographic study of human carbonic anhydrase II. *Acta Crystallographica. Section F, Structural Biology and Crystallization Communications*, *65*(Pt. 5), 495–498.

Fisher, S. Z., Kovalevsky, A. Y., Domsic, J., Mustyakimov, M., Silverman, D. N., McKenna, R., et al. (2010). Enzymes for carbon sequestration: Neutron crystallographic studies of carbonic anhydrase. *Acta Crystallographica. Section D, Biological Crystallography*, *66*(Pt. 11), 1178–1183.

Fisher, S. Z., Kovalevsky, A. Y., Mustyakimov, M., Silverman, D. N., McKenna, R., & Langan, P. (2011). Neutron structure of human carbonic anhydrase II: A hydrogen-bonded water network "switch" is observed between pH 7.8 and 10.0. *Biochemistry, 50*(44), 9421–9423.

Frost, S. C. (2014). Physiological functions of the alpha class of carbonic anhydrases. *Sub-Cellular Biochemistry, 75*, 9–30.

Groenewegen, P. P. M., & Feil, D. (1969). Molecular form factors in X-ray crystallography. *Acta Crystallographica, Section A: Crystal Physics, Diffraction, Theoretical and General Crystallography, 25*(3), 444–450.

Innocenti, A., Beyza Oztürk Sarikaya, S., Gülçin, I., & Supuran, C. T. (2010). Carbonic anhydrase inhibitors. Inhibition of mammalian isoforms I-XIV with a series of natural product polyphenols and phenolic acids. *Bioorganic & Medicinal Chemistry, 18*(6), 2159–2164.

Iverson, T. M., Alber, B. E., Kisker, C., Ferry, J. G., & Rees, D. C. (2000). A closer look at the active site of γ-class carbonic anhydrases: High-resolution crystallographic studies of the carbonic anhydrase from *Methanosarcina thermophila*. *Biochemistry, 39*(31), 9222–9231.

Jensen, E. L., Clement, R., Kosta, A., Maberly, S. C., & Gontero, B. (2019). A new widespread subclass of carbonic anhydrase in marine phytoplankton. *The ISME Journal, 13*(8), 2094–2106.

Jewell, D. A., Tu, C., Paranawithana, S. R., Tanhauser, S. M., LoGrasso, P. V., Laipis, P. J., et al. (1991). Enhancement of the catalytic properties of human carbonic anhydrase III by site-directed mutagenesis. *Biochemistry, 30*(6), 1484–1490.

Keilin, D., & Mann, T. (1940). Carbonic anhydrase. Purification and nature of the enzyme. *The Biochemical Journal, 34*(8–9), 1163–1176.

Khalifah, R. G., Strader, D. J., Bryant, S. H., & Gibson, S. M. (1977). Carbon-13 nuclear magnetic resonance probe of active-site ionizations in human carbonic anhydrase B. *Biochemistry, 16*(10), 2241–2247.

Kikutani, S., Nakajima, K., Nagasato, C., Tsuji, Y., Miyatake, A., & Matsuda, Y. (2016). Thylakoid luminal θ-carbonic anhydrase critical for growth and photosynthesis in the marine diatom *Phaeodactylum tricornutum*. *Proceedings of the National Academy of Sciences, 113*(35), 9828–9833.

Kim, C. U., Song, H., Avvaru, B. S., Gruner, S. M., Park, S., & McKenna, R. (2016). Tracking solvent and protein movement during CO_2 release in carbonic anhydrase II crystals. *Proceedings of the National Academy of Sciences of the United States of America, 113*(19), 5257–5262.

Kisker, C., Schindelin, H., Alber, B. E., Ferry, J. G., & Rees, D. C. (1996). A left-hand beta-helix revealed by the crystal structure of a carbonic anhydrase from the archaeon Methanosarcina thermophila. *The EMBO Journal, 15*(10), 2323–2330.

Koruza, K., Lafumat, B., Végvári, Á., Knecht, W., & Fisher, S. Z. (2018). Deuteration of human carbonic anhydrase for neutron crystallography: Cell culture media, protein thermostability, and crystallization behavior. *Archives of Biochemistry and Biophysics, 645*, 26–33.

Koruza, K., Mahon, B. P., Blakeley, M. P., Ostermann, A., Schrader, T. E., McKenna, R., et al. (2019). Using neutron crystallography to elucidate the basis of selective inhibition of carbonic anhydrase by saccharin and a derivative. *Journal of Structural Biology, 205*(2), 147–154.

Kovalevsky, A., Aggarwal, M., Velazquez, H., Cuneo, M. J., Blakeley, M. P., Weiss, K. L., et al. (2018). "To be or not to be" protonated: Atomic details of human carbonic anhydrase-clinical drug complexes by neutron crystallography and simulation. *Structure (London, England: 1993), 26*(3), 383–390.e3.

Krishnamurthy, V. M., Kaufman, G. K., Urbach, A. R., Gitlin, I., Gudiksen, K. L., Weibel, D. B., et al. (2008). Carbonic anhydrase as a model for biophysical and physical-organic studies of proteins and protein–ligand binding. *Chemical Reviews, 108*(3), 946–1051.

Lomelino, C. L., Andring, J. T., & McKenna, R. (2018a). Crystallography and its impact on carbonic anhydrase research. *International Journal of Medicinal Chemistry, 2018*, https://doi.org/10.1155/2018/9419521.

Lomelino, C. L., Andring, J. T., & McKenna, R. (2018b). Structural insights into the catalytic mechanism of carbonic anhydrase. In *Carbonic anhydrases: Biochemistry, mechanism of action and therapeutic applications* (p. 39): Nova Science Publishers, Incorporated.

Lomelino, C. L., Supuran, C. T., & McKenna, R. (2016). Non-classical inhibition of carbonic anhydrase. *International Journal of Molecular Sciences, 17*(7), https://doi.org/10.3390/ijms17071150.

Maupin, C. M., & Voth, G. A. (2007). Preferred orientations of His64 in human carbonic anhydrase II. *Biochemistry, 46*(11), 2938–2947.

McPherson, A. (1982). *Preparation and analysis of protein crystals.* John Wiley & Sons.

Meilleur, F., Munshi, P., Robertson, L., Stoica, A. D., Crow, L., Kovalevsky, A., et al. (2013). The IMAGINE instrument: First neutron protein structure and new capabilities for neutron macromolecular crystallography. *Acta Crystallographica Section D: Biological Crystallography, 69*(10), 2157–2160.

Meldrum, N. U., & Roughton, F. J. W. (1933). Carbonic anhydrase. Its preparation and properties. *The Journal of Physiology, 80*(2), 113–142.

Merz, K. M. (1991). Carbon dioxide binding to human carbonic anhydrase II. *Journal of the American Chemical Society, 113*(2), 406–411.

Michalczyk, R., Unkefer, C. J., Bacik, J.-P., Schrader, T. E., Ostermann, A., Kovalevsky, A. Y., et al. (2015). Joint neutron crystallographic and NMR solution studies of Tyr residue ionization and hydrogen bonding: Implications for enzyme-mediated proton transfer. *Proceedings of the National Academy of Sciences of the United States of America, 112*(18), 5673–5678.

Miller, W. H., Dessert, A. M., & Roblin, R. O. (1950). Heterocyclic Sulfonamides as carbonic anhydrase Inhibitors. *Journal of the American Chemical Society, 72*(11), 4893–4896.

Mitsuhashi, S., Mizushima, T., Yamashita, E., Yamamoto, M., Kumasaka, T., Moriyama, H., et al. (2000). X-ray structure of β-carbonic anhydrase from the red alga, Porphyridium purpureum, reveals a novel catalytic site for CO_2 hydration. *Journal of Biological Chemistry, 275*(8), 5521–5526.

Nair, S. K., Ludwig, P. A., & Christianson, D. W. (1994). Two-site binding of phenol in the active site of human carbonic anhydrase II: Structural implications for substrate association. *Journal of the American Chemical Society, 116*(8), 3659–3660.

National Institute of Standards and Technology. (2013). *Neutron scattering lengths and cross sections.* Retrieved October 3, 2019, from https://www.ncnr.nist.gov/resources/n-lengths/.

Norvell, J. C., Nunes, A. C., & Schoenborn, B. P. (1975). Neutron diffraction analysis of myoglobin: Structure of the carbon monoxide derivative. *Science (New York, N.Y.), 190*(4214), 568–570.

O'Dell, W. B., Bodenheimer, A. M., & Meilleur, F. (2016). Neutron protein crystallography: A complementary tool for locating hydrogens in proteins. *Archives of Biochemistry and Biophysics, 602*, 48–60.

Oksanen, E., Chen, J. C.-H., & Fisher, S. Z. (2017). Neutron crystallography for the study of hydrogen bonds in macromolecules. *Molecules: A Journal of Synthetic Chemistry and Natural Product Chemistry, 22*(4), 3–4.

Pettersen, E. F., Goddard, T. D., Huang, C. C., Couch, G. S., Greenblatt, D. M., Meng, E. C., et al. (2004). UCSF Chimera—A visualization system for exploratory research and analysis. *Journal of Computational Chemistry*, *25*(13), 1605–1612.

Pinard, M. A., Boone, C. D., Rife, B. D., Supuran, C. T., & McKenna, R. (2013). Structural study of interaction between brinzolamide and dorzolamide inhibition of human carbonic anhydrases. *Bioorganic & Medicinal Chemistry*, *21*(22), 7210–7215.

PyMOL. (2006). *The PyMOL Molecular Graphics System*. version 1.6. Schrödinger, LLC.

Roberts, S. B., Lane, T. W., & Morel, F. M. M. (1997). Carbonic Anhydrase in the Marine Diatom *Thalassiosira Weissflogii* (bacillariophyceae). *Journal of Phycology*, *33*(5), 845–850.

Rowlett, R. S. (2010). Structure and catalytic mechanism of the β-carbonic anhydrases. *Biochimica et Biophysica Acta (BBA)—Proteins and Proteomics*, *1804*(2), 362–373.

Rowlett, R. S. (2014). Structure and catalytic mechanism of β-carbonic anhydrases. In *Subcellular biochemistry. carbonic anhydrase: Mechanism, regulation, links to disease, and industrial applications* (pp. 53–76), Netherlands: Springer.

Sechi, M., Innocenti, A., Pala, N., Rogolino, D., Carcelli, M., Scozzafava, A., et al. (2012). Inhibition of α-class cytosolic human carbonic anhydrases I, II, IX and XII, and β-class fungal enzymes by carboxylic acids and their derivatives: New isoform-I selective nanomolar inhibitors. *Bioorganic & Medicinal Chemistry Letters*, *22*(18), 5801–5806.

Shu, F., Ramakrishnan, V., & Schoenborn, B. P. (2000). Enhanced visibility of hydrogen atoms by neutron crystallography on fully deuterated myoglobin. *Proceedings of the National Academy of Sciences of the United States of America*, *97*(8), 3872–3877.

Silverman, D. N., & Lindskog, S. (1988). The catalytic mechanism of carbonic anhydrase: Implications of a rate-limiting protolysis of water. *Accounts of Chemical Research*, *21*(1), 30–36.

Silverman, D. N., & Vincent, S. H. (1983). Proton transfer in the catalytic mechanism of carbonic anhydrase. *CRC Critical Reviews in Biochemistry*, *14*(3), 207–255.

Sippel, K. H., Robbins, A. H., Domsic, J., Genis, C., Agbandje-McKenna, M., & McKenna, R. (2009). High-resolution structure of human carbonic anhydrase II complexed with acetazolamide reveals insights into inhibitor drug design. *Acta Crystallographica Section F: Structural Biology and Crystallization Communications*, *65*(Pt. 10), 992–995.

Steiner, H., Jonsson, B. H., & Lindskog, S. (1975). The catalytic mechanism of carbonic anhydrase. Hydrogen-isotope effects on the kinetic parameters of the human C isoenzyme. *European Journal of Biochemistry*, *59*(1), 253–259.

Supuran, C. T. (2011). Bacterial carbonic anhydrases as drug targets: Toward novel antibiotics? *Frontiers in Pharmacology*, *2*.

Supuran, C. T. (2016). How many carbonic anhydrase inhibition mechanisms exist? *Journal of Enzyme Inhibition and Medicinal Chemistry*, *31*(3), 345–360.

Tanhauser, S. M., Jewell, D. A., Tu, C. K., Silverman, D. N., & Laipis, P. J. (1992). A T7 expression vector optimized for site-directed mutagenesis using oligodeoxyribonucleotide cassettes. *Gene*, *117*(1), 113–117.

Tripp, B. C., & Ferry, J. G. (2000). A structure – function study of a proton transport pathway in the γ-class carbonic anhydrase from *Methanosarcina thermophila*. *Biochemistry*, *39*(31), 9232–9240.

Viparelli, F., Monti, S. M., De Simone, G., Innocenti, A., Scozzafava, A., Xu, Y., et al. (2010). Inhibition of the R1 fragment of the cadmium-containing zeta-class carbonic anhydrase from the diatom *Thalassiosira weissflogii* with anions. *Bioorganic & Medicinal Chemistry Letters*, *20*(16), 4745–4748.

Vullo, D., De Luca, V., Scozzafava, A., Carginale, V., Rossi, M., Supuran, C. T., et al. (2012). Anion inhibition studies of the fastest carbonic anhydrase (CA) known, the extremo-CA from the bacterium Sulfurihydrogenibium azorense. *Bioorganic & Medicinal Chemistry Letters*, *22*(23), 7142–7145.

West, D., Kim, C. U., Tu, C., Robbins, A. H., Gruner, S. M., Silverman, D. N., et al. (2012). Structural and kinetic effects on changes in the CO(2) binding pocket of human carbonic anhydrase II. *Biochemistry*, *51*(45), 9156–9163.
Xu, Y., Feng, L., Jeffrey, P. D., Shi, Y., & Morel, F. M. M. (2008). Structure and metal exchange in the cadmium carbonic anhydrase of marine diatoms. *Nature*, *452*(7183), 56–61.
Xue, Y., Vidgren, J., Svensson, L. A., Liljas, A., Jonsson, B. H., & Lindskog, S. (1993). Crystallographic analysis of Thr-200—>His human carbonic anhydrase II and its complex with the substrate, $HCO3^-$. *Proteins*, *15*(1), 80–87.
Yan, X., & Maier, C. S. (2009). Hydrogen/deuterium exchange mass spectrometry. *Methods in Molecular Biology (Clifton, N.J.)*, *492*, 255–271.
Zimmerman, S. A., Tomb, J.-F., & Ferry, J. G. (2010). Characterization of CamH from *Methanosarcina thermophila*, founding member of a subclass of the γ class of carbonic anhydrases. *Journal of Bacteriology*, *192*(5), 1353–1360.

Further reading

Evans, P. (2006). Scaling and assessment of data quality. *Acta Crystallographica Section D: Biological Crystallography*, *62*(1), 72–82.
Otwinowski, Z., & Minor, W. (1997). [20] Processing of X-ray diffraction data collected in oscillation mode. In *Macromolecular crystallography Part A.Methods in enzymology*: *Vol. 276*. (pp. 307–326). Academic Press.

CHAPTER FOURTEEN

Protein kinase A in the neutron beam: Insights for catalysis from directly observing protons

Oksana Gerlits[a,*], Kevin L. Weiss[b], Matthew P. Blakeley[c], Gianluigi Veglia[d,e], Susan S. Taylor[f], Andrey Kovalevsky[b,*]

[a]Department of Natural Sciences, Tennessee Wesleyan University, Athens, TN, United States
[b]Neutron Scattering Division, Oak Ridge National Laboratory, Oak Ridge, TN, United States
[c]Large Scale Structures Group, Institut Laue–Langevin, Grenoble, France
[d]Department of Chemistry, University of Minnesota, Minneapolis, MN, United States
[e]Department of Biochemistry, Molecular Biology and Biophysics, University of Minnesota, Minneapolis, MN, United States
[f]Department of Chemistry and Biochemistry, University of California at San Diego, La Jolla, CA, United States
*Corresponding authors: e-mail address: ogerlits@tnwesleyan.edu; kovalevskyay@ornl.gov

Contents

Abstract

Protein kinases transmit chemical signals by phosphorylating substrate proteins, thus regulating a multitude of cellular processes. cAMP-dependent protein kinase (PKA), a prototypical enzyme for the whole kinase family, has been the focus of research for several decades, however, the details of the chemical mechanism of phosphoryl group transfer have remained unknown. We used neutron crystallography to map key proton sites and hydrogen bonding interactions in the PKA catalytic subunit (PKAc) in a product complex containing adenosine diphosphate (ADP) and the phosphorylated high affinity protein kinase substrate (pPKS) peptide. To improve neutron diffraction, we deuterated PKAc allowing us to use very small crystals. In the product complex, the phosphoryl

ISSN 0076-6879
https://doi.org/10.1016/bs.mie.2019.12.003

group of pPKS is protonated whereas the catalytic Asp166 is not. H/D exchange analysis of the main-chain amides and comparison with the NMR analysis of PKAc with inhibitor peptide complex revealed exchangeable amides that may distinguish the catalytic and inhibited states.

1. Introduction

Protein kinases regulate many cellular processes by transferring the γ-phosphoryl group of adenosine triphosphate (ATP), which acts as a chemical signal, to serine, threonine, tyrosine and histidine residues of various proteins. Approximately 1.7% of human genes encode protein kinases, pointing to the crucial biological significance of these signaling proteins (Manning, Whyte, Martinez, Hunter, & Sudarsanam, 2002). Dysfunctional protein kinases are implicated in many diseases and have become major therapeutic targets for the design and development of anti-cancer drugs (Klaeger et al., 2017). The cAMP-dependent protein kinase is the most studied enzyme of the kinase superfamily and has become a paradigm for all other kinases (Adams, 2001). The catalytic domain of PKA—PKAc—consists of small N-terminal and large C-terminal lobes connected by a small linker region. The nucleotide binding site is located between the two lobes, while protein substrates sit on the ledge of the C-lobe. The phosphoryl transfer reaction requires binding of one or two divalent metal ions to the PKAc active site, and the physiological metal is magnesium (i.e., Mg^{2+}) (Taylor, Zhang, Steichen, Keshwani, & Kornev, 2013).

X-ray crystallographic (Akamine et al., 2003; Bastidas et al., 2013; Das et al., 2015; Gerlits et al., 2014, 2015; Gerlits, Waltman, Taylor, Langan, & Kovalevsky, 2013; Madhusudan, Akamine, Xuong, & Taylor, 2002), NMR spectroscopic (Jin et al., 2012; Masterson et al., 2010, 2011; Srivastava et al., 2014) and biomolecular simulation (Cheng, Zhang, & McCammon, 2005; Perez-Gallegos, Garcia-Viloca, Gonzalez-Lafont, & Lluch, 2015, 2017; Valiev, Yang, Adams, Taylor, & Weare, 2007) studies have provided us with a good understanding of how PKAc functions. During the catalytic cycle, PKAc undergoes several conformational changes. It adopts an open conformation in the apo-form before ATP or protein (or peptide) substrate binds. It transitions to an intermediate, partially closed, state in binary complexes with the nucleotide or substrate. Finally, PKAc assumes a fully closed conformational state in the ternary complex, when both the nucleotide and substrate are present at the active site. Phosphoryl transfer occurs only in the closed conformation of the enzyme, whereas the MgADP product release is the rate-limiting step. The phosphoryl transfer chemical reaction involves several proton transfers, whose pathways can only be understood if protonation states of the active site residues, substrate and product functional groups are known. Among those are the catalytic residue Asp166, hydroxyl of the substrate serine, and the phosphate group on the phosphorylated product. Thus, understanding the mechanistic details of the chemical step remained challenging, because locations of protons have not been experimentally determined.

Neutron diffraction provides a means to directly observe hydrogen atoms and accurately determine their positions in proteins or other biological macromolecules (Niimura & Podjarny, 2011). Because, unlike X-rays, neutrons are scattered by atomic nuclei, instead of electron clouds, their scattering power does not depend on the number of electrons in an atom. Therefore, the lightest element hydrogen scatters neutrons nearly as well as do other atoms, such as carbon, oxygen and nitrogen that are commonly present in protein molecules, although the neutron scattering length for hydrogen is negative. Unambiguous determination of hydrogen atom positions then allows us to visualize hydrogen bonds and other non-covalent interactions, which is usually not possible even in ultra-high resolution X-ray structures (Oksanen, Chen, & Fisher, 2017). Moreover, the cold neutrons with wavelengths of ~2–5 Å used to perform crystallographic measurements do not cause radiation damage to protein crystals, thus permitting data collection at room temperature to afford radiation damage-free crystal structures at close-to-physiological conditions (Blakeley, Hasnain, & Antonyuk, 2015). One of the significant drawbacks of using neutrons is the very strong incoherent scattering by hydrogen (protium) that only contributes to background in the diffraction images, reducing the signal-to-noise ratio. To overcome this and increase the signal-to-noise ratio of the neutron diffraction from protein crystals, hydrogens are replaced with the heavy isotope deuterium. The neutron scattering length of deuterium is positive and its incoherent scattering cross section is very small. Hence, deuterium atoms appear in the neutron scattering length density maps as peaks, as do other atoms, unlike hydrogen atoms that are observed as troughs in the maps. Labile hydrogen atoms on OH, NH and SH chemical groups with pK_a values below 20 can be replaced using hydrogen/deuterium (H/D) exchange with the heavy water (D_2O), these make up ~15–20% of the hydrogen content in a protein. Non-labile hydrogen atoms on CH bonds, having pK_a values considerably higher than 20, cannot be displaced by deuterium from D_2O. Deuterium can be incorporated into the CH bonds by expressing protein in the deuterated media made with D_2O and a deuterated carbon source to produce perdeuterated protein as discussed below but this is currently limited to protein production in bacterial and yeast systems.

We have carried out a neutron diffraction experiment at room-temperature to obtain an atomic-level view of PKAc trapped in a ternary complex containing products adenosine diphosphate (ADP), a phosphorylated protein kinase substrate peptide (pPKS) and two metal ions (Fig. 1) and gained unprecedented insight into the phosphoryl transfer reaction. We experimentally pinpointed locations of protons (observed as deuterium, D, atoms) in the active site of PKAc and throughout the protein allowing us to revise the reaction mechanism, and to visualize hydrogen bonds enabling us to compare to those previously inferred from X-ray structures. Furthermore, we identified H/D exchange patterns on the main-chain amides gaining insight into the dynamics of PKAc (Gerlits et al., 2019).

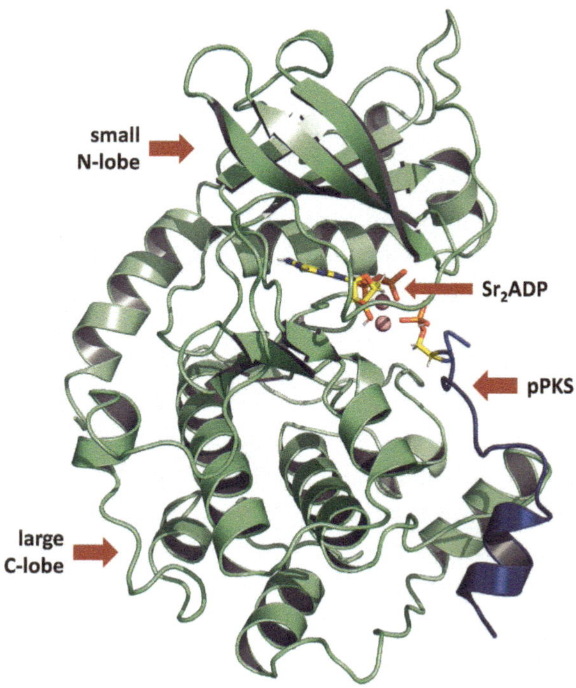

Fig. 1 Overall structure of PKAc:Sr2ADP:pPKS complex in cartoon representation. pPKS peptide product is colored blue. $pP21_{PKS}$ and ADP are shown as sticks, and two Sr^{2+} ions are pink spheres.

2. Protein sample preparation and crystal growth

Sample preparation is not an easy task for neutron crystallographic experiments. The fundamental reasons are: (1) the fluxes from neutron sources are many orders of magnitude weaker than synchrotrons and (2) the probability of neutron scattering by a nucleus is substantially lower than the probability of X-ray photon scattering from an atom. Thus, for a successful neutron diffraction measurement, production of considerably larger crystals than those necessary for X-ray diffraction is required. But, the outcomes and insights into the biological function this technique offers very well justify the effort (Langan & Chen, 2013; Niimura, Takimoto-Kamimura, & Tanaka, 2016; O'Dell, Bodenheimer, & Meilleur, 2016).

2.1 Partial protein deuteration vs perdeuteration

Although both hydrogen and deuterium atoms scatter neutrons strongly, as described above, it is advisable to deuterate the protein sample (replace H by D) for a neutron diffraction experiment. Although full protein deuteration can be very beneficial for the crystallographic experiment, it has several drawbacks. First, protein expression in a fully deuterated media is very costly, as it requires all the components of the expression media including the carbon source to be deuterated. Second, in many cases perdeuteration significantly decreases the yield of the expressed protein. Lastly, it may adversely affect the size of crystals; fully deuterated proteins may, as we have observed for PKAc, not produce large enough crystals for neutron diffraction experiment. We discovered that partial protein deuteration is effective in providing us with significant amounts of the protein and in leading to the growth of sufficiently large crystals. In our study PKAc was expressed in 80% (v/v) D_2O using minimal medium and hydrogenous glycerol as a sole carbon source for the *Escherichia coli* growth. With this partially deuterated sample of PKAc in hand, we were able to grow crystals of ~0.1 mm^3 in volume (Fig. 2) and to successfully collect a 2.5 Å room temperature quasi-Laue neutron diffraction dataset.

The His_6-tagged recombinant mouse PKAc subunit was partially deuterated using the following protocol. In the initial step, BL21 (DE3) competent cells containing the PKAc plasmid (carbenicillin resistance, 100 μg/mL) were grown in LB medium made with H_2O until OD_{600} reached ~2. The cells were spun down and resuspended in the *Enfors* minimal medium made with 80% D_2O. 3 L of fresh minimal medium containing 80% D_2O

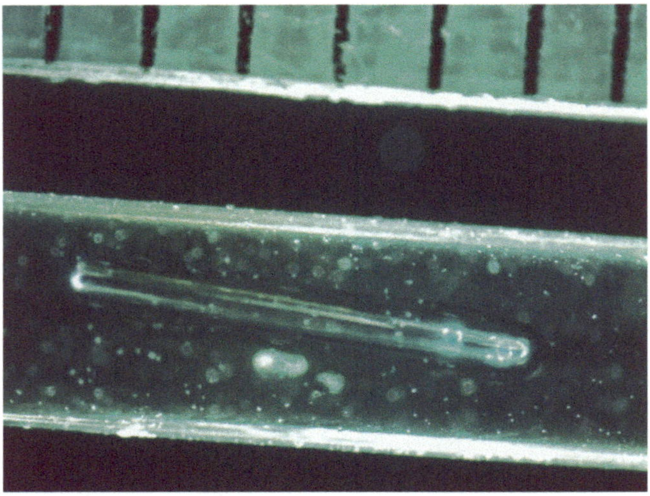

Fig. 2 Crystal of deuterated PKAc:Sr2ADP:pPKS complex of ~0.1 mm^3 volume mounted in a 2 mm inner diameter quartz capillary and used for neutron diffraction data collection. Notice that the crystal is several millimeters long, but small in the other two dimensions.

was inoculated with the 100 mL of this pre-culture and grown in a bioreactor (BioFlo 3000, New Brunswick Scientific). Because bioreactors allow *E. coli* to grow to very high cell densities due to automatic maintenance of cell growth parameters and feeding, our culture was allowed to reach OD_{600} of ~10 at 35 °C before it was induced with 1 mM isopropyl-β-D-thiogalactopyranoside (IPTG) for 20 h at 18 °C to express PKAc. The sole carbon source for the bacterial growth and protein expression was hydrogenous glycerol. The cell suspension was then spun down at 6000 g at 4 °C for 30 min to yield ~90 g of total wet cell mass. All subsequent protein purification steps were performed using buffer solutions made with H_2O. The enzyme was purified by Ni-affinity chromatography, concentrated to ~10 mg/mL, and loaded on to a gel filtration column (HiLoad 16/60 Superdex 75) equilibrated with 50 mM TRIS, 20 mM $SrCl_2$, 250 mM NaCl and 2 mM dithiothreitol (DTT). The purified partially deuterated protein, henceforth called dPKAc, was again concentrated to 10 mg/mL and used for crystallization.

2.2 Crystallization

Protein crystallization is not always a straightforward process, and it is especially tedious when preparing crystals necessary for a neutron crystallographic experiment. This technique requires crystals of orders of magnitude larger than those used at synchrotrons. If a crystal of $\sim$0.001 mm^3 (100 μm on each side) would be considered quite large and would diffract X-rays well in most cases, it would have to be at least 100 times larger than that to be suitable for neutron diffraction. Therefore, crystal growth conditions must be optimized much more carefully, and crystallization approaches may have to be altered (for example, much larger crystallization drops are usually required to grow neutron quality crystals). Also, initially a variety of conditions needs to be tested during crystallization trials on hydrogenous protein and in aqueous solutions before attempting work with the deuterated samples. Variables such as protein concentration, type of complexes (reactants vs products or intermediate analogs) and ligand, ligand-to-protein ratio, pH, precipitant type and concentration must be adjusted/selected to find those which will produce crystals of the desired size.

During such trials with hydrogenous PKAc we learned that, although we can successfully grow crystals of several PKAc complexes, trapping all stages of the catalytic cycle (Gerlits et al., 2014, 2015, 2013), it was the complex of hydrogenous PKAc with the products of phosphorylation reaction that crystallized the best. The procedure has been described in detail previously in Gerlits et al. (2013). In brief, the reaction mixture was prepared by addition of the solid reagents, the nucleotide ATP, and the 20-residue peptide substrate, PKS, to the solution of the enzyme, PKAc, in presence of $\sim$20 mM Mg_2Cl. In the formed ternary complex, the phosphoryl transfer from ATP to the reactive serine residue of PKS catalyzed by the enzyme generated the reaction products, ADP and the phosphorylated substrate (pPKS). The mixture was incubated for 30 min at 4 °C to ensure completion of the reaction, filtered and combined at 1:1 ratio with the crystallization solution. The PKAc:ATP:PKS molar ratio, in these experiments, was kept at 1:10:10 with a PKAc concentration of $\sim$10 mg/mL. We also exploited effects of different metal ions, Na^+, K^+, Ca^{2+}, Sr^{2+} and Ba^{2+} in place of the physiological Mg^{2+} and observed that the complex with Sr^{2+} reliably produced the largest crystals. Subsequent X-ray structures of the hydrogenous PKAc complexes, with different metals including Mg^{2+} obtained through the same procedure, revealed both phosphoryl transfer products, ADP and pPKS, captured at the PKAc active site including two metal ions in all studied complexes. Thereby we confirmed the formation of the product complexes with every studied metal ion. Once the proper system capable of producing crystals of the desired size was identified for hydrogenous PKAc, the conditions were tuned to obtain similar results for the deuterated sample.

Purification and crystallization of the deuterated sample were carried out in H_2O-based solutions. The product complex was prepared in a similar manner to that for hydrogenous protein prior to crystallization. Specifically, the dPKAc:ADP:pPKS complex was formed by adding sodium ATP and PKS in solid form to the solution of dPKAc in the molar ratio of 1:10:10. The reaction mixture was incubated at 4 °C for 30–40 min, allowing the phosphoryl transfer reaction to proceed to completion, filtered through 0.2 μm VWR spin filters and combined at a 1:1 ratio (v/v) with the crystallization solution containing 19.5% w/v PEG6000, 0.1 M MES (pH 5), 5 mM DTT and 50 mM $SrCl_2$. The resulting dPKAc:Sr_2ADP:pPKS complex produced several crystals of ~0.1 mm^3 in volume in the same crystallization drop. The crystals were mounted in quartz capillaries containing the reservoir solution made with 100% D_2O to allow for H/D vapor exchange of the labile H atoms (Fig. 2). The H/D exchange was allowed to proceed for several weeks prior to neutron diffraction data collection. It is important to note that in some cases H/D exchange can be accomplished directly in the crystallization drop through replacement of the reservoir solution with the solution of the same composition but prepared using 100% D_2O, which ultimately replaces H_2O with D_2O in the crystallization drops. In addition, crystals can be directly grown from solutions in D_2O. To our surprise, neither method was successful for the crystals of our dPKAc complexes. We noticed that the crystal quality quickly deteriorated when we replaced the reservoir solution with the one made in D_2O, indicating that the dPKAc:Sr_2ADP:pPKS crystals did not tolerate D_2O. Moreover, only tiny needles could be grown when we attempted crystallization of our dPKAc complexes in D_2O solutions.

3. Structural analysis

This section will focus on specific insights gained from the neutron structure of dPKAc product complex that illustrate the unique contribution of neutron crystallography to the understanding of protein function that otherwise would be impossible to infer from X-ray diffraction. Specifically, the previously unknown protonation states of several residues, assessment of hydrogen bonding interactions predicted from X-ray structures and applicability of H/D exchange patterns observed for main-chain amides to protein dynamics will be described. The neutron diffraction from a dPKAc:Sr_2ADP:pPKS crystal was first tested at room temperature using the IMAGINE beamline (Meilleur, Coates, Cuneo, Kovalevsky, & Myles, 2018; Meilleur et al., 2013) at the

High Flux Isotope Reactor (Oak Ridge National Laboratory, USA) and subsequently the complete dataset was collected on the LADI-III diffractometer (Blakeley et al., 2010) at the Institut Laue Langevin (France).

3.1 Observing important protons for PKAc function

The direct observation of protonation states on the side chains of ionizable amino acids in a protein structure is the most obvious application of neutron crystallography. Although a good number of hydrogen atom locations can be reliably predicted from an X-ray structure on the basis of the observed distances between the adjacent heavy atoms of neighboring residues, water molecules, ligands, etc., the protonation states of some chemical groups usually remain hard to determine unequivocally. Ironically, very often those residues, functional groups on ligands and water species are the most important for the biological function of the macromolecule, and therefore are of the most interest for a researcher.

Proteins kinases have a conserved fold and, consequently, their structures and active sites are very similar. PKA has become a prototypical kinase enzyme and the structural and functional knowledge gained for PKA has been applied to understand structure-function relationships of other kinases. The proton distribution at the site of the phosphoryl transfer has always been debated. Specifically, the protonation states of the conserved catalytic residue Asp166, which is the closest ionizable residue to the OH group of a substrate that could act as a catalytic base, and of the phosphate group of the peptide product have remained unknown. In the X-ray structure of the PKAc product complex we can observe that the carboxyl group of Asp166 and the phosphorylated Ser21 of the substrate pPKS ($pS21_{PKS}$) participate in numerous interactions having distances suggestive of numerous hydrogen bonds. The $pS21_{PKS}$ phosphate makes five such interactions—three with water molecules, one with its own main-chain amide and one with hydroxyl group of Ser53. Asp166 has four interactions—three with water molecules and one with the Thr201 side chain (Fig. 3A). However, this does not answer the important questions such as what the Asp166 and $pS21_{PKS}$ protonation states are, and which interactions are true hydrogen bonds, because these structural details cannot be inferred with confidence from the X-ray data. We cannot unequivocally distinguish between hydrogen bond donors and acceptors in these networks. Indeed, each one of the two contributors in a given interaction can act as either or both. Because the position of hydrogen atoms on the hydroxyl groups of Ser53 and Thr201 and orientation of water molecules in the networks

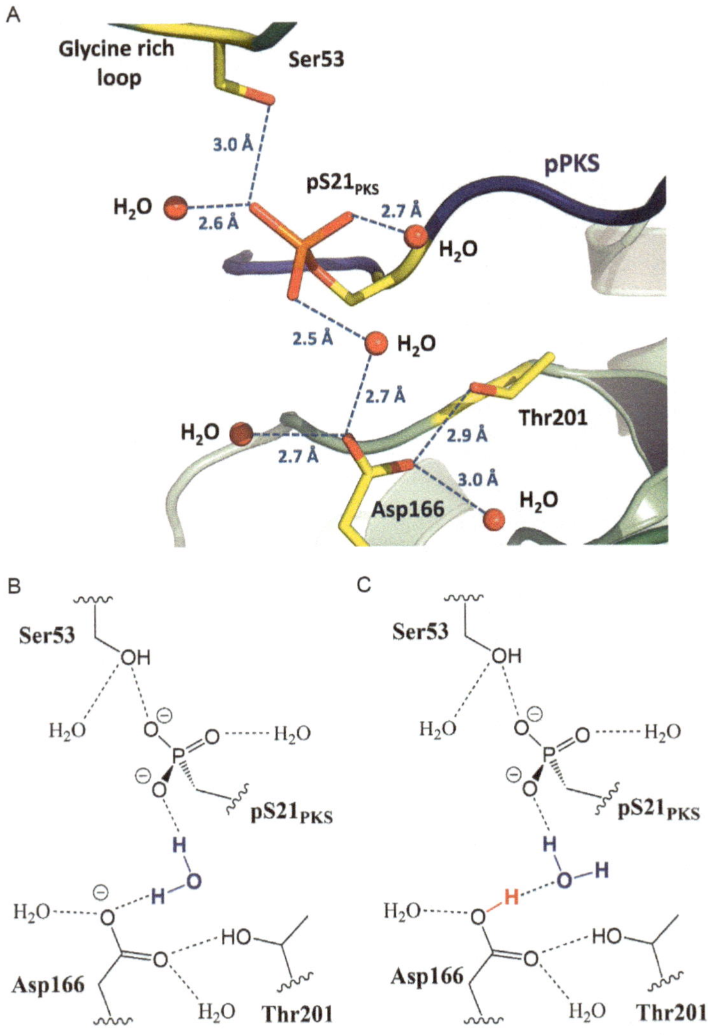

Fig. 3 (A) Possible hydrogen bond interactions made by the phosphate of $pS21_{PKS}$ and Asp166 based on the O…O distances. (B and C) Chemical diagram of the same part of the active site depicted in (A) demonstrating differences in hydrogen bonding caused by lack or presence of a proton on Asp166. Importantly, several additional chemical diagrams are possible if the phosphate of $pS21_{PKS}$ is protonated because the proton can be on any of the three oxygen atoms.

are unknown, at least two plausible diagrams can be sketched for each of the groups that differ in the specific locations of known hydrogen atoms for the observed contacts. This is illustrated in Fig. 3B and C for Asp166, one diagram can describe the situation when Asp166 is deprotonated, whereas the other depicts Asp166 being protonated in the same environment. Something similar, only more complicated, could be drawn for $pS21_{PKS}$; this functional group can have more than one protonation state. Only by using neutron crystallography, we were able to unambiguously determine the hydrogen atom positions.

The neutron structure of the dPKAc product complex provides a complete picture with experimentally determined positions of hydrogen atoms (observed as D) and water (D_2O) orientations. The neutron data clearly show that in the product-bound state Asp166 is deprotonated and the $pS21_{PKS}$ phosphoryl group is monoprotonated (Fig. 4). Analysis of these observations generated conclusions that are not obvious without seeing the observed protons: (1) the protonation of the phosphoryl group on the peptide product probably induces its rotation toward the bulk solvent noted previously from comparison of X-ray structures of the product and the pseudo-Michaelis complexes (Gerlits et al., 2015); (2) this rotation, very

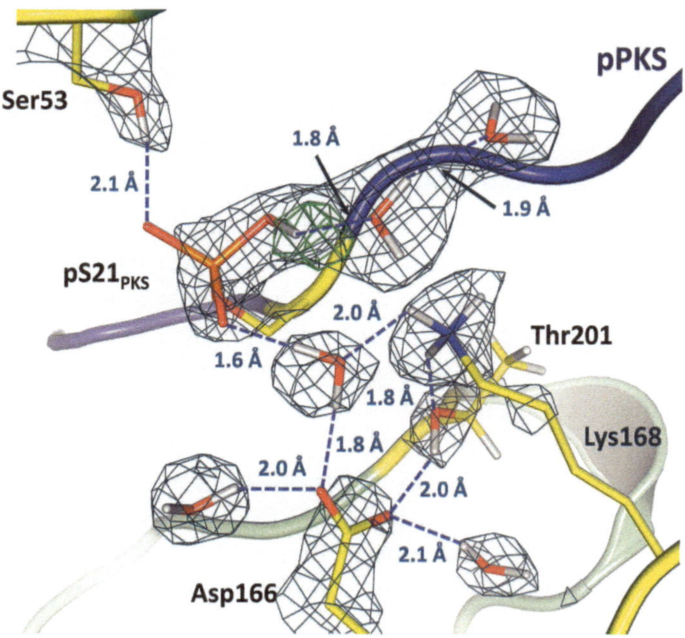

Fig. 4 Neutron scattering length density maps, protonation states and hydrogen bonding interactions visualized for $pS21_{PKS}$, Asp166, other active site residues and the associated water molecules. The $2F_O - F_C$ map is contoured at 1.5 σ level, and the omit difference $F_O - F_C$ map directly showing protonation of the phosphate of $P21_{PKS}$ is contoured at 3 σ level. Hydrogen bonds are depicted as blue dashed lines and the D...O distances are in Å.

possibly, triggers early stages of the product peptide release. Having visualized the protonation states of these chemical groups in the PKAc active site, we can further suggest the possible mechanism of the proton transfer events occurring during the phosphoryl transfer reaction. The OH group of the substrate Ser21 residue should be protonated when the PKS binds to PKAc. The hydroxyl hydrogen can then be transferred to the Asp166 carboxylic group during the phosphoryl transfer step. Thus, Asp166 acts a general base during this reaction step. Once the phosphoryl group is transferred to PKS, the protonated Asp166 donates the proton to an oxygen of the $pS21_{PKS}$ phosphate, now acting as a general acid. The protonated $pS21_{PKS}$ side chain later rotates away from the metal ion to the position detected in our crystal structures. Therefore, it appears that Asp166 may have a dual role in the PKAc catalysis—it can first act as a general base to abstract a proton from the substrate hydroxyl in sync with the nucleophilic attack on the ATP γ-phosphorus, but after the phosphoryl transfer chemical step has occurred it can change its role to become a general acid donating its proton to the $pS21_{PKS}$ phosphate.

ATP binds in an exquisitely tailored nucleotide binding site that accommodates three parts of the molecule—the adenine group, the ribose sugar ring and the phosphates tail (Fig. 5). There are also two metal ion binding sites, with the cations coordinating with both the phosphates and PKAc active site residues. The adenine group is located in a sterically tight subsite, interacting with Ala70, Val57 and Leu73 through hydrophobic C–H…π contacts. Its exocyclic N6 amino group makes a single hydrogen bond with the main chain carbonyl of Glu121, which is conserved in all kinases. The phosphates tail stretches over to the phosphoryl transfer site, interacting with the conserved protonated Lys72 and deprotonated Glu91. Importantly, we found that interactions made by the ribose are in agreement with this complex being a product complex, poised for product release. Both 2′- and 3′-OH groups of the sugar ring do not participate in strong hydrogen bonding. The D atom of 2′-OH is rotated away from the carboxylic oxygen of Glu127, even though the O…O separation of 2.7 Å implies hydrogen bonding. Moreover, 3′-OH forms only a weak hydrogen bond with the main-chain carbonyl of Glu170.

The catalytic cycle of PKAc is highly regulated and dynamic (Nolen, Taylor, & Ghosh, 2004). In the active form, which is facilitated by autophosphorylation of the activation loop Thr197 (pThr197), the kinase catalytic core fluctuates between open (apo-PKAc), intermediate (nucleotide bound PKAc) and closed (nucleotide and substrate bound)

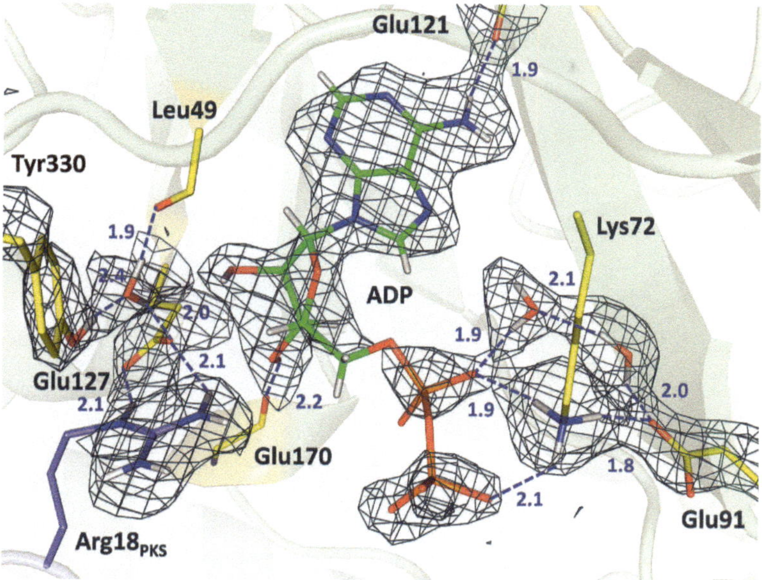

Fig. 5 Neutron scattering length density map, protonation states and hydrogen bonding interactions visualized for ADP, other active site residues and the associated water molecules. The $2F_O - F_C$ map is contoured at 1.5 σ level. Hydrogen bonds are depicted as blue dashed lines and the D…O distances are in Å.

conformations (Johnson, Akamine, Radzio-Andzelm, Madhusudan, & Taylor, 2001). The phosphorylation reaction only happens in the closed state. To release the products of the reaction PKAc transitions back to the open conformation and the catalytic cycle is repeated. A contact between pThr197 in the activation loop and His87 at the tip of the αC-helix is one of the key interactions that regulates the active site mechanics (Fig. 6). This contact is always present in the structures of PKAc in the intermediate or closed conformations, but the interaction does not form in the open conformation (Akamine et al., 2003). The interaction of His87 with the phosphorylation site of the activation loop has been shown to act as an anchor that slows down opening of the active site cleft and consequently the catalytic cycle. We found that His87 is doubly protonated, carrying a positive charge which stabilizes the negative charge on the deprotonated phosphate of pThr197. It is likely that in the open conformation of apo-PKAc His87 would be not protonated and have a neutral charge. Thus, His87 is a crucial part of the switch mechanism that slows down catalysis and makes the release of ADP the rate-limiting step. Of note is the lack of neutron scattering length density on the phosphate, which is canceled

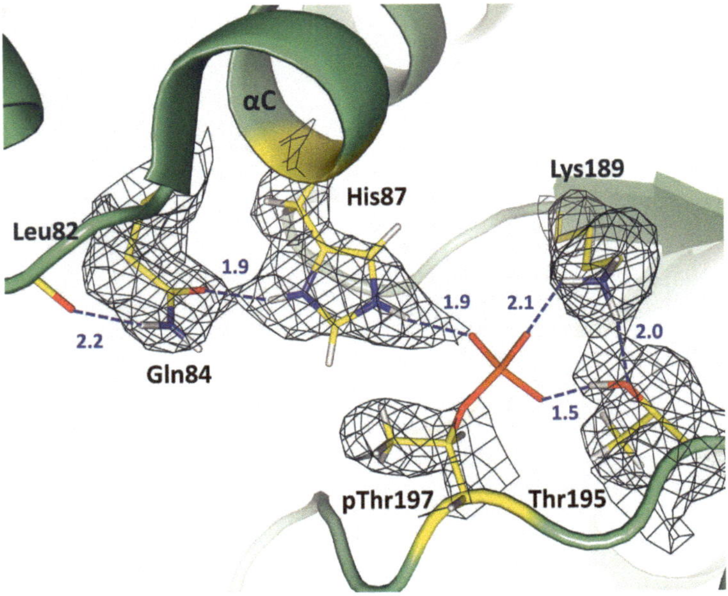

Fig. 6 Neutron scattering length density map, protonation states and hydrogen bonding interactions visualized for pThr197, His87 and other nearby residues. The $2F_O - F_C$ map is contoured at 1.5 σ level. Hydrogen bonds are depicted as blue dashed lines and the D...O distances are in Å.

due to the smaller scattering length for P relative to other atoms and low resolution of the dataset. This can be compared with the observation of the strong neutron scattering length density on the phosphate of $pS21_{PKS}$, which may be related to its deuteration. Another important observation afforded by our neutron structure is that Cys199 (in the activation segment) is protonated and is a thiol. In the apo-PKAc, however, it is highly reactive and can be easily chemically modified leading to loss of the kinase activity (Jimenez, Kupfer, Gani, & Shaltiel, 1982). In this case, Cys199 is likely deprotonated (thiolate), but protonated Cys199 (thiol) is less reactive than thiolate, ensuring catalytic activity of PKAc is not blocked by chemical modification.

3.2 Dynamic features through H/D exchange pattern and comparison to NMR

Apart from the obvious implications related to protein function, the ability to visualize H/D exchange patterns can be used to determine certain aspects of protein dynamics. This is normally done by examining the H/D exchange levels of the main chain amides, extracting information about the protein molecule regions that are more flexible to allow exchange or more rigid where exchange is very slow. The crystal of dPKAc:Sr_2ADP:pPKS complex was allowed to H/D exchange for several weeks in order to ensure significant H/D exchange occurred for the sites with rates of $>10^{-6}\,s^{-1}$, and also taking into account that exchange rate decreases at acidic pH (Connelly, Bai, Jeng, & Englander, 1993). The H/D exchange pattern in the dPKAc:Sr_2ADP:pPKS complex is depicted in Fig. 7. We consider sites to be fully exchanged when the refined D atom occupancies are in the range of 0.61–1.0, partially exchanged D atoms have occupancies in the range of 0.0–0.60, whereas when the D occupancy is refined to be −0.56 to 0.0 the main chain amide is considered non-exchanged. Because the neutron scattering length of H is −0.56 times the scattering length of D, the amount of D (%D) at each position can be calculated according to the following equation: $\%D=[(D_{occup.}+0.56)/1.56]*100$. Thus, the amount of D is >3/4 for fully exchanged, between 1/3 and 3/4 for partially exchanged and <1/3 for non-exchanged amide sites. dPKAc has ~53% of the amide protons fully exchanged, ~20% of the amide protons partially exchanged, and ~27% sites with no exchange (excluding proline residues that lack exchangeable protons) (Fig. 7). Clearly, only about half of the main chain amides underwent complete H/D exchange, whereas the other half exchanged poorly, indicating that the enzyme is rather solvent protected. This is not surprising taking into account that dPKAc:Sr_2ADP:pPKS is a stable ternary complex with the high-affinity peptide substrate pPKS. The main-chain amides in the large

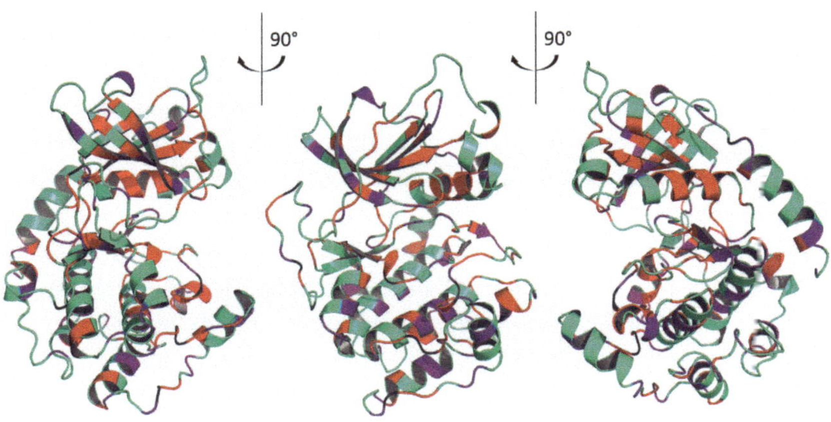

Fig. 7 H/D exchange pattern for the main-chain amide protons in the deuterated PKAc. Fully exchanged sites are colored cyan green, partially exchanged amides are purple and non-exchanged amides are red. The pPKS and Sr_2ADP are omitted for clarity.

C-lobe appear to be H/D exchanged to a high degree, with the helices and loops containing mostly fully exchanged sites. But, in the small N-lobe the extended β-sheet and the αC-helix pushing at it from below are mostly not exchanged, suggesting that they are less dynamic. The low level of H/D exchange in the β-sheet is not surprising, because this secondary structure element is greatly stabilized by strong hydrogen bonds between amides of one β-strand and carbonyls of the opposite β-strand. Helices, however, frequently contain well-exchanged amides. Therefore, it appears that αC-helix is less dynamic than other helices in dPKAc, probably because it is tightly packed against the β-sheet.

It is also instructive to compare the H/D exchange of the main-chain amides obtained from our neutron structure of dPKAc:Sr_2ADP:pPKS complex with H/D fractionation factors, φ, measured by the NMR for PKAc:Mg_2ADP:PKI complex (where PKI is an inhibitor peptide identical to PKS, except for a single Ser-to-Ala substitution at the phosphorylation site 21 (Li, Srivastava, Kim, Taylor, & Veglia, 2015)). φ values determine the strength of hydrogen bonds, with $\varphi < 0.5$ indicative of stronger hydrogen bonds and higher φ values are associated with weaker hydrogen bonds. The φ values should correlate with H/D exchange levels because stronger hydrogen bonds are less accessible and therefore accumulate less deuterium compared to the weaker and more accessible ones. To perform this analysis, we grouped the main-chain amides into two groups: (a) slow-exchanging amides with D occupancies below 0.5 ($\varphi < 0.5$) and (b) fast-exchanging amides with D occupancies above 0.5 ($\varphi > 0.5$). We then focused on the slow-exchanging sites, because they indicate reduced dynamics of the specific regions. In the NMR dataset, there are 55 residues that belong to the category of slow-exchanging amides for the PKAc:Mg_2ADP:PKI complex, with most of them confined to the small N-lobe, in striking agreement with the H/D exchange pattern observed in the neutron structure. Interestingly, though, about 1/3 of these amides are found fully exchanged in our neutron structure of the dPKAc:Sr_2ADP:pPKS complex. The φ values for the same amides had increased values in the NMR datasets for apo-PKAc or Mg_2ADP binary complex, which correlates with the PKAc transitioning from closed (ternary) to semiopen (binary) and open (apo) conformations, with weakened hydrogen bonds and increased dynamics (Li et al., 2015; Masterson et al., 2010). We concluded that the differences in the H/D exchange pattern, indicative of altered dynamics, observed in the neutron structure arise from our complex representing products of the phosphoryl transfer reaction rather than reactants.

3.3 Analysis of D incorporation into C–H bonds

An interesting aspect for further analysis is how much D is incorporated into the non-exchangeable C–H bonds of dPKAc by the *E. coli* metabolic pathways. This is especially interesting given the fact that the enzyme was expressed in 80% D_2O, with fully hydrogenous glycerol as a sole carbon source. Table 1 presents this analysis. To make this analysis manageable and comprehensible we first averaged occupancies of all D atoms of the C–D bonds for each amino acid residue and treated the whole residue as a single

Table 1 Incorporation of D into the C–H bonds.

Amino acid	% No incorporation	% Partial incorporation	% Full incorporation	Number of residues
Hydrophobic				
Gly	20	50	30	20
Ala	26	64	10	19
Val	11	84	5	19
Ile	5	76	19	21
Leu	0	81	19	32
Pro	0	69	31	13
Met	0	100	0	6
Average	*9*	*75*	*16*	
Hydrophilic				
Ser	47	27	26	15
Cys	0	50	50	2
Thr	13	53	34	15
Asp	21	53	26	19
Asn	7	53	40	15
Glu	0	75	25	24
Gln	7	50	43	14
Arg	7	73	20	15
Lys	13	78	9	32
Average	*13*	*57*	*30*	

Continued

Table 1 Incorporation of D into the C–H bonds.—cont'd

Amino acid	% No incorporation	% Partial incorporation	% Full incorporation	Number of residues
Aromatic				
His	13	63	24	8
Tyr	23	77	0	13
Phe	19	73	8	26
Trp	17	83	0	6
Average	*18*	*74*	*8*	
Overall Average	*13*	*69*	*18*	

Percentage values indicate how many residues of the same type relative to their total number have a particular level of D incorporation. The level of D incorporation for each residue is calculated by averaging the occupancies of D atoms bound to C atoms.

entity. Thus, we identified each residue as having no D incorporation, partial D incorporation or full D incorporation. On average, ~70% of all amino acid residues in dPKAc have partial incorporation of D into their C–H bonds, ~18% with full D incorporation and ~13% with no D incorporation. Interestingly, ¾ of hydrophobic and aromatic residues have partial D incorporation, whereas hydrophilic residues have higher levels of D incorporation. All leucine, proline and methionine residues at least partially incorporate D, with no methionine residues having either no or full D incorporation. On the other hand, none of the tyrosine and trypotophan residues have full D incorporation. In fact, aromatic residues show the poorest levels of D incorporation into the C–H bonds. Hydrophilic residues have most residues with full D incorporation (~30%) compared to 16% of hydrophobic and only 8% of aromatic residues. It is quite remarkable that, in spite of expressing dPKAc in a solution containing 20% H_2O and 100% hydrogenous glycerol, nearly one-fifth of amino acid residues have full incorporation of D into the non-exchangeable C–H bonds, with five residues containing ≥94% of D (one each of glycine, proline, threonine, asparagine and glutamine), reflecting the complexity of metabolic processes in the bacterial cell.

4. Summary

Macromolecular neutron crystallography is a powerful technique capable of providing mechanistic details of enzyme function that are hard, or impossible, to obtain with other biophysical methods. Some challenges, including large crystal growth, protein deuteration, etc. mostly stemming from the weak fluxes of neutron beams, need to be overcome to elevate neutron crystallography to a level of a mainstream structural biology technique. It is nevertheless advisable to consider neutron crystallography in many cases because, if successful, it can answer long-standing biochemical questions. By exploring different protein deuteration and crystal growth methodologies and by careful manipulation of the complex chemical composition, we were able to obtain a neutron structure of PKAc trapped in a product complex with bound Sr_2ADP and phosphorylated PKS peptide. We directly observed protonation of the phosphate group transferred to the serine residue of PKS and determined that the catalytic Asp166 is deprotonated. We also visualized numerous hydrogen bonding interactions by correctly orienting O–D and N–D groups and D_2O molecules, and by determining the protonation states of ionizable residues, such as histidine. Analysis of the main-chain amide H/D exchange levels gave us an opportunity to look into the dynamics of PKAc. Comparison of the H/D exchange with the previous NMR measurements provided further understanding of how dynamics might be altered in the product complex relative to the substrate complex.

Acknowledgments

The neutron diffraction experiments were performed on IMAGINE beamline at Oak Ridge National Laboratory's High Flux Isotope Reactor sponsored by the Scientific User Facilities Division, Office of Basic Energy Sciences, U.S. Department of Energy (DOE), and on beamline LADI-III at the Institut Laue Langevin. Protein deuteration, purification and crystallization were carried out at ORNL's Center for Structural Molecular Biology (CSMB) supported by the DOE Office of Biological and Environmental Research (BER), using facilities supported by the Scientific User Facilities Division, Office of Basic Energy Sciences (BES). This work was supported by the NIH grants GM19301 and GM100310, by a UCOP grant, and by the DOE BES.

References

Adams, J. A. (2001). Kinetic and catalytic mechanisms of protein kinases. *Chemical Reviews*, *101*, 2271–2290.

Akamine, P., Madhusudan, Wu, J., Xuong, N.-H., Ten Eyck, L. F., & Taylor, S. S. (2003). Dynamic features of cAMP-dependent protein kinase revealed by apoenzyme crystal structure. *Journal of Molecular Biology*, *327*, 159–171.

Bastidas, A. C., Deal, M. S., Steichen, J. M., Guo, Y., Wu, J., & Taylor, S. S. (2013). Phosphoryl transfer by protein kinase A is captured in a crystal lattice. *Journal of the American Chemical Society*, *135*, 4788–4798.

Blakeley, M. P., Hasnain, S. S., & Antonyuk, S. V. (2015). Sub-atomic resolution X-ray crystallography and neutron crystallography: Promise, challenges and potential. *IUCrJ*, *2*(Pt. 4), 464–474.

Blakeley, M. P., Teixeira, S. C., Petit-Haertlein, I., Hazemann, I., Mitschler, A., Haertlein, M., et al. (2010). Neutron macromolecular crystallography with LADI-III. *Acta Crystallographica. Section D, Biological Crystallography*, *66*(Pt. 11), 1198–1205.

Cheng, Y., Zhang, Y., & McCammon, J. A. (2005). How does the cAMP-dependent protein kinase catalyze the phosphorylation reaction: An ab initio QM/MM study. *Journal of the American Chemical Society*, *127*, 1553–1562.

Connelly, G. P., Bai, Y., Jeng, M. F., & Englander, S. W. (1993). Isotope effects in peptide group hydrogen exchange. *Proteins*, *17*, 87–92.

Das, A., Gerlits, O., Parks, J. M., Langan, P., Kovalevsky, A., & Heller, W. T. (2015). Protein kinase A catalytic subunit primed for action: Time-lapse crystallography of Michaelis complex formation. *Structure*, *23*, 2331–2340.

Gerlits, O., Das, A., Keshwani, M. M., Taylor, S., Waltman, M. J., Langan, P., et al. (2014). Metal-free cAMP-dependent protein kinase can catalyze phosphoryl transfer. *Biochemistry*, *53*, 3179–3186.

Gerlits, O., Tian, J., Das, A., Langan, P., Heller, W. T., & Kovalevsky, A. (2015). Phosphoryl transfer reaction snapshots in crystals. Insights into the mechanism of protein kinase A catalytic subunit. *The Journal of Biological Chemistry*, *290*, 15538–15548.

Gerlits, O. O., Waltman, M. J., Taylor, S., Langan, P., & Kovalevsky, A. (2013). Insights into the phosphoryl transfer catalyzed by cAMP-dependent protein kinase: An X-ray crystallographic study of complexes with various metals and peptide substrate SP20. *Biochemistry*, *52*, 3721–3727.

Gerlits, O., Weiss, K. L., Blakeley, M. P., Veglia, G., Taylor, S. S., & Kovalevsky, A. (2019). Zooming in on protons: Neutron structure of protein kinase A trapped in a product complex. *Science Advances*, *5*, eaav0482.

Jimenez, J. S., Kupfer, A., Gani, V., & Shaltiel, S. (1982). Salt-induced conformational changes in the catalytic subunit of adenosine cyclic 3′,5′-phosphate dependent kinase. Use for establishing a connection between one sulfhydryl group and the γ-P subsite in the ATP site of this subunit. *Biochemistry*, *21*, 1623–1630.

Jin, Y., Cliff, M. J., Baxter, N. J., Dannatt, H. R. W., Hounslow, A. M., Bowler, M. W., et al. (2012). Charge-balanced metal fluoride complexes for protein kinase A with adenosine diphosphate and substrate peptide SP20. *Angewandte Chemie, International Edition*, *51*, 12242–12245.

Johnson, D. A., Akamine, P., Radzio-Andzelm, E., Madhusudan, M., & Taylor, S. S. (2001). Dynamics of cAMP-dependent protein kinase. *Chemical Reviews*, *101*, 2243–2270.

Klaeger, S., et al. (2017). The target landscape of clinical kinase drugs. *Science*, *358*, eaan4368.

Langan, P., & Chen, J. C.-H. (2013). Seeing the chemistry in biology with neutron crystallography. *Physical Chemistry Chemical Physics*, *15*, 13705–13712.

Li, G. C., Srivastava, A. K., Kim, J., Taylor, S. S., & Veglia, G. (2015). Mapping the hydrogen bond networks in the catalytic subunit of protein kinase A using H/D fractionation factors. *Biochemistry*, *54*, 4042–4049.

Madhusudan, Akamine, P., Xuong, N.-H., & Taylor, S. S. (2002). Crystal structure of a transition state mimic of the catalytic subunit of cAMP-dependent protein kinase. *Nature Structural Biology*, *9*, 273–277.

Manning, G., Whyte, D. B., Martinez, R., Hunter, T., & Sudarsanam, S. (2002). The protein kinase complement of the human genome. *Science*, *298*, 1912–1934.

Masterson, L. R., Cheng, C., Yu, T., Tonelli, M., Kornev, A., Taylor, S. S., et al. (2010). Dynamics connect substrate recognition to catalysis in protein kinase A. *Nature Chemical Biology*, *6*, 821–828.

Masterson, L. R., Shi, L., Metcalfe, E., Gao, J., Taylor, S. S., & Veglia, G. (2011). Dynamically committed, uncommitted, and quenched states encoded in protein kinase A revealed by NMR spectroscopy. *Proceedings of the National Academy of Sciences of the United States of America*, *108*, 6969–6974.

Meilleur, F., Coates, L., Cuneo, M. J., Kovalevsky, A., & Myles, D. A. A. (2018). The neutron macromolecular crystallography instruments at Oak Ridge National Laboratory: Advances, challenges, and opportunities. *Crystals, 8*, 388.

Meilleur, F., Munshi, P., Robertson, L., Stoica, A. D., Crow, L., Kovalevsky, A., et al. (2013). The IMAGINE instrument: First neutron protein structure and new capabilities for neutron macromolecular crystallography. *Acta Crystallographica. Section D, Biological Crystallography, 69*, 2157–2160.

Niimura, N., & Podjarny, A. (2011). *Neutron protein crystallography.* Oxford University Press. p. 250.

Niimura, N., Takimoto-Kamimura, M., & Tanaka, I. (2016). Neutron diffraction in studies of protein dynamics and functions, application of. In *Encyclopedia of analytical chemistry: Applications, theory and instrumentation* (pp. 1–30). John Wiley and Sons, Ltd.

Nolen, B., Taylor, S., & Ghosh, G. (2004). Regulation of protein kinases: Controlling activity through activation segment conformation. *Molecular cell, 15*, 661–675.

O'Dell, W. B., Bodenheimer, A. M., & Meilleur, F. (2016). Neutron protein crystallography: A complementary tool for locating hydrogens in proteins. *Archives of Biochemistry and Biophysics, 602*, 48–60.

Oksanen, E., Chen, J. C.-H., & Fisher, Z. (2017). Neutron crystallography for the study of hydrogen bonds in macromolecules. *Molecules, 22*, 596.

Perez-Gallegos, A., Garcia-Viloca, M., Gonzalez-Lafont, A., & Lluch, J. M. (2015) SP20 phosphorylation reaction catalyzed by protein kinase A: QM/MM calculations based on recently determined crystallographic structures. *ACS Catalysis, 5*, 4897–4912.

Perez-Gallegos, A., Garcia-Viloca, M., Gonzalez-Lafont, A., & Lluch, J. M. (2017). Understanding how cAMP-dependent protein kinase can catalyze phosphoryl transfer in the presence of Ca^{2+} and Sr^{2+}: A QM/MM study. *Physical Chemistry Chemical Physics, 19*, 10377–10394.

Srivastava, A. K., McDonald, L. R., Cembran, A., Kim, J., Masterson, L. R., McClendon, C. L., et al. (2014). Synchronous opening and closing motions are essential for cAMP-dependent protein kinase A signaling. *Structure, 22*, 1735–1743.

Taylor, S. S., Zhang, P., Steichen, J. M., Keshwani, M. M., & Kornev, A. P. (2013). PKA: Lessons learned after twenty years. *Biochimica et Biophysica Acta, 1834*, 1271–1278.

Valiev, M., Yang, J., Adams, J. A., Taylor, S. S., & Weare, J. H. (2007). Phosphorylation reaction in cAPK protein kinase-free energy quantum mechanical/molecular mechanics simulations. *The Journal of Physical Chemistry. B, 111*, 13455–13464.

CHAPTER FIFTEEN

Pyridoxal 5′-phosphate dependent reactions: Analyzing the mechanism of aspartate aminotransferase

Timothy C. Mueser[a,]*, Victoria Drago[a], Andrey Kovalevsky[b], Steven Dajnowicz[a,b,†]

[a]Department of Chemistry and Biochemistry, University of Toledo, Toledo, OH, United States
[b]Neutron Scattering Division, Oak Ridge National Laboratory, Oak Ridge, TN, United States
*Corresponding author: e-mail address: timothy.mueser@utoledo.edu

Contents

† Current address: Schrödinger. 120 West 45th Street, New York, NY.

Methods in Enzymology, Volume 634
ISSN 0076-6879
https://doi.org/10.1016/bs.mie.2020.01.009

Abstract

Enzyme catalysis is the primary activity in energy and information metabolism and enzyme cofactors are key to the catalytic ability of most enzymes. Pyridoxal 5′-phosphate (PLP) cofactor, derived from Vitamin B6, is widely distributed in nature and has significant latitude in catalytic diversity. X-ray crystallography has revealed the structures of diverse PLP dependent enzymes from multiple families. But these structures are incomplete, lacking the positions of protons essential for understanding enzymatic mechanisms. Here, we review the diversity of PLP and discuss the use of neutron crystallography and joint X-ray/neutron refinement of Fold Type I aspartate aminotransferase to visualize the positions of protons in both the internal and external aldimine forms. Strategies used to prepare extremely large crystals required for neutron diffraction and the approach to data refinement including the PLP cofactor are discussed. The observed positions of protons, including one located in a previously unknown low-barrier hydrogen bond, have been used to create more accurate models for computational analysis. The results revealed a new mechanism for the transaminase reaction where hyperconjugation is key to reducing the energy barrier which finally provides a clear explanation of the Dunathan alignment.

1. Introduction

Enzymatic activity defines the work accomplished by living organisms; the transformation of chemicals into energy and biomolecules. The Enzyme Commission of the International Union of Biochemistry (https://www.qmul.ac.uk/sbcs/iubmb) has categorized the protein enzymes into six classes based on the types of chemical transformation: oxidoreductases, transferases, hydrolases, lyases, isomerases, and ligases (Dixon & Webb, 1979). Another significant type of classification is the absence or presence of cofactors. Proteins can function by using side chain moieties alone or the apoenzyme may require an additional inorganic or organic component to create a holoenzyme. Inorganic cofactors include metal ions of magnesium, zinc, cobalt, iron, etc. Organic cofactors, small molecules referred to as coenzymes, are mostly synthesized by microbes and ingested as vitamins. The cofactors are categorized by the enzyme functionality, in which they assist. For example, niacin (NAD) and the flavonoids (FAD and FMN) sequester electrons, coenzyme A activates and sequesters acyl groups, and vitamin B_6 helps sequester toxic ammonia. While these categorizations are critical for pedagogy, the fundamental comprehension of life processes lies in the mechanisms of the chemical transformation; the modulation of electron density, the stabilization or destabilization of particular reaction intermediates and transition states and, most importantly for this book, the contribution to these phenomena through the arrangement of protons.

1.1 The pyridoxal 5′-phosphate cofactor

In this chapter, we summarize in context our recent neutron crystallography studies of pyridoxal 5′-phosphate (PLP)-dependent enzymes. Fig. 1 shows the PLP cofactor in both aldimine forms with atom designations. PLP, derived from pyridoxamine or vitamin B_6, is one of the most widely distributed coenzymes in nature. In human metabolism, the pyridoxamine precursor of microbial origin is converted to PLP in two steps: (1) the reduction of C4′ methylamine to an aldehyde by an NADH dependent reductase and (2) the phosphorylation of the C5′ primary alcohol by an ATP-dependent kinase. In the enzyme active site, the C4′ formyl group of PLP reacts with the ε-amine of a lysine residue to form a Schiff base, a reversible covalent attachment referred to as the *internal aldimine*. During most reactions, the Schiff base shifts attachment to a substrate amino group, referred to as the *external aldimine*.

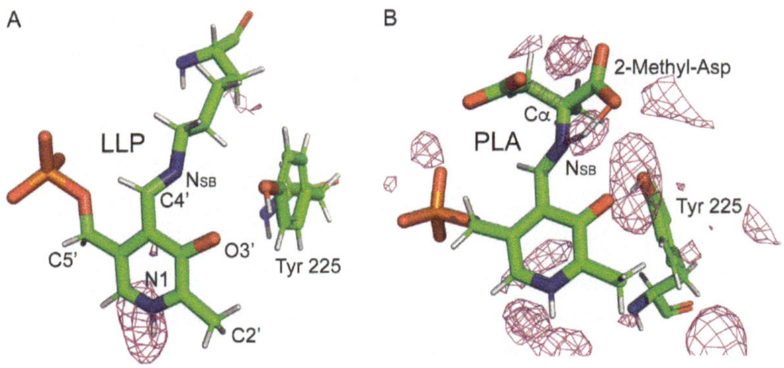

Fig. 1 The cofactors in the two active sites of porcine cytosolic aspartate aminotransferase are shown (*re* face forward). Neutron density difference maps (magenta) were created by deleting the active site exchangeable hydrogens in the final model. In this structure, the protein was perdeuterated, the PLP was hydrogenated, and the crystal was D_2O-exchanged over several months. (A) Pyridoxal phosphate in the internal aldimine form (LLP) Schiff base coupled to the ε-amine of Lys 258. Neither the Schiff base nitrogen (NSB) or the PLP phenol (O3′) show protonation. The pyridine nitrogen (N1) appears protonated. (B) Pyridoxal phosphate in the external aldimine form Schiff base coupled to the substrate inhibitor 2-methyl-aspartate (PLA). The OH of Tyr 225 has become deuterated, with the deuteron positioned in a low-barrier hydrogen bond (LBHB) between N_{SB} and the substrate carboxyl group.

A recent genomics study attributes 4% of total enzyme activity to PLP with extensive catalytic diversity exemplified by involvement in 5 of the 6 enzyme classes (Percudani & Peracchi, 2003). Of the 140 PLP-dependent enzyme activities recognized, the majority are transferases and lyases with smaller numbers of isomerases, hydrolases, and oxidoreductases (Percudani & Peracchi, 2003). A query for the keyword "pyridoxal phosphate" retrieves over ~17,000 entries in SciFinder (A Chemical Abstracts Service, American Chemical Society) and over 1500 entries in the RCSB PDB. PLP is widespread in both nature and in scientific inquiry. This chapter is not intended to be an extensive review of the PLP-dependent enzymes but instead describe the potential for neutron diffraction analysis of these diverse reactions.

1.2 The fold types and active sites of PLP-dependent enzymes

The reaction diversity of PLP-dependent enzymes is well documented. With the advent of structural genomics, Jansonius, in the late 1990s, suggested the five "fold types" of PLP-dependent enzymes (Jansonius, 1998), which are shown in Fig. 2A with their active sites shown in Fig. 2B. While each fold type has distinctive characteristics, common features exist. The cofactor binding sites are located in the loop regions atop similar core αβα domains and the active sites are covered by a diverse set of capping domains.

Starting with Fold Type I, each 412-residue subunit of aspartate aminotransferase homodimer (AAT, EC 2.6.1.1, PDB: 2AAT (Smith, Almo, Toney, & Ringe, 1989) and 5TOQ (Dajnowicz, Parks, et al., 2017)) has a seven-stranded parallel beta sheet surrounded by seven α-helices that comprise the cofactor binding domain. Lys 258 forms the internal aldimine with PLP, known as the LLP moiety. The small domain over the active site has residues from the N- and C-termini. AAT reacts with L-aspartate releasing oxaloacetate and converting PLP to the pyridoxamine, known as the PMP moiety. In the second part of the transferase reaction, AAT PMP reacts with α-ketoglutarate transferring the sequestered amine to the substrate creating L-glutamate.

For Fold Type II, the β-subunit of tryptophan synthase αββα tetramer (TS, EC 4.2.1.20, PDB: 1BKS; Rhee, Parris, Ahmed, Miles, & Davies, 1996) is 394 residues, in which a six-stranded parallel beta sheet surrounded by seven α-helices comprise the cofactor binding domain. The small domain covering the active site is formed by a segment of residues near the

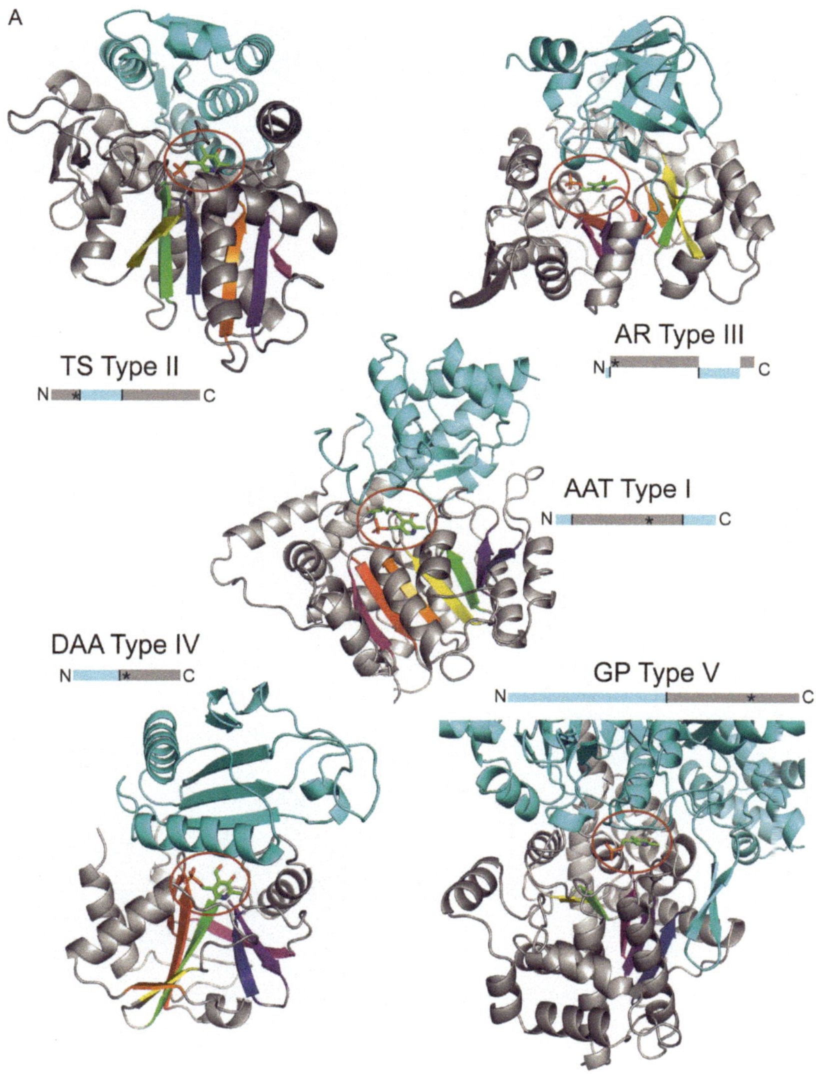

Fig. 2 (A) Shown are ribbon diagrams of representatives of the five main fold types of PLP-dependent enzymes; the α/β PLP binding domain (gray) with β-strands color coded by sequence (blue to red), the active site circled, and the capping domain (cyan) The bar below each name indicates the position with the sequence of the core and capping domains. Fold Type I—aspartate aminotransferase (AAT) family, PDB: 2AAT (PDB: 5TOQ); Fold Type II—tryptophan synthase (TS) family, PDB: 1BKS; Fold Type III—alanine racemase (AR) family, PDB: 1SFT; Fold Type IV—D-amino acid aminotransferase (DAA) family, PDB: 1DAA; Fold Type V—glycogen phosphorylase family (GP), PDB: 1PYG. *(Continued)*

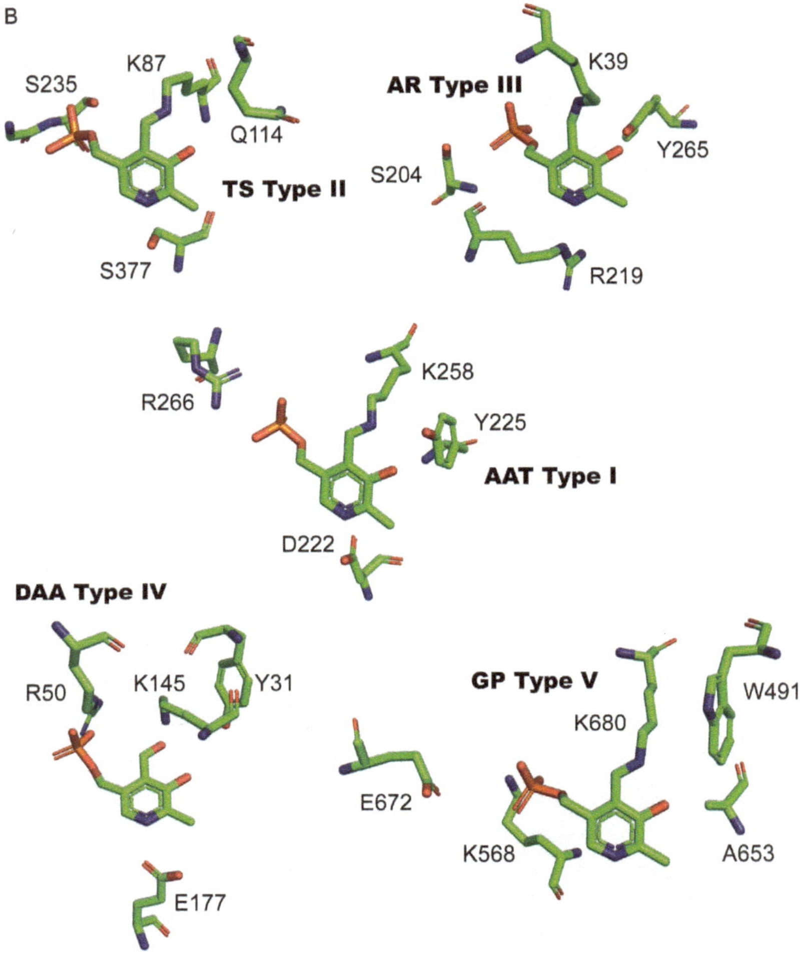

Fig. 2—Cont'd (B) The active site residues of the representative fold types.

N-termini with Lys 87 forming LLP. TS β-subunit reacts with serine and the subsequent release of water by β-elimination creates an amino-acrylate. An indole ring arrives through an internal channel between the α and β subunits to react with the PLP bound amino-acrylate, generating tryptophan.

The N-terminal core subdomain of the 383-residue Fold Type III alanine racemase homodimer (AR, EC 5.1.1.1, PDB: 1SFT; Shaw, Petsko, & Ringe, 1997) is a seven-stranded β-barrel with Lys 39 forming LLP. Interestingly, the small domain covering the active site is formed by a

C-terminal region of the second monomer in a domain swapping arrangement. The alanine-PLP external aldimine undergoes loss and gain of protonation at the Cα position resulting in a racemic mixture.

The 277-residue D-amino acid aminotransferase Fold Type IV homodimer (DAA, EC 2.6.1.21, PDB: 1DAA; Sugio, Petsko, Manning, Soda, & Ringe, 1995) has a C-terminal six-stranded β-barrel core domain with an N-terminal small domain and Lys 145 forming the LLP. Similar to the Fold Type I reaction, DAA maintains the original chirality at the Cα position during transamination.

The 841-residue Fold Type V glycogen phosphorylase homodimer (GP, EC 2.4.1.1, PDB: 1PYG; Sprang, Withers, Goldsmith, Fletterick, & Madsen, 1991) is much larger than the other fold types. The C-terminal cofactor binding domain has the familiar αβα core with Lys 680 forming LLP. The N-terminal cap domain is significantly larger than in the other fold types and the action of the PLP is unique. Instead of forming an external aldimine, the GP enzyme releases inorganic phosphate from glucose-1-phosphate utilizing the LLP C5′ phosphate.

Located on the top face of the αβα cofactor binding domain, the PLP-dependent active sites have five distinct points of contact, the N1 pyridine nitrogen, the O3′ phenolic oxygen, the Schiff base (SB), the C5′ phosphate, and a buttress for the pyridine ring. In Type I AAT, the *re* face of PLP lies flat against Trp 140 with the *si* face downward against Ala 224. The PLP inserts sideways into Type II TS active site with the *re* face against Glu 356 and the *si* face against His 86. In the Type III AR, the *re* face is downward suspended over a pocket while the *si* face is up and interacting with His 166 and Tyr 265. The PLP position in Type IV DAA and Type V GP are similar to Type I AAT with the *re* face up and the *si* face down. The Type IV DAA has an exposed *re* face while Type V GP is wedged between Tyr 648 and Arg 138 on the *si* and *re* faces, respectively.

The PLP cofactor acts as an electron sink to stabilize the withdrawal of electrons to induce bond cleavage and rearrangements. The protonation states of the N1, O3′, and SB enforce the diversity of the enzyme activity. The residues attached to these positions influence both p*K*a and electron resonance. For the N1 position, Type I AAT and Type IV DAA have acidic groups in hydrogen bonding position promoting protonation of N1. The protonated N1 is required for the formation of the "quinonoid intermediate" proposed in many of the PLP mechanisms, and discussed in more detail in the next section. The N1 of Type II TS is hydrogen bonded to Ser 377 and coupled to Arg 219 in Type III AR suggesting N1 remains deprotonated during these transformations. Residues interacting with the SB and O3′ positions also have varying influence. Types I AAT, III AR, and IV DAA have the phenolic group of tyrosine in close proximity. Type II TS has Gln 114 nearby but this is turned away, and Type V GP has Trp 491 and Ala 653 creating a rigid wall.

The visualization of the PLP-dependent enzymes using X-ray crystallography has enabled a wealth of structural information, but they are missing approximately 30% of the atoms due to the inability to accurately determine the positions of protons. The structures can be populated with protons by chemical inference, known as "riding hydrogens," but this approach cannot accurately position the protons of titratable groups. For the PLP cofactor, several methods are employed to determine the positions of protons. The earliest studies utilized UV/vis spectroscopy to reveal multiple stable and transient electronic states (Clark, Jansonius, & Mehler, 1993a, 1993b; Metzler, Cahill, & Metzler, 1980). More recently, solid-state NMR and computation have added significant credence to the proposed spectroscopic states (Chan-Huot et al., 2013; Mueller, 2017; Toney, 2011). In this chapter, we discuss the use of neutron crystallography and joint refinement with X-ray diffraction data to provide detailed snapshots of complete atomic structures (Dajnowicz, Blakeley, Keen, Kovalevsky, & Mueser, 2017; Dajnowicz, Johnston, et al., 2017). Initially, we assumed that this work would discern between established mechanisms, but instead we found protons in unpredicted positions providing new mechanistic interpretation.

2. Catalytic mechanisms and protonation states

The pyridoxal 5′-phosphate (PLP) cofactor participates in a myriad of chemical transformations promoted by the electrophilicity of the cofactor. PLP acts as an "electron sink," withdrawing electron density from bonds within substrates into the conjugated system of the cofactor. To initiate a PLP-dependent reaction, a substrate would bind to the "open" form of the enzyme near the internal aldimine (LLP) as the "Michalis Complex," with a binding affinity described by the substrate's K_M. The presence of a substrate initiates a conformational change where the capping domain covers the active site, resulting in a "closed" structure with the substrate isolated from solvent. With exception of Type V GP, a geminal diamine intermediate is formed between a deprotonated substrate amine and the internal aldimine Schiff base (SB). In many reactions, an active site residue (or the PLP cofactor itself) acting as a Lewis base is thought to deprotonate the substrate amine to initiate the coupling to PLP. This is not a simple task, especially for the α-amino acids where the protonated amine is stabilized within an internal zwitterion. Through bond rearrangement, the SB moves to the substrate creating the external aldimine which promotes the various reactions. Most reactions are thought to involve the heterolytic cleavage of a C—H bond primarily through the creation of a "fleeting carbanion," a step with a very high energy barrier (Toney, 2011, 2019).

2.1 Transamination by type I aspartate aminotransferase (AAT)

Aspartate aminotransferases are essential for nitrogen metabolism in respiring cells, e.g., the mitochondrial isoform (GOT2) funnels nitrogen into urea biosynthesis and the cytosolic isoform (GOT1) is involved in maintaining glutamate levels (NCBI Gene, https://www.ncbi.nlm.nih.gov/gene). In the first half of the reaction, aspartate binds to PLP and the enzyme releases oxaloacetate with the cofactor sequestering the amine as pyridoxamine (PMP) (Fig. 3A) (Hayashi & Kagamiyama, 1995; Toney, 2014). The transaminase reaction is considered energetically neutral but a remarkable high-energy step occurs in which the ε-amino group of Lys 258, released from LLP, acts as a weak Lewis base to abstract the proton from the Cα-H bond to generate a carbanion. This apparently untenable reaction can only occur because of the electrophilicity of PLP, which will be discussed in further detail below.

In crystal structures of PLP dependent enzymes with LLP internal aldimines, the dihedral angle between the pyridine ring and the SB (C3-C4-C4′-N_{SB}) is 45° out of plane, an alignment where there is reduced resonance overlap between the pyridine ring and the Schiff base. This has been considered an example of protein induced strain (Hayashi et al., 2003; Toney, 2011) but is more likely caused by the lesser resonance stabilization that promotes planarity (Dajnowicz, Johnston, et al., 2017), Proposed mechanisms for AAT involve switching from LLP to the external aldimine through a gem-diamine intermediate with the release of a deprotonated LLP lysine and N_{SB} becoming protonated. A hydrogen bond should then form between the protonated N_{SB} and O3′, and the SB becomes co-planar with the PLP. The protonation of the N1 pyridine nitrogen ($pK_a \sim 5.8$ in solution) is required for the putative quinonoid intermediate. Upon activation by substrate, the formation of an O3′ phenoxy anion ($pK_a \sim 9$) is coupled to the protonation of N_{SB} ($pK_a \sim 7$ internal ~9

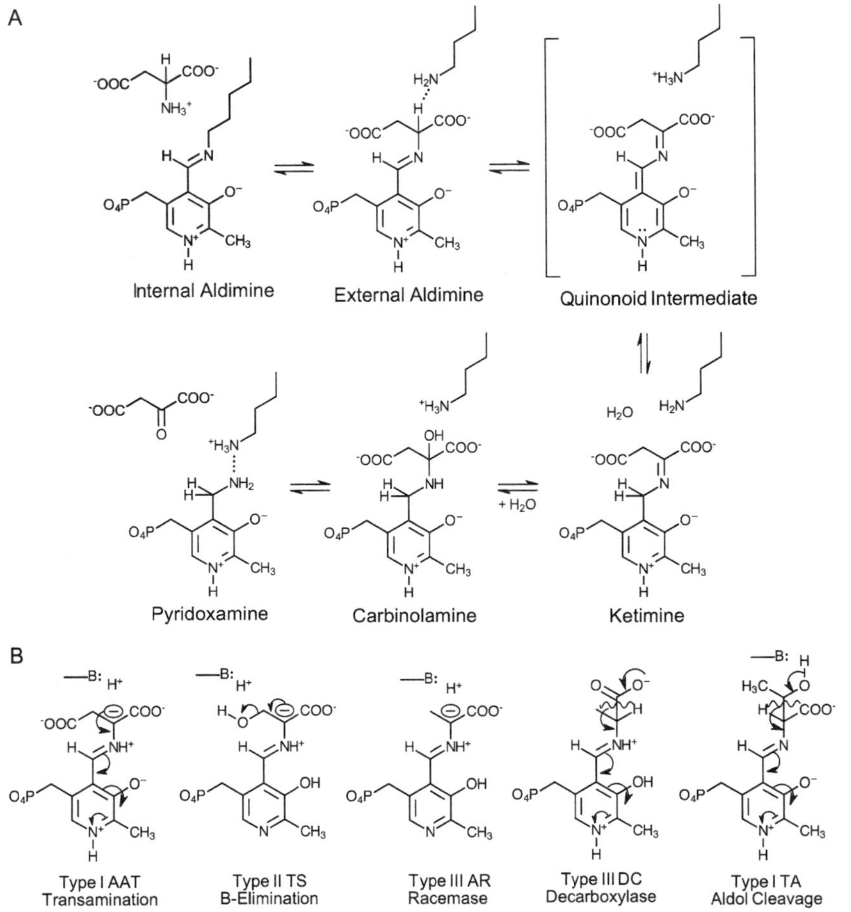

Fig. 3 (A) The mechanism of transamination, catalyzed by PLP, is shown. The internal aldimine Schiff base transfers to the α-amine of the Michalis complex substrate through a gem-diamine intermediate (not shown). The deprotonated ε-amine of Lys 258, released from the internal aldimine, accepts the proton from Cα. The carbanion created by the heterolytic cleavage of the C—H bond is stabilized by resonance structures (via hyperconjugation to antibonding orbitals) of the PLP cofactor exemplified by the quinonoid intermediate. A 1,3-proton shift protonates C4′ (ketimine) which allows water to hydrolyze (carbinolamine) and release the α-keto-acid product yielding pyridoxamine. (B) The PLP cofactor acts as an "electron sink" to create a delay between electronic transitions, which occur in the femtosecond time frame, and atomic rearrangements, which occur in the picosecond time frame. In Type I transamination, Type II β-elimination, and Type III racemase reactions, the heterolytic cleavage of the Cα-H bond requires similar stabilization but the resonance structures involving PLP are difficult to conceive when N1 is deprotonated. Bond cleavage in Type III decarboxylase and Type I threonine aldolase have electronic shifts involving PLP in which N1 is protonated.

Michaelis, and ~10 external) (Hayashi & Kagamiyama, 1997; Sharif et al., 2007). As Lys 258 removes the Cα proton, the protonated N1 allows for the formation of a "quinonoid intermediate," a resonance form in which the Cα-H σ bond becomes the π bond of N_{SB}-Cα with a shifting of the SB double bond to between C4′ and C4. This enhanced electron withdrawing capabilities of a N1 protonated PLP cofactor increases the negative charge on C4′. Lys 258 donates the abstracted proton to the C4′ position, a 1,3-proton shift, forming the "ketimine intermediate" which disrupts the conjugation between N_{SB} and PLP. The deprotonated Lys 258 then activates water for addition to the Cα position, the "carbinolamine intermediate," which promotes hydrolysis and release of oxaloacetate.

We now have a clearer picture of the AAT mechanism. AAT is a homodimer and crystal contacts keep one domain open while the second domain can react with substrate, a fortuitous result which allows visualization of both the internal and external aldimines in the same data set. In the neutron crystal structure of AAT, we find the LLP N1 is protonated while the O3′ and N_{SB} are deprotonated (Dajnowicz, Blakeley, et al., 2017; Dajnowicz, Johnston, et al., 2017) (Fig. 1A). The external aldimine, captured by introducing the inhibitor 2-methyl-aspartate (PLA) into the crystal, shows N1 protonated, O3′ deprotonated, and a proton positioned in a low-barrier hydrogen bond (LBHB) between N_{SB} and the α-carboxyl group of the aspartate substrate (Fig. 1B). The Lys released by the formation of the external aldimine is deprotonated and in position for Cα-H abstraction as postulated in most mechanisms. The new result, visualized for the first time by neutron crystallography, is that the hydrogen bond between O3′ and N_{SB}, which is presumed necessary for PLP-SB planarity, does not appear to exist as suggested in nearly all proposed mechanisms. Instead, the SB-PLP planarity is primarily promoted by the protonation of N1 (Dajnowicz, Blakeley, et al., 2017; Dajnowicz, Johnston, et al., 2017). Perdeuterated AAT was purified and crystallized using hydrogenated solvents. The crystals were vapor exchanged with D_2O to repopulate the exchangeable hydrogens. Tyr 225 OH is hydrogen bonded to PLP O3′. A positive peak for D appears in this bond within the PLA moiety but, curiously, the LLP does not appear to have exchanged. In another new observation, a neighbor to the N1 position, His 189, deprotonated in LLP, becomes protonated when PLA forms; a switching of local charge near N1 which may affect the electron distribution within the PLP ring.

2.2 The role of hyperconjugation and protonation states the Dunathan alignment

As expected, the rate limiting step in the removal of amine from aspartate is the abstraction of the Cα-H proton by Lys 258 (Fig. 3B). In all PLP mechanisms, bond cleavage is governed by the Dunathan rule that states that the bond that is cleaved is coupled to and perpendicular to the plane of the conjugated system (Dunathan, 1966). In the AAT transaminase reaction, the Cα hydrogen is aligned perpendicular to the PLP ring, parallel to the p-orbitals. The formation of a carbanion at Cα is stabilized by hyperconjugation between the Cα σ bond and the N_{SB} π^* orbitals (Dajnowicz et al., 2018; Toney, 2011). This is energetically feasible when the SB double bond is in resonance with PLP and can move to the C4′-C4 position when the carbanion forms, a change that promotes the 1,3-proton shift.

2.3 β-Elimination, racemization, decarboxylation, and aldol cleavage

In Type II TS β subunit Ser-PLP external aldimine, the Cα proton is perpendicular to the pyridine ring but the N1 position is deprotonated (Fig. 3B). The Cα carbanion that forms is not stabilized by the quinonoid intermediate due to the deprotonation of N1 in TS and therefore does not promote deamination. Instead, water is released by β-elimination from Ser 337 creating an amino-acrylate capable of C—C bond formation with indole. In Type III racemase, the N1 position is also likely deprotonated, enforced by the neighboring arginine residue, Arg 219 (Fig. 2B). The Cα carbanion formation does not lead to electron delocalization, with electrons remaining in Cα position to switch chirality by adding a proton from the other face. In the Type III decarboxylase reaction, N1 is protonated allowing bond shifting, the Cα-COO bond is in Dunathan alignment, and CO_2 is released as a consequence. Significant diversity also exists within the diverse fold types of PLP-dependent enzymes. In Type I threonine aldolase, the N1 position is likely protonated allowing bond rearrangement. However, the Cα-Cβ bond is in Dunathan alignment and retro-aldol cleavage releases acetaldehyde. In Type II threonine synthase, a phosphate helps promote gamma elimination. The lack of N1 resonance forms can be used to rationalize β-elimination and racemization but it does not explain how the carbanion can form in the first place; the energy barrier will remain high without significant delocalization of the transient carbanion. Alternate resonance forms have been proposed but not shown experimentally

(Ortega-Castro et al., 2010). Using neutron crystallography to visualize the protonation states of the various PLP-dependent enzymes will allow us to propose alternate resonance states and better understand the mechanisms used by this ubiquitous cofactor.

3. Preparation of large AAT crystals for neutron diffraction

The vast majority of crystals now used in X-ray crystallography are not suitable for neutron crystallography. The low flux of neutron sources and the low scattering efficiencies have to be overcome by maximizing the number of scattering centers. The main indicators for feasibility are (1) a propensity to form very large volume crystals (>1.0 mm^3) with minimal defects, (2) a crystal form with small unit cell dimensions, (3) a crystal form with high symmetry is preferable, (4) the ability to express a fully perdeuterated protein, and (5) most importantly, the crystals need to diffract X-rays well (typically higher than 1.5 Å). PLP-dependent enzymes have been studied extensively and the early X-ray studies were dependent on very large crystals. When planning our first neutron crystallography studies, we recalled these early projects and were able to recreate the cytosolic aspartate aminotransferase (AAT) obtained from pig heart initiated in the Metzler lab (Arnone et al., 1981; Metzler et al., 1978).

3.1 Expression of recombinant AAT

An *E. coli* codon optimized gene for *S. scrofa* wild-type AAT was synthesized (GeneScript) using the protein sequence from PDB: 1AJR, inserted into a Gateway vector (which included an N-terminal polyhistidine tag, a TEV protease site, and streptomycin resistance) and transformed into a T7 Express Lac^{IQ} *E. coli* expression host. Single colonies from LB agar plates were tested for time dependent IPTG induced expression in LB + strep liquid cultures (50 mL LB with 25 μg/mL streptomycin, induced at OD_{600} = 0.4–0.6, 37 °C for 3 h or 22 °C for 12 h), and the best expressers stored in −80 °C glycerol stocks. If necessary, high levels of expression can be maintained/recovered by either using BLR DE3 (recA-) cells or by IPTG streak testing 20–30 individual colonies from an LB + strep plate on an LB + strep IPTG agar plate. For IPTG streak testing, IPTG (100 μL, 10 mM) is spread on half of an LB + strep agar plate and allowed to absorb (~1 h). Single colonies from an LB + strep plate were picked using a sterile wooden toothpick and straight line streaked across the LB only/LB +IPTG boundary. Typically, high expression streaks exhibited normal growth on the LB-only side of the plate and reduced growth on the LB-IPTG side.

Large-scale expression of AAT was achieved through controlled growth of cells in a bioreactor (3 L, BIOFLO 110, New Brunswick Scientific) with enriched LB media. Standard LB media was supplemented with Enfors minimal media: 53 mM $(NH_4)_2SO_4$, 37 mM Na_2HPO_4, 11.8 mM KH_2PO_4, 2.2 mM $(NH_4)_2$-H-Citrate, 54.3 mM glycerol, 0.02% $MgSO_4 \cdot 7H_2O$ (w,v), and 1 × trace salts (3.4 μM $CaCl_2 \cdot 2H_2O$, 0.75 μM $CoCl_2$, 0.64 μM $CuSO_4 \cdot 5H_2O$, 62 μM $FeCl_3 \cdot 6H_2O$, 0.67 μM $MnSO_4 \cdot H_2O$, 60 μM $Na_2EDTA \cdot 2H_2O$, 0.63 μM $ZnSO_4 \cdot 7H_2O$). The OD_{600} was monitored periodically and expression induced with 1 mM IPTG at $OD_{600} = 1.5$ at 22 °C and cells grown for 12 h. During induction, glycerol was fed into the bioreactor at a rate of 0.3 mL/min. The harvested cells were pelleted by centrifugation at 4000 rpm, weighed, and stored at −80 °C until further use.

3.2 Adaptation to minimal media and perdeuteration of AAT

Perdeuteration of proteins is discussed elsewhere. Here, we outline our experiences and preferences. The method of perdeuteration is very similar to isotopic replacement for NMR studies using LeMaster's media, a method adapted for selenomethionine incorporation for MAD phasing (Hendrickson, Horton, & Lemaster, 1990). For these studies, cells expressing wild-type AAT were stepwise adapted from LB/Enfors to Enfors only minimal media in H_2O using serial 1:100 inoculation and growth (50 mL, 100% LB/Enfors, 50% LB + 100%, then 100% Enfors) with expression levels determined at each step. Multiple inoculations at a single step were sometimes necessary. Alternatively, M9 salts with glycerol can be used. The minimal media adapted cells were then stepwise adapted (1:100 inoculum) to D_2O-minimal media (10 mL, 25%, 50%, 75%, 100% D_2O) with each step monitored for expression levels and multiple adaptations at a single step when necessary. Large scale expression was then completed (1 L Enfors:D_2O with H-glycerol, at the ORNL perdeuteration facility). If necessary, the D_2O minimal media can be enriched with deuterated supplements (amino acids, etc.) to reduce the biosynthetic requirements of the cells (Haertlein et al., 2016).

3.3 Purification of recombinant hydrogenated and perdeuterated AAT

Frozen cells containing either H-AAT or D-AAT were thawed in a lysis buffer (5 mL Buffer A per g cells; Buffer A: 50 mM $NaHPO_4$ pH 7.5, 250 mM NaCl, 10 mM imidazole), lysed by sonication, clarified by centrifugation (12,000 × *g*), and purified using Ni-NTA (Qiagen) (10 cv linear gradient, Buffer A to Buffer B (Buffer A plus 500 mM imidazole)). TEV protease (1% w/w) was added to the pooled fractions which were then dialyzed (12 h, 50 mM Tris-HCl pH 7.5, 0.1 mM EDTA, 1 mM DTT, 4 °C) and reapplied to the regenerated Ni-NTA column, collecting the tagless protein in the flow-through. AAT is buffer exchanged (Buffer C, 40 mM sodium acetate pH 5.4, 2 mM PLP), concentrated to 25 mg/mL and filtered (0.22 μm centrifugal filter).

3.4 Crystallization of recombinant H-AAT and D-AAT

In the original work by Metzler and Arnone, small plate-crystals of native AAT, purified from porcine heart muscle, were grown in hanging drop (20 mg/mL protein, 40 mM sodium acetate pH 5.4, 2 mM PLP, and 8% w/v polyethylene glycol 6000) (Arnone et al., 1977). These conditions can be used to coax growth of very large crystals by macroseeding in the same solution (5–10 mm^3, C.C. Hyde and P. Rogers, personal communication). Crystals of recombinant AAT were grown in identical conditions as described for the native protein. The new crystals have the same space group and cell dimensions but surprisingly, the external crystal morphology is different, large hexagonal rods instead of thin plates (Dajnowicz, Parks, et al., 2017). Fortuitously, the new crystal form does not require macroseeding and is likely the result of our introduction of surface mutations (the protein sequence derived from the PDB deposition has three Asp/Asn differences compared to the sequence derived from the gene). Two methods for growing large crystals were tested, large-volume sitting drop vapor diffusion (9-well siliconized glass depression plates in sandwich boxes, Hampton Research) and large-diameter capillary dialysis (unit gravity controls for the Toledo Crystallization Box, SpaceX CRS-15, CRS-18, International Space Station microgravity crystal growth, to be described elsewhere). Polydispersity of the final AAT solution (typically ~12%) was measured using by dynamic light scattering. Large-scale sitting drop vapor diffusion (25, 50, and 75 μL drops, 1:1 protein:well, 40 mM sodium acetate pH 5.4, 2 mM PLP, and 8% w/v polyethylene glycol 6000) was performed for both hydrogenated and perdeuterated protein. Crystals for neutron data collection were placed in fresh mother liquor to which the various substrates and inhibitors were added to trap different stages of catalysis. Thus far, several grams of hydrogenated AAT and approximately a hundred milligrams of perdeuterated AAT have been prepared. Crystal production was an overwhelming success but only two crystals so far have been coaxed to neutron diffraction quality, one perdeuterated AAT crystal grown using depression plates and one hydrogenated AAT crystal grown by capillary dialysis.

3.5 Crystal mounting and deuterium exchange

Crystals were prepared for room-temperature neutron diffraction studies by drift mounting in sealed quartz capillaries (2 mm dia., CV2024-Q, VitroCom, Mountain Lakes, NJ). The capillary was sealed at one end, secured to a mounting pin with wax, and placed in an upright position. Mother liquor was added to the capillary with a 10 mm air gap between the bottom liquid and the crystal mounting position then filled slightly beyond the top of the tube. Using a large, sturdy cryoloop, the crystal was transferred to the capillary and allowed to settle at the air gap interface. Using a syringe with a blunt tipped needle, the top solution was pulled off without disturbing the crystal and a paper wick was used to remove excess liquid from around the crystal. The capillary was then filled with mother liquor, leaving another 10 mm air gap above the crystal. To complete H/D exchange of the labile hydrogens, vapor diffusion was performed by mounting in capillaries with D_2O-based mother liquor. H/D exchange was performed for a minimum of 1 month prior to data collection.

4. Data collection, data processing, and structure refinement for AAT

Joint refinement of neutron and X-ray diffraction data is essentially a special case of isomorphous replacement, and the same care should be taken to ensure the data sets are isomorphous. Smaller, frozen crystals can be used for high-resolution X-ray structures but large, ambient temperature crystals should be used for X-ray data in joint refinement. Preferably, an X-ray data set is collected on the same crystal after neutron diffraction has been completed, using as low a dose as is compatible with the required data. The ambient temperature X-ray data is then used to position all heavy (i.e., non-hydrogen) atoms, using the high-resolution structure as a starting model.

The structural waters are reevaluated against the lower resolution ambient temperature data. Joint refinement then allows visualization of H/D positions using (Fo–Fc) difference maps. Compared to X-ray data, neutron data has a poor signal-to-noise ratio and is mostly incomplete, which may introduce Fourier artifacts; care must be taken to not overinterpret. In our limited experience, deuterium peaks best appear in difference maps at the exchangeable positions within hydrogenated proteins and best appear in the (2Fo–fc) model maps of perdeuterated proteins. The presence of hydrogen, with a negative scattering length, can cause local damping of scattering intensity which may result in noisier maps. This makes interpretation a bit more challenging but chemical restraints are employed to minimize the disparity. The most challenging part of joint refinement is the bookkeeping. In our AAT homodimer, components that have identical connectivity in X-ray refinement can be different with the neutron data, i.e., ligand protonation, variable protonation of histidines and lysines, etc. The water structure also becomes more challenging, the spherical electron density of the central oxygen seen by X-rays can be seen as a bent ellipsoid for D_2O with neutrons. Multiple cycles of water rebuilding dominate the early stages of joint refinement. An additional complexity is partial occupancy. Due to the very high cost of deuterated solvents, perdeuterated proteins are often purified and crystallized in hydrogenated solutions. The crystals are then vapor exchanged with D_2O which leads to partial exchange. Currently, the choice for joint refinement is between nCNS and Phenix (see chapter "Implementation of the riding hydrogen model in *CCTBX* to support the next generation of X-ray and neutron joint refinement in *Phenix*" by Liebschner et al.). To date, 68 neutron structures have been submitted to the Protein Data Bank. Approximately 52% of those structures have been refined in PHENIX and 34% refined with the nCNS, a patch created for CNS. Here, we have chosen to use nCNS to build on our personal experience and familiarity with XPlor/CNS built up over many years.

4.1 Neutron and X-ray data collection for joint refinement

Preliminary neutron diffraction experiments were carried out on beamline CG-4D (IMAGINE) located at the High Flux Isotope Reactor (Oak Ridge National Laboratory). A full quasi-Laue data set was collected on LADI-III beamline at the Institut Laue Langevin (ILL). A modified version of Daresbury Laboratory LAUE suite program, LAUEGEN, was used to process the data. LSCALE was used to scale and normalize the data based on the intensities of symmetry-equivalent reflections measured at different wavelengths. The data were merged using the CCP4 program, SCALA (Winn et al., 2011). The ambient X-ray data were collected on a Rigaku HighFlux HomeLab instrument fitted with a MicroMax-007 HF X-ray generator, Osmic VariMax optics, and a R-Axis IV++ image plate detector. The data were integrated and scaled using HKL3000. The refined X-ray structure was solved by molecular replacement using PDB: 1AJS (Rhee et al., 1997) as the search model and refined with SHELX-2016.

4.2 Reflection, coordinate, topology, and parameter files

The program nCNS uses topology (protein-allhdg_d_new.top) and parameter (protein-allhdg_d_new.param) files modified to include deuterium to define atomic geometry and nonbonded energy terms, respectively. Reflection files for the X-ray and neutron data were converted from mtz to ASCII XPLOR format using the ccp4 program, "Mtz2various (export)." R-free flags were set independently for both data sets using either CCP4 "uniqueify (import)" or nCNS "make_cv.inp." Chain ID A and B can be used in X-ray refinement to distinguish the two chains of the homodimer. However, duplicate residue numbers cause problems when writing patches to distinguish protonation states. The residues were therefore renumbered; Chain A residues 0001–0412 and Chain B residues 1001–1412. Dual conformers introduced during high-resolution X-ray refinement were removed for nCNS joint refinement. All waters were renamed from "HOH" to "DOD" and "HETATM" record types to "ATOM."

New topology and parameter files need to be created for the different forms of PLP. The ligand models were built using GaussView Version 6.1 (Semichem Inc., Shawnee Mission, KS) (comparable programs such as Maestro (Schrödinger, New York, NY) or Spartan (Wavefunction, Inc., Irvine, CA) will work as well). Hydrogens (or deuteriums) were manually added to the model as appropriate, with ionizable groups initially left unprotonated to be viewed in difference maps generated during joint-XN refinement. The ligands were optimized in GaussView SMD/B3LYP-D3BJ/Def2-SVP level of theory. During the geometry optimization, the backbone atoms and phosphate group of the internal aldimine were constrained to the atomic positions given by the X-ray structure. For the external aldimine, the carboxyl oxygens and the phosphate group were constrained. The optimized geometry from GaussView, in PDB file format, was input into XPLO2D and the AUTOdic option used to generate the cofactor ligand topology and parameter files. The exchangeable hydrogens in the output PDB were changed to deuteriums. In the topology file, the exchanged hydrogens were changed to deuterium and the masses of the remaining atoms were confirmed. All the dihedral angles were reviewed and constraints were placed/removed as necessary. The LLP internal aldimine is incorporated into the model using ligands.link located in the nCNS library.

4.3 Joint X-ray/neutron refinement of AAT using nCNS

The input file "generate_easy.inp" was used to create the molecular topology file (.mtf) and new PDB to be used in the joint refinement. Before using the input file, patches were added to account for the *cis*-prolines in the structure. The geometry of the output PDB was checked with "model_stats.inp," which lists geometry violations such as bond lengths, bond and dihedral angles. Any major geometry violations were examined in a molecular viewer such as Coot (Emsley, Lohkamp, Scott, & Cowtan, 2010). Occasionally, hydrogens are added with poor geometry. If unable to manually correct this in Coot, the hydrogens were deleted from the side chain and generate was run again. A rigid body refinement was carried out to align the two structures with input file "xn_rigid.inp." The rigid body refinement was only performed in the first cycle of refinements, unlike the other steps which are repeated several times. Next, a positional refinement was carried out using "xn_minimize.inp" to minimize the potential energy between the structures. Individual B-factors were refined using the input file "xn_bindividual.inp" and individual occupancies were refined using "xn_qindividual.inp." The rigid body, minimization, B-factor, and occupancy refinements used a joint maximum likelihood target using amplitudes.

The template, "model_maps.inp," was used to create input files to generate Fo–Fc and 2Fo–Fc maps. The maps were opened in Coot along with the coordinates from the individual occupancy refinement. The Fo–Fc maps are initially viewed at 3 σ and the 2Fo–Fc maps at 1 σ. Each residue in the structure was examined to reorient hydrogens/deuteriums on ionizable side chains to their correct positions. Generate was rerun and joint energy minimization, b-individual, and q-individual refinements were performed. New maps were generated and visualized in Coot. The nuclear density noticeably improved with the correct positioning of the ionizable side chains. Each water is examined and the deuteriums are rotated around the fixed oxygen into the observed difference density peak to produce correct hydrogen bonding. Patches using residue numbers are written to delete hydrogens/deuteriums on side chains that appear to be deprotonated according to the nuclear density. After the waters were repositioned and any necessary patches were added to the input file, generate, minimization, b-individual, and q-individual, map generation, and model rebuilding steps were repeated until satisfactory.

5. Using accurate protonation states for computational chemistry methods

Computational chemistry methods, including the well-known quantum mechanical (QM) and molecular mechanical (MM) hybrid method (QM/MM) (van der Kamp & Mulholland, 2013), have been used to study enzymatic function for several decades. The countless computational studies have provided deep understanding of active site interactions and even long-range forces that aid in the drug design, enzyme design, catalysis, and reaction mechanism (Gerlits et al., 2016). This section introduces how neutron crystallography can assist in structure preparation for computational methods and highlight the fundamental insights in PLP-dependent catalysis.

5.1 The use of neutron crystallography in computational chemistry

The intrinsically complex chemical environment within an enzyme have been optimized by nature to modulate the pK_a of key active site residues, enhancing catalysis and functionality of the enzyme. Specifically, the chemical environment can drastically shift the pK_a of ionizable groups, switching the protonation equilibria compared to a solvent environment. The extended interaction networks within the protein environment makes it a challenging problem to predict the protonation states of ionizable groups. Modeling the correct charge distribution for either QM or MM methods is critical to obtain physically meaningful results. Therefore, accurate determination of the protonation states is needed. Using neutron structures as input models for computational chemistry methods is one of most direct means for obtaining accurate protonation states. However, it is important to note that significant progress has been made for predicting pK_a values within proteins (Alexov et al., 2011; Bas, Rogers, & Jensen, 2008; Nielsen, Gunner, & Garcia-Moreno, 2011).

Before the neutron structure of AAT (PDB: 5VJZ) was obtained, investigation of an intricate hydrogen bond network, located at the edge of PLP opposite to the reactive Schiff base, provided evidence that the extended network of interactions influences the pK_a of PLP-N1 and ultimately the rate of catalysis (Dajnowicz, Parks, et al., 2017). To gain insight how this extended hydrogen bond network, which involves residues Asp 222, His 143, Thr 139, and His 189, four mutants were produced: single mutants D222T and H143L, and double mutants H143L:H189L and H143F: H189. Moreover, quantum chemical cluster calculations on internal aldimine models support the hypothesis that, in addition to Asp 222, the local environment produced by His 143, Thr 139 and His 189 increases the pK_a of the pyridinyl nitrogen, stabilizing the pyridinium form of PLP. Since experimental hydrogen positions were not yet obtained, and due to the complexity of the extended interaction network near PLP-N1, four permutations of the internal aldimine each with different His 143/His 189 protonation profiles were optimized to compare with high-resolution X-ray structure of AAT (PDB: 5TOQ). In-depth analysis of the hydrogen bond distances, RMSD values, and relative residue conformation of the four permutations indicated the His 143-neutral/His 189-neutral configuration was the best fit for later reaction enthalpy QM calculations. Afterward, the neutron structure of AAT showed that both His 143 and His 189 are neutral in the internal aldimine, validating our QM models, and showcasing the benefit of using neutron structure as input for computational chemistry methods.

5.2 Insight into PLP-dependent catalysis and chemical mechanism of aspartate aminotransferase

In AAT, one of the partially rate limiting steps during the transformation of aspartic acid to oxaloacetic is the Cα-H bond cleavage of the external aldimine. As stated above, the external aldimine is a pivotal point in the chemical mechanism of PLP-dependent enzymes, and electronic level detail of this key intermediate may provide an understanding of the catalytic versatility. Thus, obtaining accurate charge distribution of the active site is needed to study the physical determinates that drive the Dunathan stereo-electronic alignment, transition state stabilization, and/or ground state destabilization. In the case of the external aldimine, the neutron structured provided an additional level of detail which may have been overlooked otherwise, and would result in the incorrect charge distribution of the active site for our computational models. Specifically, the neutron structure shows that His 189 becomes positively charged in the external aldimine, and we hypothesize that this a response to counterbalance the active site.

Starting from the neutron structure as a guide, the MD simulation was conducted to analyze key interactions that assist in bond alignment, while quantum cluster calculations were performed to obtain accurate reaction and orbital energies of external aldimine Cα deprotonation (Dajnowicz et al., 2018). To model the native external aldimine state of AAT for classical MD simulations, the 1-deazaPLP X-ray structure (PDB: 3QPG; Griswold & Toney, 2011) was used to reposition Lys 258 toward the Cα and the Cα-methyl group was replaced with a hydrogen atom (Dajnowicz et al., 2018). Consequently, Lys 258 was shifted back to the active position. It is important to note that neutron structure showed a low barrier hydrogen bond (LBHB) between N_{SB} and the C-terminal carboxylate group of the external aldimine, which was not considered due the potential need to accurately describe nuclear quantum effects. We are continuing to work toward obtaining accurate models of the LBHB in AAT and to gain insight into its role in catalysis and possible nuclear quantum effect. For the QM model of the external aldimine, a quantum mechanical cluster approach (Siegbahn & Himo, 2011) was used to estimate the activation energy of the Cα proton transfer to Lys 258 in AAT. The QM model was generated from an MD snapshot near the end of the 180-ns simulation (Dajnowicz et al., 2018). Specifically, the quantum cluster model included the external aldimine, the side chains of 12 active site residues (Tyr 70, Asp 222, His 143, His 189, Asn 194, Tyr 225, Ser 255, Ser 257, Lys 258, Arg 266, Arg 292, Arg 386), and 1 crystallographic water molecule. Natural bond orbital (NBO) analysis and second-order perturbative orbital interaction energy analysis was performed to gain sight into stabilization energy of the external aldimine through a intramolecular charge-transfer interaction (Fig. 4) (Glendening et al., 2019).

In the external aldimine of AAT, the closed state of the enzyme, Arg 386 and Arg 292 anchor the substrate through salt bridges, orientating and holding the Cα-H bond perpendicular to the π-conjugated system of PLP. The results from our MD simulation show that the Cα-H bond is nearly perpendicular throughout the 180-ns simulation and samples a narrow distribution. This pre-organization and alignment of the external aldimine of AAT is driven by the closing of the small domain, orientating Arg 386 in a position to form a salt bridge with the carboxylate group of the substrate. This stereoelectronic Dunathan alignment has been observed in many crystal structures of PLP-dependent enzymes (Toney, 2011). However, no detailed physical interpretation has been offered to connect the bond alignment to the catalytic power of PLP-dependent enzymes.

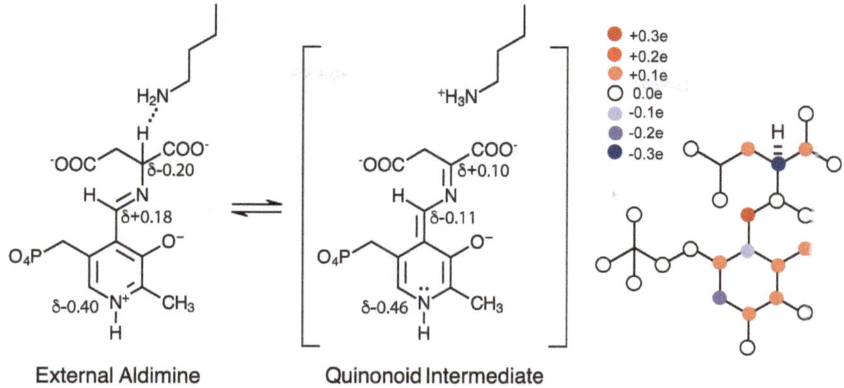

Fig. 4 With all atoms positioned properly, including hydrogens, quantum cluster analysis can be reliably used to analyze the molecular orbitals associated with a chemical transformation. For the transamination reaction, heterolytic cleavage of the Cα–H was calculated and the QM models of the reactant (external aldimine, left) and the product (quinonoid intermediate, middle) were analyzed using density function theory with natural bond order (NBO) analysis (Glendening, Landis, & Weinhold, 2019). Instead of formal Lewis charges depicted in resonance structures, we can display the differences in partial natural charge (right). The redistribution of the "carbanion" has the carbon nearest neighbors in the substrate and the PLP C2, C3, C5 and O3′ all gaining 0.1 electrons. The Cα loses 0.3 electrons while the C4′ gains 0.3 electrons, which correlates to the 1,3-proton shift. The N1 pyridine nitrogen gains only 0.06 electrons but is depicted as changing from +1 to 0 in the resonance drawings.

Our QM calculations demonstrate that the $\sigma \rightarrow \pi^*$ hyperconjugation interaction has a significant stabilizing effect on the reactants, providing evidence that the ground state destabilization mechanism is not invoked in the stereoelectronic Dunathan alignment. However, we calculated that the presence of the $\sigma \rightarrow \pi^*$ hyperconjugation stabilizes the transition state ~4.5 times more compared to the reactants by delocalizing the electrons from the Cα-H sigma orbital to the π^* of the Schiff base. Our NBO analysis on external aldimine models of AAT provides evidence that hyperconjugation enhances PLP-dependent catalysis through transition state stabilization and is manifested in several types of PLP-dependent enzymes where the Dunathan alignment occurs. After the Cα-H proton is transferred to Lys 258, the covalently linked PLP-substrate forms the planar quinonoid intermediate. The calculated enthalpy for the proton transfer reaction of the external aldimine is $-0.1\,\text{kcal}\,\text{mol}^{-1}$. Therefore, the quinonoid intermediate is isoenergetic with respect to the external aldimine, suggesting that the quinonoid intermediate is stable. This agrees with its detection by UV–vis spectroscopy (Toney, 2014).

The protonated pyridine N1 in AAT enhances charge delocalization and the natural population charge analysis on our QM cluster models this hypothesis. The partial charges on the Cα and C4′ atoms are −0.19 and 0.21 in the external aldimine, respectively, and 0.09 and −0.18 in the quinonoid intermediate, respectively. Thus, the electrons on Cα are delocalized throughout the PLP cofactor, suggesting that the carbanion intermediate is not stable in AAT. In addition, the computed charge distribution of the quinonoid intermediate can help explain the following reaction mechanism. Since C4′ has a partial negative charge in the quinonoid intermediate, it can act as a nucleophile, assisting in the 1,3-proton transfer to form the ketamine intermediate.

In contrast, the previous natural population charge analysis of tryptophan synthase, a PLP-dependent enzyme conducting β-elimination reactions, shows the opposite charge distribution compared to AAT, which disfavors the 1,3-proton transfer seen in the transamination mechanism (Caulkins et al., 2016). In addition, the fleeting carbanionic intermediate in TS has a partial negative charge on the Cα carbon, in contrast to the quinonoid intermediate for AAT. Thus, the protonated N1 in PLP-dependent enzymes assist in charge redistribution and, in AAT, stabilizes the quinonoid resonance structure and promotes the 1,3-proton transfer. However, in TS, N1 is deprotonated, disfavoring quinonoid resonance structure, localizing the negative charge on the Cα carbon, promoting Cα-Cβ bond breakage, and driving the substrates toward the β-elimination reaction path. In summary, PLP-dependent enzymes exploit electronic properties of the vitamin B_6-derived cofactor to promote chemistry by orbital alignment, caging the substrate into a reactive conformation, and prompting various reaction path through selective protonation of the PLP cofactor.

References

Alexov, E., Mehler, E. L., Baker, N., Baptista, A. M., Huang, Y., Milletti, F., et al. (2011). Progress in the prediction of pK(a) values in proteins. *Proteins*, *79*, 3260–3275.

Arnone, A., Briley, P. D., Rogers, P. H., Hyde, C. C., Metzler, C. M., & Metzler, D. E. (1981). Crystallographic studies showing domain movement in cytoplasmic aspartate-aminotransferase. *Federation Proceedings*, *40*, 1757.

Arnone, A., Rogers, P. H., Schmidt, J., Han, C. N., Harris, C. M., & Metzler, D. E. (1977). Preliminary crystallographic study of aspartate: 2-oxoglutarate aminotransferase from pig heart. *Journal of Molecular Biology*, *112*, 509–513.

Bas, D. C., Rogers, D. M., & Jensen, J. H. (2008). Very fast prediction and rationalization of pK(a) values for protein-ligand complexes. *Proteins*, *73*, 765–783.

Caulkins, B. G., Young, R. P., Kudla, R. A., Yang, C., Bittbauer, T. J., Bastin, B., et al. (2016). NMR crystallography of a carbanionic intermediate in tryptophan synthase: Chemical structure, tautomerization, and reaction specificity. *Journal of the American Chemical Society*, *138*, 15214–15226.

Chan-Huot, M., Dos, A., Zander, R., Sharif, S., Tolstoy, P. M., Compton, S., et al. (2013). NMR studies of protonation and hydrogen bond states of internal Aldimines of pyridoxal 5'-phosphate Acid-Base in alanine racemase, aspartate aminotransferase, and poly-L-lysine. *Journal of the American Chemical Society, 135*, 18160–18175.

Clark, P. A., Jansonius, J. N., & Mehler, E. L. (1993a). Semiempirical calculations of the electronic absorption-spectrum of mitochondrial aspartate-aminotransferase 2. The 2-methylaspartate complex. *Journal of the American Chemical Society, 115*, 9789–9793.

Clark, P. A., Jansonius, J. N., & Mehler, E. L. (1993b). Semiempirical calculations of the electronic absorption-spectra of vitamin-B6 derivatives in the active-site of mitochondrial aspartate-aminotransferase. *Journal of the American Chemical Society, 115*, 1894–1902.

Dajnowicz, S., Blakeley, M., Keen, D., Kovalevsky, A., & Mueser, T. (2017). Direct observation of protonation states in a PLP-dependent enzyme by neutron crystallography. *Acta Crystallographica Section A, 73*, A26.

Dajnowicz, S., Johnston, R. C., Parks, J. M., Blakeley, M. P., Keen, D. A., Weiss, K. L., et al. (2017). Direct visualization of critical hydrogen atoms in a pyridoxal 5'-phosphate enzyme. *Nature Communications, 8*.

Dajnowicz, S., Parks, J. M., Hu, X. C., Gesler, K., Kovalevsky, A. Y., & Mueser, T. C. (2017). Direct evidence that an extended hydrogen-bonding network influences activation of pyridoxal 5-phosphate in aspartate aminotransferase. *The Journal of Biological Chemistry, 292*, 5970–5980.

Dajnowicz, S., Parks, J. M., Hu, X. C., Johnston, R. C., Kovalevsky, A. Y., & Mueser, T. C. (2018). Hyperconjugation promotes catalysis in a pyridoxal 5'-phosphate-dependent enzyme. *ACS Catalysis, 8*, 6733–6737.

Dixon, M., & Webb, E. C. (1979). *Enzymes* (3rd ed.). Academic Press.

Dunathan, H. C. (1966). Conformation and reaction specificity in pyridoxal phosphate enzymes. *Proceedings of the National Academy of Sciences of the United States of America, 55*, 712–716.

Emsley, P., Lohkamp, B., Scott, W. G., & Cowtan, K. (2010). Features and development of Coot. *Acta Crystallographica Section D-Biological Crystallography, 66*, 486–501.

Gerlits, O., Wymore, T., Das, A., Shen, C. H., Parks, J. M., Smith, J. C., et al. (2016). Long-range electrostatics-induced two-proton transfer captured by neutron crystallography in an enzyme catalytic site. *Angewandte Chemie (International ed. in English), 55*, 4924–4927.

Glendening, E. D., Landis, C. R., & Weinhold, F. (2019). NBO 7.0: New vistas in localized and delocalized chemical bonding theory. *Journal of Computational Chemistry, 40*, 2234–2241.

Griswold, W. R., & Toney, M. D. (2011). Role of the pyridine nitrogen in pyridoxal 5'-phosphate catalysis: Activity of three classes of PLP enzymes reconstituted with deazapyridoxal 5'-phosphate. *Journal of the American Chemical Society, 133*, 14823–14830.

Haertlein, M., Moulin, M., Devos, J. M., Laux, V., Dunne, O., & Forsyth, V. T. (2016). Biomolecular Deuteration for neutron structural biology and dynamics. *Methods in Enzymology, 566*, 113–157.

Hayashi, H., & Kagamiyama, H. (1995). Reaction of aspartate-aminotransferase with L-Erythro-3-Hydroxyaspartate—Involvement of Tyr70 in stabilization of the catalytic intermediates. *Biochemistry, 34*, 9413–9423.

Hayashi, H., & Kagamiyama, H. (1997). Transient-state kinetics of the reaction of aspartate aminotransferase with aspartate at low pH reveals dual routes in the enzyme-substrate association process. *Biochemistry, 36*, 13558–13569.

Hayashi, H., Mizuguchi, H., Miyahara, I., Islam, M. M., Ikushiro, H., Nakajima, Y., et al. (2003). Strain and catalysis in aspartate aminotransferase. *Biochimica et Biophysica Acta, 1647*, 103–109.

Hendrickson, W. A., Horton, J. R., & Lemaster, D. M. (1990). Selenomethionyl proteins produced for analysis by multiwavelength anomalous diffraction (Mad)—A vehicle for direct determination of 3-dimensional structure. *The EMBO Journal*, *9*, 1665–1672.

Jansonius, J. N. (1998). Structure, evolution and action of vitamin B-6-dependent enzymes. *Current Opinion in Structural Biology*, *8*, 759–769.

Metzler, C. M., Cahill, A., & Metzler, D. E. (1980). Equilibria and absorption-spectra of Schiff-bases. *Journal of the American Chemical Society*, *102*, 6075–6082.

Metzler, C. M., Metzler, D. E., Martin, D. S., Newman, R., Arnone, A., & Rogers, P. (1978). Crystalline enzyme-substrate complexes of aspartate-aminotransferase. *The Journal of Biological Chemistry*, *253*, 5251–5254.

Mueller, L. J. (2017). NMR crystallography in the enzyme active site of tryptophan synthase. *Acta Crystallographica Section A*, *73*, A57.

Nielsen, J. E., Gunner, M. R., & Garcia-Moreno, E. B. (2011). The pK(a) cooperative: A collaborative effort to advance structure-based calculations of pK(a) values and electrostatic effects in proteins. *Proteins*, *79*, 3249–3259.

Ortega-Castro, J., Adrover, M., Frau, J., Salva, A., Donoso, J., & Munoz, F. (2010). DFT studies on Schiff base formation of vitamin B6 analogues. Reaction between a pyridoxamine-analogue and carbonyl compounds. *The Journal of Physical Chemistry A*, *114*, 4634–4640.

Percudani, R., & Peracchi, A. (2003). A genomic overview of pyridoxal-phosphate-dependent enzymes. *EMBO Reports*, *4*, 850–854.

Rhee, S., Parris, K. D., Ahmed, S. A., Miles, E. W., & Davies, D. R. (1996). Exchange of K+ or Cs+ for Na+ induces local and long-range changes in the three-dimensional structure of the tryptophan synthase alpha(2)beta(2) complex. *Biochemistry*, *35*, 4211–4221.

Rhee, S., Silva, M. M., Hyde, C. C., Rogers, P. H., Metzler, C. M., Metzler, D. E., et al. (1997). Refinement and comparisons of the crystal structures of pig cytosolic aspartate aminotransferase and its complex with 2-methylaspartate. *The Journal of Biological Chemistry*, *272*, 17293–17302.

Sharif, S., Huot, M. C., Tolstoy, P. M., Toney, M. D., Jonsson, K. H. M., & Limbach, H. H. (2007). N-15 nuclear magnetic resonance studies of acid-base properties of pyridoxal-5'-phosphate aldimines in aqueous solution. *The Journal of Physical Chemistry B*, *111*, 3869–3876.

Shaw, J. P., Petsko, G. A., & Ringe, D. (1997). Determination of the structure of alanine racemase from Bacillus stearothermophilus at 1.9-angstrom resolution. *Biochemistry*, *36*, 1329–1342.

Siegbahn, P. E. M., & Himo, F. (2011). The quantum chemical cluster approach for modeling enzyme reactions. *WIREs Computational Molecular Science*, *1*, 323–336.

Smith, D. L., Almo, S. C., Toney, M. D., & Ringe, D. (1989). 2.8 ÅA-resolution crystal-structure of an active-site mutant of aspartate-aminotransferase from Escherichia-coli. *Biochemistry*, *28*, 8161–8167.

Sprang, S. R., Withers, S. G., Goldsmith, E. J., Fletterick, R. J., & Madsen, N. B. (1991). Structural basis for the activation of glycogen phosphorylase-B by adenosine-monophosphate. *Science*, *254*, 1367–1371.

Sugio, S., Petsko, G. A., Manning, J. M., Soda, K., & Ringe, D. (1995). Crystal-structure of a D-amino-acid aminotransferase—How the protein controls stereoselectivity. *Biochemistry*, *34*, 9661–9669.

Toney, M. D. (2011). Controlling reaction specificity in pyridoxal phosphate enzymes. *Biochimica et biophysica acta-Proteins and Proteomics*, *1814*, 1407–1418.

Toney, M. D. (2014). Aspartate aminotransferase: An old dog teaches new tricks. *Archives of Biochemistry and Biophysics*, *544*, 119–127.

Toney, M. D. (2019). Carbon acidity in enzyme active sites. *Frontiers in Bioengineering and Biotechnology*, 7.

van der Kamp, M. W., & Mulholland, A. J. (2013). Combined quantum mechanics/molecular mechanics (QM/MM) methods in computational enzymology. *Biochemistry*, *52*, 2708–2728.

Winn, M. D., Ballard, C. C., Cowtan, K. D., Dodson, E. J., Emsley, P., Evans, P. R., et al. (2011). Overview of the CCP4 suite and current developments. *Acta crystallographica Section D, Biological Crystallography*, *67*, 235–242.

CHAPTER SIXTEEN

The role of hydrogen atoms in redox catalysis by the flavoenzyme cholesterol oxidase

Tatiana Kerber, Alice Vrielink*
School of Molecular Sciences, University of Western Australia, Perth, WA, Australia
*Corresponding author: e-mail address: alice.vrielink@uwa.edu.au

Contents

Abstract

Flavoenzymes comprise a large class of proteins that carry out a diverse range of important redox chemistry. Although X-ray crystal structures of many flavoenzymes have been determined, there are still unresolved questions regarding the actual oxidation state of the flavin cofactors in these structures due to photoreduction by the ionizing radiation of the X-ray beam during the diffraction experiment. Additionally, the ability to visualize hydrogen atoms in X-ray structures is difficult due to the weak scattering capability of these atoms. Since hydrogen atoms affect the electrostatic nature of enzyme active sites and play important roles in the chemistry of key amino acid residues, visualizing the precise positions of these atoms provides a more detailed understanding of their role in enzyme catalysis. Single crystal neutron diffraction is an alternative method to structure determination, circumventing problems associated with photoreduction of the sample thus providing a clearer view of the structural features of a flavoenzyme in different redox states. Additionally, the larger neutron scattering factors for hydrogen and deuterium atoms enables one to visualize these atoms much more easily than from X-ray scattering measurements. In this chapter we give an overview of neutron and X-ray crystallography studies on the flavoenzyme, cholesterol oxidase and how the observations of unusual hydrogen atom positions provide insight into the redox chemistry of the flavin cofactor.

Methods in Enzymology, Volume 634
ISSN 0076-6879
https://doi.org/10.1016/bs.mie.2019.12.004

1. Introduction

Flavoenzymes are proteins that contain a flavin cofactor, either flavin adenine dinucleotide (FAD) or flavin mononucleotide (FMN). These enzymes are important to many biological processes such as energy production, biodegradation, chromatin remodeling, protein folding and neural development (Joosten & vanBerkel, 2007). Since their discovery a vast number of flavoenzymes have been isolated from different organisms ranging from microbes to mammals. The flavin cofactor forms the reactive center for the redox chemistry and can be used to catalyze a wide variety of chemical and biological reactions (Massey, 2000; Miura, 2001).

The versatility of flavoenzymes derives from the presence of the FAD or FMN cofactor bound to the enzyme which can accept and donate electrons through the isoalloxazine ring moiety (Hefti, Vervoort, & van Berkel, 2003). The cofactors are able to exist in three different oxidation states: an oxidized quinone, a semiquinone and a hydroquinone. The semiquinone and hydroquinone forms can additionally assume protonated or deprotonated states (Fig. 1). Numerous studies on the spectroscopic and electrochemical properties of flavins have been carried out over the years to understand the unique chemistry and reaction mechanisms of different flavoenzymes (Batie & Kamin, 1981; Fraaije & Mattevi, 2000; Ghisla & Massey, 1989; Malinowski, Barber, Whitaker, & Smith, 2007; Masuda & Bauer, 2002; Nakaminami, Kuwabata, & Yoneyama, 1997; Senda, Senda, Kimura, & Ishida, 2009). The different redox states of the flavin cofactors have characteristic absorption profiles near 450 nm (Fig. 2); in the oxidized state flavin containing proteins display a signature yellow color, have an absorption maximum at approximately 450 nm and, under physiological conditions, the isoalloxazine moiety is present in a neutral form. A one electron reduction step (to generate a semiquinone), results in a blue (neutral protonated) or red (anionic) colored sample, depending on the protonation state of the isoalloxazine moiety and generates absorption maxima at 350 and 490 nm. Upon full reduction (2 electron) to the hydroquinone, the flavin containing protein adopts a nearly colorless form with an absorption maximum at 350 nm. The hydroquinone can also exist in both neutral and anionic states under different conditions.

There are thousands of entries for FMN and FAD-containing proteins in the Protein Data Bank (PDB) for which X-ray crystal structures have been solved. Although X-ray diffraction has helped to reveal many structural

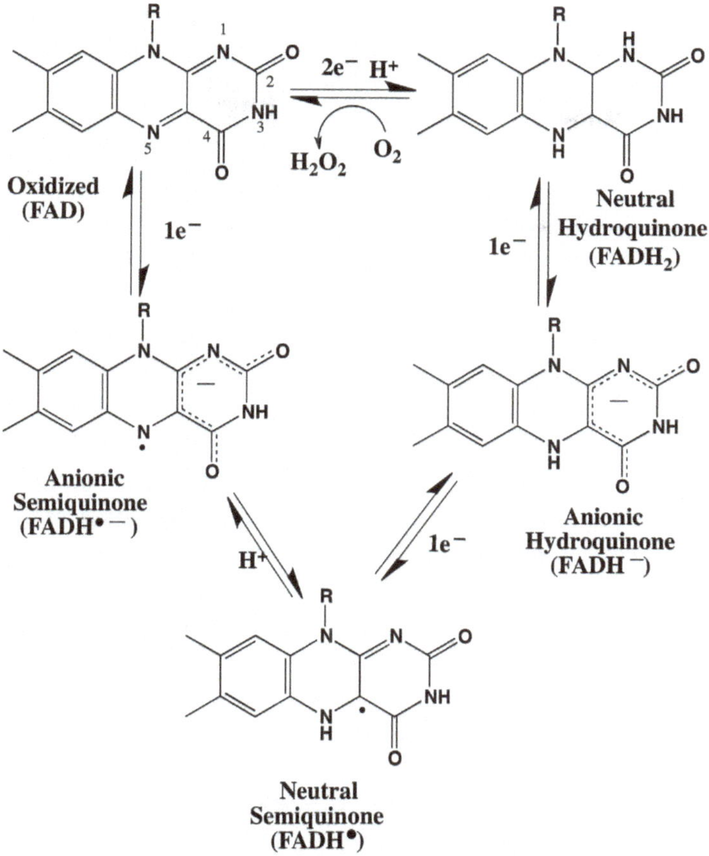

Fig. 1 The isoalloxazine moiety of the flavin cofactor in different redox states of flavins.

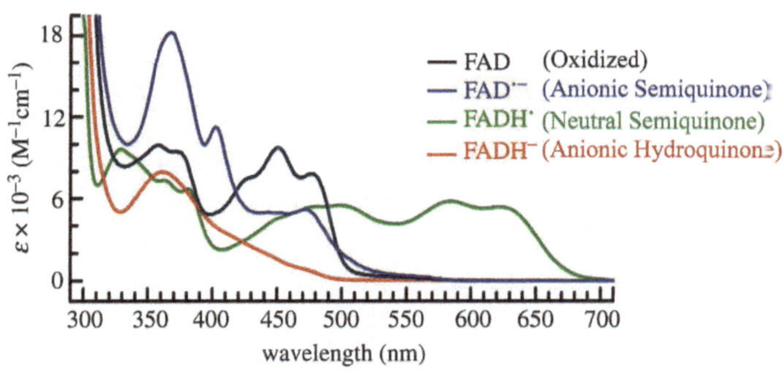

Fig. 2 Absorption spectra of FAD in different redox states. *Adapted from Evans, E. W., Dodson, C. A., Maeda, K., Biskup, T., Wedge, C. J., & Timmel, C. R. (2013). Magnetic field effects in flavoproteins and related systems.* Interface Focus, *3(5), 20130037. doi:10.1098/rsfs. 2013.0037.*

features of flavoenzymes providing insights into enzyme mechanism as well as atomic details regarding redox-dependent conformational changes (Bourgeois & Royant, 2005; Fritchie & Johnston, 1975; Guo et al., 2012; Kuo, Dunn, & Fritchie, 1974; Ledwidge et al., 2005; Lennon, Williams, & Ludwig, 1999; Li et al., 2011; Li, Vrielink, Brick, & Blow, 1993; Trus, Wells, Johnstone, Fritchie, & Marsh, 1971; Winkler et al., 2008), the technique has a number of limitations. First, the ionizing radiation of the X-ray beam can result in significant radiation damage to biological macromolecules. This is particularly serious at synchrotron facilities where the intensity of the beam enhances the potential for significant radiation damage due to the high radiation dosage on the crystal sample (Blakeley, Hasnain, & Antonyuk, 2015; Carugo & Djinovic Carugo, 2005; Garman, 2010; Ravelli, Leiros, Pan, Caffrey, & McSweeney, 2003). Additionally, the X-ray beam can cause radiation induced photo reduction of the cofactor present in the protein (Berglund et al., 2002; Rohr, Hersleth, & Andersson, 2010). There are a number of examples where attempts to elucidate different redox states of an enzyme have been hampered by the synchrotron beam induced photo reduction producing either fully reduced cofactors or mixed redox states of the enzyme (Berglund et al., 2002; Berkholz, Faber, Savvides, & Karplus, 2008; Kort, Komori, Adachi, Miki, & Eker, 2004). Thus, radiation from the X-ray beam itself can influence the flavin redox state and consequently, the flavin structure as well as the protein structure surrounding the flavin, and this can confound interpretations regarding the catalytic mechanisms for redox chemistry. This issue raises questions concerning the actual oxidation states of flavin cofactors in reported crystal structures of flavoenzymes deposited in the PDB (Kwon, Smith, Raven, & Moody, 2017).

2. Cholesterol oxidase

Cholesterol oxidase (ChOx, EC 1.1.3.6) is a bacterial flavoenzyme that catalyzes the oxidation and isomerization of cholesterol to cholest-4-en-3-one (Fig. 3). It contains an FAD cofactor, which plays a key role in the structural features of the protein and the enzyme activity. Two forms of ChOx have been characterized, one found where the FAD cofactor is covalently linked to the enzyme, and another where the cofactor is non-covalently bound to the enzyme (see Vrielink & Ghisla, 2009 for a review). The crystal structures of both forms of the enzyme have been determined (Coulombe, Yue, Ghisla, & Vrielink, 2001; Yue, Kass, Sampson, & Vrielink, 1999) and kinetic

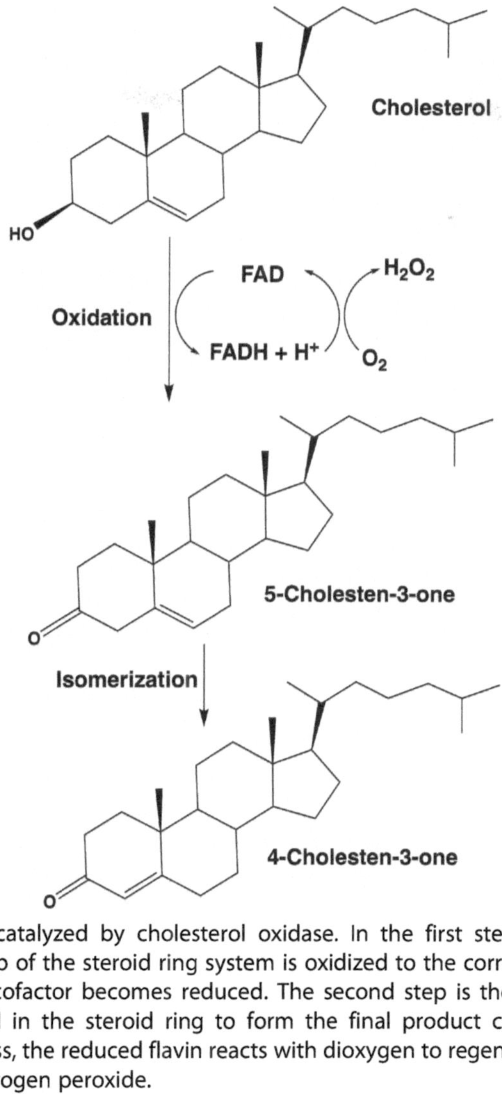

Fig. 3 Reaction catalyzed by cholesterol oxidase. In the first step (oxidation), the 3β-hydroxyl group of the steroid ring system is oxidized to the corresponding ketone while the flavin cofactor becomes reduced. The second step is the isomerization of the double bond in the steroid ring to form the final product cholest-4-en-3-one. During the process, the reduced flavin reacts with dioxygen to regenerate the oxidized enzyme and hydrogen peroxide.

and redox studies carried out (Gadda et al., 1997; Kass & Sampson, 1995, 1998a, 1998b; Lyubimov, Chen, Brammer, Sampson, & Vrielink, 2009; Lyubimov, Heard, Tang, Sampson, & Vrielink, 2007; Piubelli et al., 2008; Pollegioni, Wels, Pilone, & Ghisla, 1999; Yin, Liu, Anderson, & Sampson, 2002; Yin, Sampson, Vrielink, & Lario, 2001; Yue et al., 1999). These studies have shown that both the active site topology and thus the catalytic chemistry and the redox potentials are significantly different for each form of the enzyme.

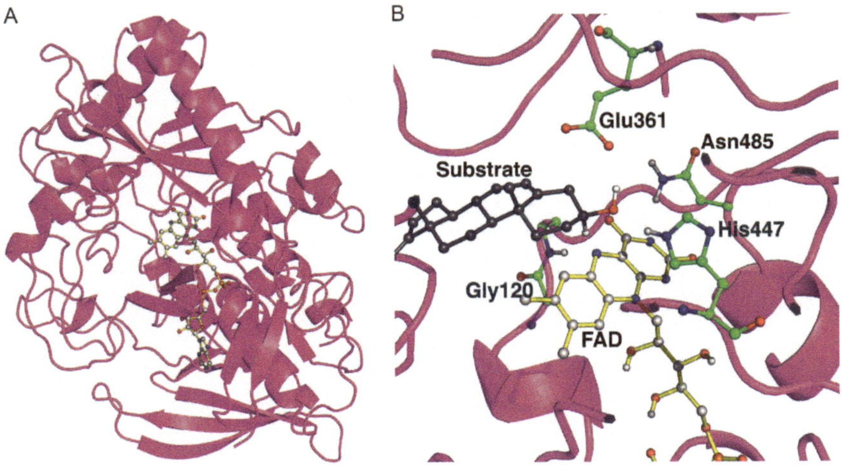

Fig. 4 The structure of ChOx containing the FAD cofactor non-covalently bound to the enzyme. (A) Secondary structure representation of the enzyme (PDB file: 1b4v) (B) Active site of the enzyme and its key residues. The structure includes a model for the cholesterol substrate shown in gray bonds. Amino acids implicated as important for catalysis are shown in ball and stick representation. The FAD cofactor is shown in with yellow bonds and the amino acids important for catalysis are shown in green bonds.

In the case of the non-covalently bound form of the enzyme X-ray crystal structures have been determined to sub-Ångstrom resolution and at different pH values (Lario, Sampson, & Vrielink, 2003; Lyubimov, Lario, Moustafa, & Vrielink, 2006). This work has provided unprecedented views of the enzyme and given enticing information on the roles of specific residues in the catalytic mechanism of flavin mediated redox catalysis. The three-dimensional structure and the active site of the non-covalently bound form of ChOx are presented in Fig. 4. However, despite the atomic resolution structures obtained from X-ray diffraction studies, there are still unresolved questions regarding the contribution of hydrogen atoms in the flavin-assisted redox catalysis and the structural differences between different redox states of the enzyme. These studies emphasize the importance of visualizing the precise positions of hydrogen atoms in different redox states of the enzyme.

3. Neutron crystallography studies of cholesterol oxidase

Although X-ray crystallography has provided an unprecedented wealth of information about protein structure the inherent inability to visualize the positions of hydrogen atoms in medium resolution structures results in an incomplete view of the structure needed for a full understanding of the unique chemistry undertaken by these macromolecules. Even when using X-ray diffraction data to atomic resolution (>1.2 Å), hydrogen atoms are difficult to see and often one can only estimate their positions. Furthermore, due to the lack of electrons, X-ray crystallography is unable to provide visualization of protons. Hydrogen atoms play a critical role in the structure and function of macromolecules, and particularly enzymes. They affect the electrostatic nature of enzyme active sites and can alter the chemistry of key amino acid residues. Active site hydrogen atoms and those that are more mobile due to movements during the catalytic reaction tend to be difficult to visualize at convincing levels in the electron density maps.

As discussed in the chapter "Fundamentals of neutron crystallography in structural biology" by Helliwell, and elsewhere in this volume, the advantage of neutron crystallography over X-ray crystallography is the ability to visualize hydrogen atoms as a result of the strong scattering power of hydrogen by neutrons thus allowing easier detection of these small atoms in nuclear density maps even at medium resolution (~2.0–2.5 Å). This signal is further amplified, resulting in strong positive peaks in nuclear density maps, when deuterium is used in place of hydrogen. The ability to visualize hydrogen atoms within enzyme structures has been elegantly shown for a number of neutron structures including aldose reductase, elastase, carbonic anhydrase, xylose isomerase, HIV protease, and β lactamase (Adachi et al., 2009; Blakeley et al., 2008; Fisher, Aggarwal, Kovalevsky, Silverman, & McKenna, 2012; Hazemann et al., 2005; Katz et al., 2006; Kurihara et al., 2004; Meilleur, Snell, van der Woerd, Judge, & Myles, 2006) and the results from these studies have enhanced an understanding of their catalytic mechanisms. As enzyme chemistry often involves rearrangements of hydrogen atoms between amino acid side chains of the protein, cofactors and bound ligands during the catalytic cycle, a detailed knowledge of the exact positions of these atoms is critical to fully understand their role in enzyme mechanism. In the context of redox enzymes such as ChOx, during the process of substrate oxidation and flavin reduction, the enzyme abstracts a hydroxyl proton and transfers a hydride from the substrate to the cofactor. A complete understanding of the redox mechanism of ChOx would require a detailed view of the location of the hydrogen atoms in different redox states. This would enable identification of structural features such as low barrier hydrogen bonds.

In addition to better visualization of hydrogen atoms, neutron crystallography is an ideal method for studying redox enzymes, including flavoenzymes, as the potential for X-ray induced reduction chemistry during the data collection is eliminated thus giving a clearer interpretation of the structural results for different redox states of the protein. Atomic resolution X-ray structures of ChOx have suggested that specific hydrogen atoms near the flavin ring system adopt unusual positions (Lyubimov et al., 2006). For example, electron density maps suggest that a Cβ hydrogen of Asn119 located near to N5 of the isoalloxazine ring system adopts different positions as a function of pH (Fig. 5). Atomic resolution structures of the enzyme at different pHs have been proposed to represent different states of the FAD cofactor. A high pH structure (pH 9.0) corresponds to a negatively charged FAD anion mimicking the anionic reduced form of the cofactor. Furthermore, crystals exposed to X-rays at high pH are colorless, suggesting a reduced cofactor state, while crystals at low pH remain yellow in color indicative of an oxidized flavin cofactor (Fig. 5). These observations regarding crystal color correlated with redox states of the cofactor as well as density features around suggest that stabilization of the charged cofactor state may be conferred by an aliphatic–π interaction with a side chain Cβ hydrogen atom of Asn119 and the FAD. Testing of this hypothesis would be facilitated by neutron diffraction studies of the enzyme in discrete well determined redox states, enabling robust visualization of hydrogen/deuterium atoms without confounding problems associated with photo-reduction. An understanding of the role of hydrogen atoms in redox stabilization for ChOx would also give a more generalized understanding of flavin assisted redox chemistry.

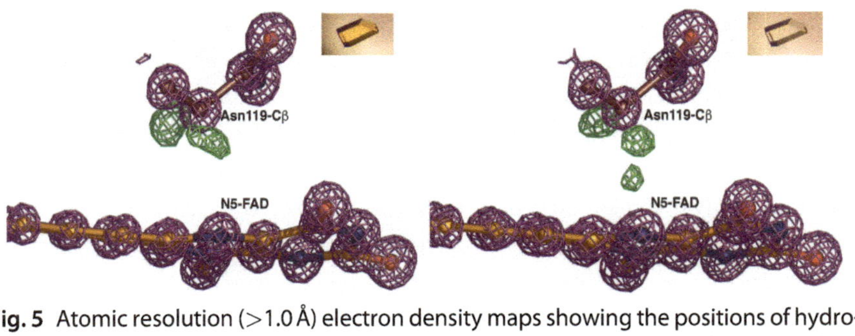

Fig. 5 Atomic resolution (>1.0 Å) electron density maps showing the positions of hydrogen atoms for the oxidized form of cholesterol oxidase (left) obtained from crystals at low pH (4.0) and a high pH (9.0) structure, which shows the flavin in an anionic, reduced form (right) (Lyubimov et al., 2006). Magenta corresponds to a 2Fo − Fc electron density map (contoured at 4σ) and green corresponds to the $F_o - F_c$ density (contoured at 2σ). Insets show the crystal color for the oxidized and reduced forms.

A limitation of neutron crystallography in the past was the need to grow large single crystals volumes (>1 mm^3) and the long data collection times due to the low flux of the neutron beam. However, significant advancements have been made in data collection methods, instrumentation and detectors (Gamble, Clauser, & Kossiakoff, 1994; Golden & Vrielink, 2014; Shu, Ramakrishnan, & Schoenborn, 2000) have helped to overcome issues with long data collection times. Additionally, the ability to obtain fully perdeuterated protein samples by recombinant methods (Golden, Attwood, Duff, Meilleur, & Vrielink, 2015; Liu, Hanson, Langan, & Viola, 2007; Meilleur, Dauvergne, Schlichting, & Myles, 2005), have alleviated the requirements for very large crystals. Typically, crystals of perdeuterated protein in the order of 0.3–0.4 mm^3 provide enough signal to obtain precise positions of hydrogen atoms in macromolecular structure not visible by X-ray diffraction. All these improvements have made neutron crystallography a more feasible technique providing shorter exposure times and requiring smaller crystal volumes. These advances enabled the first neutron study of ChOx to be undertaken to establish the structure for the fully oxidized state of the enzyme without any radiation induced photoreduction. Crystals of the protiated enzyme were grown through an iterative macroseeding method to achieve a size suitable size for neutron diffraction studies (Golden et al., 2015). Once a suitable size was reached from protiated crystallization buffers the crystals were soaked in deuterated buffers in order to replace exchangeable hydrogen atoms with deuterium atoms (Golden et al., 2017). Neutron data from a single crystal (0.4 mm^3) using the quasi Laue method were collected to 2.2 Å resolution followed by an X-ray dataset to 1.5 Å resolution on the same crystal. The data were subjected to joint neutron/X-ray refinement. The density maps showed the positions of deuterium atoms, particularly in well-ordered regions of the structure. Deuterium atoms were also observed for many water molecules which enabled establishment of their orientations. Most interestingly, the neutron residual difference density maps ($F_o - F_c$) revealed difference density near to the main chain nitrogen atom of the peptide bond linking Asn119 and Gly120 and N5 of the FAD cofactor (Fig. 6). This peak was presumed to be a deuterium atom, positioned exactly midway between the nitrogen atom of the peptide bond and N5 of FAD. The distance from the peptide bond nitrogen atom was considerably longer than an expected N–H bond length and longer than N–H bonds in other well-defined regions of the structure. The lengthened N–H bond observed in the neutron map suggests a highly unusual micro-environment around this particular peptide bond. This environment appears to be supported by a second-order interaction

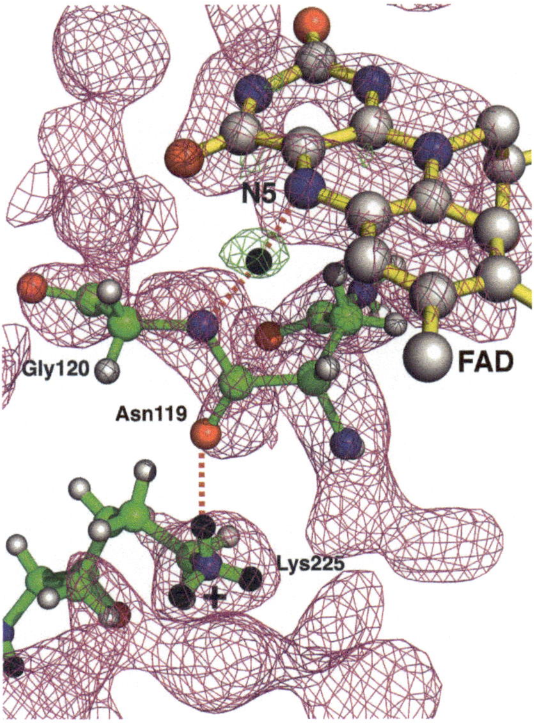

Fig. 6 Nuclear density omit map near the isoalloxazine moiety of the flavin cofactor and Lys225 showing the second order interaction with the N5 reactive center of FAD. The deuterium atom on the main chain nitrogen of Asn119 has been omitted from the model. The magenta colored map corresponds to the $2F_o - F_c$ nuclear density contoured at 1.5 σ and the green colored map corresponds to the $F_o - F_c$ nuclear density contoured at 2 σ. A deuterium atom has been positioned in the green $F_o - F_c$ nuclear density and is colored as a black sphere. The deuterium atoms on the ammonium group of Lys225 have also been colored as black spheres. The FAD is shown in yellow colored bonds and the protein is shown in green colored bonds.

involving the main chain carbonyl oxygen atom of Asn119 and the side chain ammonium group of a buried lysine residue (Lys225) situated within hydrogen bonding distance to the peptide bond oxygen atom (Fig. 6). The structural conservation of a charged side chain interacting with a carbonyl oxygen atom near to the FAD N5 atom in other members of the glucose-methanol-choline (GMC) oxidoreductase family (Cavener, 1992) has been proposed to polarize the peptide bond such that an elongated N–H bond toward the flavin reactive center is facilitated. Furthermore, this interaction is thought to play a role in directing the geometry of the orbitals

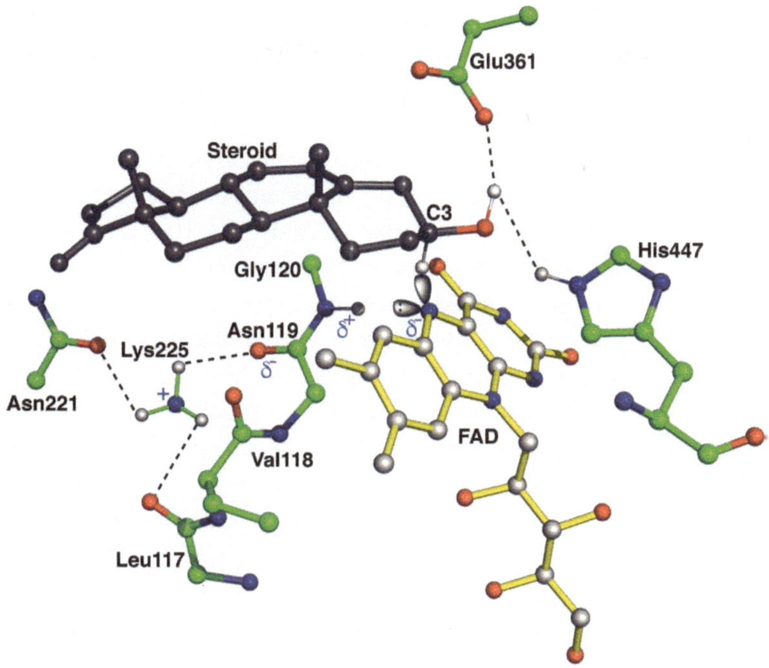

Fig. 7 Interaction between the positively charged Lys225 residue and the FAD reactive center (N5). The side-chain of Lys225 is stabilized by three hydrogen bonds to carbonyl oxygen atoms (main-chain oxygen of Leu117 and Asn119 and side chain of Asn221). The C3 hydrogen on the substrate, to be transferred as a hydride, and the alcohol hydrogen on the substrate, to be extracted by the proposed base, Glu361, are indicated on the steroid. The orbitals on the flavin N5 are shown.

on the N5 atoms of the isoalloxazine moiety of the flavin cofactor to generate a more tetrahedral orientation thus pre-organizing the flavin N5 to ideally position the hydride accepting orbital for flavin reduction by the steroid substrate (Fig. 7).

The close interaction of the main chain of the NH group of Gly120 and the flavin N5 was correlated with observations in the 1.12 Å X-ray structure of an enzyme-substrate complex where crystals had been soaked in isopropanol (a substrate mimic) under anaerobic conditions in order to trap the reduced enzyme (Golden, Karton, & Vrielink, 2014). In this structure difference electron density was observed 1.03 Å from the flavin N5 atom. In the reduced enzyme, the flavin is expected to accept a hydride from the C3 center of the cholesterol steroid substrate. The difference electron density observed in the X-ray structure was modeled as a hydride atom bound to N5 of the flavin. Interestingly the N5 adopted a tetrahedral interaction of the hydrogen atom between the main chain N of Gly120 and the flavin N5 atom and suggests that this interaction stabilizes the N5 during catalysis. This work prompted density-functional theory calculations (Golden et al., 2014, 2017) to be undertaken and showed that the hydrogen bond formed between the protein main chain and the flavin reactive center acts to stabilize the position of the hydride on the reduced flavin. Additionally, these studies showed that the conservation of the hydrogen bonded system through a second order effect from the buried Lys225 acts to increase the Gly120 N–H bond distance such that the hydrogen atom is able to lie closer to the flavin N5 atom. Further studies to investigate the effect of this positive charge at the position of Lys225 through the determination of the X-ray/neutron structures and redox potential of various Lys225 mutants are underway.

The neutron density maps from the hydrogenated/D_2O exchanged structure showed enticing examples of density providing important insights into the role of hydrogen in the redox activity of the FAD cofactor. However, a clearer neutron density map would be obtained if a completely perdeuterated protein sample were crystallized. This would give details of hydrogen positions, not only for exchangeable hydrogen atoms but also for non-exchangeable hydrogen atoms (such as the Cβ hydrogen atom on the side chain of Asn119 as discussed above), which in this case would all be present as deuteriums. Importantly these studies will also allow for maximal signal to noise ratios for neutron maps as well as studies of both oxidized and reduced states of the enzymes in the crystal state.

Toward these aims, Golden and coworkers have successfully expressed, purified and characterized the perdeuterated form of ChOx from *Streptomyces* SA-COO (Golden et al., 2015). Kinetic assays confirmed that the perdeuterated and the hydrogenated forms of the enzyme were functionally identical and crystallization studies succeeded in producing crystals from the same conditions as the hydrogenated form of the enzyme (Golden et al., 2014). High-resolution X-ray structures obtained for perdeuterated ChOx crystals also showed no significant structural differences between the two forms of the enzyme. These results indicate that perdeuterated Chox can be successfully expressed and purified with adequate yields. Also, since the perdeuteration process does not appear to adversely affect the function or structure of the protein, further neutron diffraction studies are likely to provide more complete insights regarding the role of hydrogen atoms in the redox mechanism.

4. Concluding remarks

Neutron protein crystallography has enabled determination of the precise locations of hydrogen atoms in the structure of ChOx. This has given unique insights into the role of hydrogen atoms in redox catalysis, both near to the flavin redox center as well as in a secondary shell away from the cofactor. Specifically, these insights have suggested that contacts between the protein and the flavin can modulate the redox reaction center of the cofactor to facilitate hydride transfer from the substrate to the isoalloxazine moiety through an orbital alignment mechanism. These interpretations may guide a better understanding of the precise mechanism of flavin mediated redox chemistry for this large family of enzymes. For example, the proposed model can also be applied to other GMC oxidoreductases since the positively charged Lys225 residue is conserved within the family. Future studies directed at a neutron structure of the reduced state of the enzyme may provide further details on how the reduced cofactor is stabilized by the surrounding protein. The combined approach of X-ray and neutron crystallography provides a powerful tool to obtain complementary data and delineate structural factors associated with redox state of flavoenzymes as well as other redox proteins.

References

Adachi, M., Ohhara, T., Kurihara, K., Tamada, T., Honjo, E., Okazaki, N., et al. (2009). Structure of HIV-1 protease in complex with potent inhibitor KNI-272 determined by high-resolution X-ray and neutron crystallography. *Proceedings of the National Academy of Sciences of the United States of America, 106*(12), 4641–4646. https://doi.org/10.1073/pnas.0809400106.

Batie, C. J., & Kamin, H. (1981). The relation of pH and oxidation-reduction potential to the association state of the ferredoxin.ferredoxin:NADP+ reductase complex. *The Journal of Biological Chemistry, 256*(15), 7756–7763.

Berglund, G. I., Carlsson, G. H., Smith, A. T., Szoke, H., Henriksen, A., & Hajdu, J. (2002). The catalytic pathway of horseradish peroxidase at high resolution. *Nature, 417*(6887), 463–468.

Berkholz, D. S., Faber, H. R., Savvides, S. N., & Karplus, P. A. (2008). Catalytic cycle of human glutathione reductase near 1 Å resolution. *Journal of Molecular Biology, 382*(2), 371–384. https://doi.org/10.1016/j.jmb.2008.06.083.

Blakeley, M. P., Hasnain, S. S., & Antonyuk, S. V. (2015). Sub-atomic resolution X-ray crystallography and neutron crystallography: Promise, challenges and potential. *IUCrJ, 2*(Pt. 4), 464–474. https://doi.org/10.1107/S2052252515011239.

Blakeley, M. P., Ruiz, F., Cachau, R., Hazemann, I., Meilleur, F., Mitschler, A., et al. (2008). Quantum model of catalysis based on a mobile proton revealed by subatomic X-ray and neutron diffraction studies of h-aldose reductase. *Proceedings of the National Academy of Sciences of the United States of America, 105*(6), 1844–1848. https://doi.org/10.1073/pnas.0711659105.

Bourgeois, D., & Royant, A. (2005). Advances in kinetic protein crystallography. *Current Opinion in Structural Biology*, *15*(5), 538–547. https://doi.org/10.1016/j.sbi.2005.08.002.

Carugo, O., & Djinovic Carugo, K. (2005). When X-rays modify the protein structure: Radiation damage at work. *Trends in Biochemical Sciences*, *30*(4), 213–219. https://doi.org/10.1016/j.tibs.2005.02.009.

Cavener, D. (1992). GMC oxidoreductases. A newly defined family of homologous proteins with diverse catalytic activities. *Journal of Molecular Biology*, *223*, 811–814.

Coulombe, R., Yue, K. Q., Ghisla, S., & Vrielink, A. (2001). Oxygen access to the active site of cholesterol oxidase through a narrow channel is gated by an Arg-Glu pair. *Journal of Biological Chemistry*, *276*, 30435–30441.

Fisher, S. Z., Aggarwal, M., Kovalevsky, A. Y., Silverman, D. N., & McKenna, R. (2012). Neutron diffraction of acetazolamide-bound human carbonic anhydrase II reveals atomic details of drug binding. *Journal of the American Chemical Society*, *134*(36), 14726–14729. https://doi.org/10.1021/ja3068098.

Fraaije, M. W., & Mattevi, A. (2000). Flavoenzymes: Diverse catalysts with recurrent features. *Trends in Biochemical Sciences*, *25*, 126–132.

Fritchie, C. J., & Johnston, R. M. (1975). Molecular complexes of flavins. The crystal structure of lumiflavin-bis(naphthalene-2,3-diol) trihydrate. *Acta Crystallographica. Section B*, *31*, 454–461.

Gadda, G., Wels, G., Pollegioni, L., Zucchelli, S., Ambrosius, D., Pilone, M. S., et al. (1997). Characterization of cholesterol oxidase from *Streptomyces hydroscopicus* and *Brevibacterium sterolicum*. *European Journal of Biochemistry*, *250*, 369–376.

Gamble, T. R., Clauser, K. R., & Kossiakoff, A. A. (1994). The production and X-ray structure determination of perdeuterated Staphylococcal nuclease. *Biophysical Chemistry*, *53*(1–2), 15–25.

Garman, E. F. (2010). Radiation damage in macromolecular crystallography: What is it and why should we care? *Acta Crystallographica. Section D, Biological Crystallography*, *66*(Pt. 4), 339–351. https://doi.org/10.1107/S0907444910008656.

Ghisla, S., & Massey, V. (1989). Mechanism of flavoprotein-catalysed reactions. *European Journal of Biochemistry*, *181*, 1–17.

Golden, E., Attwood, P. V., Duff, A. P., Meilleur, F., & Vrielink, A. (2015). Production and characterization of recombinant perdeuterated cholesterol oxidase. *Analytical Biochemistry*, *485*, 102–108. https://doi.org/10.1016/j.ab.2015.06.008.

Golden, E., Karton, A., & Vrielink, A. (2014). High-resolution structures of cholesterol oxidase in the reduced state provide insights into redox stabilization. *Acta Crystallographica. Section D, Biological Crystallography*, *70*(Pt. 12), 3155–3166. https://doi.org/10.1107/S139900471402286X.

Golden, E., & Vrielink, A. (2014). Looking for hydrogen atoms: Neutron crystallography provides novel insights into protein structure and function. *Australian Journal of Chemistry*, *67*, 1751–1762.

Golden, E., Yu, L. J., Meilleur, F., Blakeley, M. P., Duff, A. P., Karton, A., et al. (2017). An extended N-H bond, driven by a conserved second-order interaction, orients the flavin N5 orbital in cholesterol oxidase. *Scientific Reports*, 7, 40517. https://doi.org/10.1038/srep40517.

Guo, P. C., Ma, J. D., Jiang, Y. L., Wang, S. J., Bao, Z. Z., Yu, X. J., et al. (2012). Structure of yeast sulfhydryl oxidase erv1 reveals electron transfer of the disulfide relay system in the mitochondrial intermembrane space. *The Journal of Biological Chemistry*, *287*(42), 34961–34969. https://doi.org/10.1074/jbc.M112.394759.

Hazemann, I., Dauvergne, M. T., Blakeley, M. P., Meilleur, F., Haertlein, M., van Dorsselaer, A., et al. (2005). High-resolution neutron protein crystallography with radically small crystal volumes: Application of perdeuteration to human aldose reductase. *Acta Crystallographica. Section D, Biological Crystallography*, *61*, 1413–1417.

Hefti, M. H., Vervoort, J., & van Berkel, W. J. (2003). Deflavination and reconstitution of flavoproteins. *European Journal of Biochemistry, 270*(21), 4227–4242. https://doi.org/10.1046/j.1432-1033.2003.03802.x.

Joosten, V., & vanBerkel, W. J. H. (2007). Flavoenzymes. *Current Opinion in Chemical Biology, 11*, 195–202.

Kass, I. J., & Sampson, N. S. (1995). The isomerization catalyzed by Brevibacterium sterolicum cholesterol oxidase proceeds stereospecifically with one base. *Biochemical and Biophysical Research Communications, 206*(2), 688–693.

Kass, I. J., & Sampson, N. S. (1998a). Evaluation of the role of His447 in the reaction catalyzed by cholesterol oxidase. *Biochemistry, 37*(51), 17990–18000.

Kass, I. J., & Sampson, N. S. (1998b). The importance of Glu^{361} position in the reaction catalyzed by cholesterol oxidase. *Bioorganic and Medicinal Chemistry Letters, 8*, 2663–2668.

Katz, A. K., Li, X., Carrell, H. L., Hanson, B. L., Langan, P., Coates, L., et al. (2006). Locating active-site hydrogen atoms in D-xylose isomerase: Time-of-flight neutron diffraction. *Proceedings of the National Academy of Sciences of the United States of America, 103*(22), 8342–8347. https://doi.org/10.1073/pnas.0602598103.

Kort, R., Komori, H., Adachi, S., Miki, K., & Eker, A. (2004). DNA apophotolyase from Anacystis nidulans: 1.8 Å structure, 8-HDF reconstitution and X-ray-induced FAD reduction. *Acta Crystallographica. Section D, Biological Crystallography, 60*(Pt. 7), 1205–1213. https://doi.org/10.1107/S0907444904009321.

Kuo, M. C., Dunn, J. B. R., & Fritchie, Jnr, C. J. (1974). The crystal structure of a flavin molecular complex: 10-propylisoalloxazine-bis(naphthalene-2,3-diol). *Acta Crystallographica. Section B, 30*, 1766–1771.

Kurihara, K., Tanaka, I., Chatake, T., Adams, M. W., Jenney, F. E., Jr., Moiseeva, N., et al. (2004). Neutron crystallographic study on rubredoxin from *Pyrococcus furiosus* by BIX-3, a single-crystal diffractometer for biomacromolecules. *Proceedings of the National Academy of Sciences of the United States of America, 101*(31), 11215–11220. https://doi.org/10.1073/pnas.0403807101.

Kwon, H., Smith, O., Raven, E. L., & Moody, P. C. E. (2017). Combining X-ray and neutron crystallography with spectroscopy. *Acta Crystallographica. Section D, Structural Biology, 73*, 141–147. https://doi.org/10.1107/S2059798316016314.

Lario, P. I., Sampson, N., & Vrielink, A. (2003). Sub-atomic resolution crystal structure of cholesterol oxidase: What atomic resolution crystallography reveals about enzyme mechanism and the role of the FAD cofactor in redox activity. *Journal of Molecular Biology, 326*(5), 1635–1650.

Ledwidge, R., Patel, B., Dong, A., Fiedler, D., Falkowski, M., Zelikova, J., et al. (2005). NmerA, the metal binding domain of mercuric ion reductase, removes Hg^{2+} from proteins, delivers it to the catalytic core, and protects cells under glutathione-depleted conditions. *Biochemistry, 44*(34), 11402–11416. https://doi.org/10.1021/bi050519d.

Lennon, B. W., Williams, C. H., Jr., & Ludwig, M. L. (1999). Crystal structure of reduced thioredoxin reductase from *Escherichia coli*: Structural flexibility in the isoalloxazine ring of the flavin adenine dinucleotide cofactor. *Protein Science, 8*(11), 2366–2379.

Li, Y., Gao, Z., Hou, H., Li, L., Zhang, J., Yang, H., et al. (2011). Crystal structure and site-directed mutagenesis of a nitroalkane oxidase from Streptomyces ansochromogenes. *Biochemical and Biophysical Research Communications, 405*(3), 344–348. https://doi.org/10.1016/j.bbrc.2010.12.050.

Li, J., Vrielink, A., Brick, P., & Blow, D. M. (1993). Crystal structure of cholesterol oxidase complexed with a steroid substrate: Implications for flavin adenine dinucleotide dependent alcohol oxidases. *Biochemistry, 32*, 11507–11515.

Liu, X., Hanson, B. L., Langan, P., & Viola, R. E. (2007). The effect of deuteration on protein structure: A high-resolution comparison of hydrogenous and perdeuterated haloalkane dehalogenase. *Acta Crystallographica. Section D, Biological Crystallography*, *63*(Pt. 9), 1000–1008. https://doi.org/10.1107/S0907444907037705.

Lyubimov, A. Y., Chen, L., Brammer, L., Sampson, N. S., & Vrielink, A. (2009). A hydrogen-bonding network is important for oxidation and isomerization in the reaction catalyzed by cholesterol oxidase. *Acta Crystallographica. Section D, Biological Crystallography*, *65*, 1222–1231.

Lyubimov, A. Y., Heard, K., Tang, H., Sampson, N. S., & Vrielink, A. (2007). Distortion of flavin geometry linked to ligand binding in cholesterol oxidase. *Protein Science*, *16*, 2647–2656.

Lyubimov, A. Y., Lario, P. I., Moustafa, I., & Vrielink, A. (2006). Atomic resolution crystallography reveals how changes in pH shape the protein microenvironment. *Nature Chemical Biology*, *2*(5), 259–264.

Malinowski, E. R., Barber, M. J., Whitaker, G. T., & Smith, E. T. (2007). Factor analysis of spectroelectrochemical reduction of FAD reveals the pK(a) of the reduced state and the reduction pathway. *Journal of Chemometrics*, *21*(12), 520–528. https://doi.org/10.1002/cem.1058.

Massey, V. (2000). The chemical and biological versatility of riboflavin. *Biochemical Society Transactions*, *28*(4), 283–296.

Masuda, S., & Bauer, C. E. (2002). AppA is a blue light photoreceptor that antirepresses photosynthesis gene expression in *Rhodobacter sphaeroides*. *Cell*, *110*(5), 613–623. https://doi.org/10.1016/s0092-8674(02)00876-0.

Meilleur, F., Dauvergne, M. T., Schlichting, I., & Myles, D. A. (2005). Production and X-ray crystallographic analysis of fully deuterated cytochrome P450cam. *Acta Crystallographica. Section D, Biological Crystallography*, *61*, 539–544.

Meilleur, F., Snell, E. H., van der Woerd, M. J., Judge, R. A., & Myles, D. A. (2006). A quasi-Laue neutron crystallographic study of D-xylose isomerase. *European Biophysics Journal*, *35*(7), 601–609.

Miura, R. (2001). Versatility and specificity in flavoenzymes: Control mechanisms of flavin reactivity. *Chemical Record*, *1*(3), 183–194.

Nakaminami, T., Kuwabata, S., & Yoneyama, H. (1997). Electrochemical oxidation of cholesterol catalyzed by cholesterol oxidase with use of an artificial electron mediator. *Analytical Chemistry*, *69*(13), 2367–2372. https://doi.org/10.1021/ac960996p.

Piubelli, L., Pedotti, M., Molla, G., Feindler-Boeckh, S., Ghisla, S., Pilone, M. S., et al. (2008). On the oxygen reactivity of flavoprotein oxidases: An oxygen access tunnel and gate in brevibacterium sterolicum cholesterol oxidase. *The Journal of Biological Chemistry*, *283*(36), 24738–24747. https://doi.org/10.1074/jbc.M802321200.

Pollegioni, L., Wels, G., Pilone, M. S., & Ghisla, S. (1999). Kinetic mechanisms of cholesterol oxidase from *Streptomyces hygroscopicus* and *Brevibacterium sterolicum*. *European Journal of Biochemistry*, *263*, 1–13.

Ravelli, R. B., Leiros, H. K., Pan, B., Caffrey, M., & McSweeney, S. (2003). Specific radiation damage can be used to solve macromolecular crystal structures. *Structure*, *11*(2), 217–224.

Rohr, A. K., Hersleth, H. P., & Andersson, K. K. (2010). Tracking flavin conformations in protein crystal structures with Raman spectroscopy and QM/MM calculations. *Angewandte Chemie (International Ed. in English)*, *49*(13), 2324–2327. https://doi.org/10.1002/anie.200907143.

Senda, T., Senda, M., Kimura, S., & Ishida, T. (2009). Redox control of protein conformation in flavoproteins. *Antioxidants & Redox Signaling*, *11*(7), 1741–1766. https://doi.org/10.1089/ARS.2008.2348.

Shu, F., Ramakrishnan, V., & Schoenborn, B. P. (2000). Enhanced visibility of hydrogen atoms by neutron crystallography on fully deuterated myoglobin. *Proceedings of the National Academy of Sciences of the United States of America*, *97*(8), 3872–3877. https://doi.org/10.1073/pnas.060024697.

Trus, B. L., Wells, J. L., Johnstone, R. M., Fritchie, C. J., & Marsh, R. E. (1971). Crystal structure of the yellow 1:2 molecular complex lumiflavin–bisnaphthalene-2,3-diol. *Journal of the Chemical Society D: Chemical Communications*, *14*, 751–752.

Vrielink, A., & Ghisla, S. (2009). Cholesterol oxidase: Biochemistry and structural features *The FEBS Journal*, *276*(23), 6826–6843. https://doi.org/10.1111/j.1742-4658.2009.07377.x.

Winkler, A., Lyskowski, A., Riedl, S., Puhl, M., Kutchan, T. M., Macheroux, P., et al. (2008). A concerted mechanism for berberine bridge enzyme. *Nature Chemical Biology*, *4*(12), 739–741. https://doi.org/10.1038/nchembio.123.

Yin, Y., Liu, P., Anderson, R. G., & Sampson, N. S. (2002). Construction of a catalytically inactive cholesterol oxidase mutant: Investigation of the interplay between active site-residues glutamate 361 and histidine 447. *Archives of Biochemistry and Biophysics*, *402*(2), 235–242.

Yin, Y., Sampson, N. S., Vrielink, A., & Lario, P. I. (2001). The presence of a hydrogen bond between asparagine 485 and the pi system of FAD modulates the redox potential in the reaction catalyzed by cholesterol oxidase. *Biochemistry*, *40*(46), 13779–13787.

Yue, Q. K., Kass, I. J., Sampson, N. S., & Vrielink, A. (1999). Crystal structure determination of cholesterol oxidase from Streptomyces and structural characterization of key active site mutants. *Biochemistry*, *38*, 4277–4286.

CHAPTER SEVENTEEN

Heme peroxidase—Trapping intermediates by cryo neutron crystallography

Hanna Kwon[a,*], Tobias E. Schrader[b], Andreas Ostermann[c], Matthew P. Blakeley[d], Emma L. Raven[a], Peter C.E. Moody[e]

[a]School of Chemistry, University of Bristol, Bristol, United Kingdom
[b]Forschungszentrum Jülich GmbH, Jülich Centre for Neutron Science (JCNS) at Heinz Maier-Leibnitz Zentrum (MLZ), Garching, Germany
[c]Heinz Maier-Leibnitz Zentrum (MLZ), Technische Universität München, Garching, Germany
[d]Large Scale Structures Group, Institut Laue–Langevin, Grenoble, France
[e]Department of Molecular and Cell Biology and Leicester Institute of Structural and Chemical Biology, University of Leicester, Leicester, United Kingdom
*Corresponding author: e-mail address: hanna.kwon@bristol.ac.uk

Contents

Abstract

By combining the normal practice for X-ray crystallography of collecting diffraction data at 100K with neutron crystallography the structures of cryo-trapped enzyme intermediates have been determined, revealing the positions of the previously hidden hydrogens that are essential to a better understanding of the involved mechanism.

1. Introduction

Heme-containing enzymes moderate many redox processes *in vivo* including protection from peroxides, drug metabolism and electron transfer, *in vitro* they have many biopharmaceutical applications. The reaction of heme enzymes all proceed through high-valent ferryl (Fe^{IV}) intermediates

Methods in Enzymology, Volume 634
ISSN 0076-6879
https://doi.org/10.1016/bs.mie.2020.01.010

known as Compound I and Compound II. Understanding the chemistry of these intermediates through their structure requires a knowledge of the protonation states of the ferryl heme and the amino acid residues involved in catalysis. Properly characterizing these unstable intermediates has only been possible with neutron crystallography at cryogenic temperatures. Here, we describe how this work was done and the additional experiments that validate the interpretation of the results.

X-ray crystallography has been used to solve the structures of key examples of these enzymes (Patterson & Poulos, 1995; Pelletier & Kraut, 1992; Poulos et al., 1980) and the X-ray structures of several intermediates had been looked at by various means (Berglund et al., 2002; Bonagura et al., 2003; Fulop et al., 1994), this work provided the framework to understanding the mechanisms of enzyme reactions. However, X-ray methods have important limitations, especially when studying redox enzymes. First, X-ray are scattered by the atomic electrons and so electron-poor hydrogen atoms are difficult or impossible to locate with confidence, even at sub-Ångstrom resolution. Protonation states of bonded atoms can sometimes be deduced from the bond lengths of the connected atoms (as single bonds are longer than double bonds), this requires sufficient accuracy in atomic positions to enable unambiguous distinction. Of course, this still does not necessarily give hydrogen positions. When the environment contains a relatively electron dense atom (as is the case for heme-enzymes) there may be additional distortions in apparent positions (Fülöp, Watmough, & Ferguson, 2000). The electrophilic nature of the heme centre means that the high-valent intermediates discussed here (and in redox enzymes in general; Kwon, Smith, Raven, & Moody, 2017) are particularly susceptible to the reducing effects of X-rays. Therefore, a trapped intermediate in a crystal may change its nature during the course of data collection, further complicating the interpretation of maps. Unsurprisingly, the nature of the intermediates remained uncertain, with conflicting bond distances for the key iron oxygen distance reported. The use of low doses on multiple crystals, and experiments to exploit the photo-reduction induced changes chemical changes provided some consensus, but not confidence.

Fortunately, neutron crystallography can place hydrogen (as deuterium) at modest resolutions and the cold or thermal neutrons used do not perturb the chemistry. Neutrons are scattered by the nuclei rather than the electrons and so this is independent of atomic number. Thus deuterium has a similar scattering cross section as compared to the heavier atoms (e.g., carbon, nitrogen, oxygen) that form proteins, and can be observed equally well.

However, neutron crystallography has its own problems. The relatively (to X-rays) weak neutron flux available at the limited number of suitable reactors or spallation sources means that obtaining crystal diffraction data requires long experiments on very large crystals on scarce resources. The data recorded are often weak and lacking in completeness. This in turn means that the quality of maps (and their interpretation) may be limited, this may be partially overcome by incorporating X-ray data into joint refinements. Replacing hydrogen with deuterium might also not be trivial. Historians might compare the speed of data collection for neutron crystallography to what was done some 50 years ago with X-rays.

In this article, examples of the heme peroxidases will be discussed, showing how spectroscopic monitoring of intermediate formation and stability has allowed the cryo-trapping of the intermediates, Compound I (Casadei et al., 2014) and Compound II (Kwon et al., 2016). This has allowed the determination of these structures using combined neutron and X-ray crystallography methods. These studies have shown the nature of the ferryl heme and active site in these intermediates, revising the accepted enzyme mechanism.

2. Cryo neutron crystallography

Neutron crystallography is the only technique where diffraction data can be collected at room temperature without radiation damage to the samples. Despite being routine for X-ray work, the collection of neutron data at cryogenic temperatures is usually not required to extend the lifetime of the crystals as neutron beams do not cause any measurable damage to crystals. Nonetheless, the ability to collect data at cryogenic temperatures has several potential advantages, although it also presents some other challenges. As the temperature is reduced the thermal motion of atoms is also reduced, thereby improving the Bragg diffraction of crystals which leads to improvement in nuclear density definition or the potential for reducing the size requirements of crystals. In cryogenic conditions less stable proteins will not degrade and there is no risk of the crystals drying out. The widespread use of cryogenic environments for protein X-ray crystallography has resulted in robust cooling technology and standard mounting. However, the relatively large size of crystals means that necessarily the cooling will take longer than for smaller crystals, increasing the risk of crystal damage through the stresses on contraction or that vitrification may be slow enough to form ice crystals. The long data collection times also increases the chances of ice formation on the exterior of the sample. In the context of the work described here, cryo-crystallography allows the recording of diffraction data from labile intermediates.

Using neutron cryo-crystallography, we were able to determine the structures of the Compound I of cytochrome *c* peroxidase (C*c*P) and Compound II of ascorbate peroxidase (APX), respectively, resolving the previously ambiguous interpretations of their structures and provoking a re-assessment of the catalytic steps.

3. Heme peroxidases

Heme peroxidases are widespread in biology. They have played key roles in expanding our understanding of structural and mechanistic biochemistry which are used to develop rapid reaction methods. Heme peroxidases share a common catalytic cycle and the general processes involve the catalytic pathway that are well established. They catalyse reduction of peroxide to water using high oxidation state intermediates called compound I (CI) and compound II (CII) (Scheme 1). A key challenge has been to clarify the precise nature of the ferryl heme species as the literature reports conflicting evidence (Groves, 2014; Groves & Boaz, 2014; Karlin, 2010; Sligar, 2010).

1) $Fe^{III} + H_2O_2 \rightarrow Fe^{IV}\text{-O } P\cdot + H_2O$ (Compound I)
2) $Fe^{IV}\text{-O } P\cdot + S \rightarrow Fe^{IV}\text{-O } P + S\cdot$ (Compound II)
3) $Fe^{IV}\text{-O } P + S\cdot \rightarrow Fe^{III} + H_2O + S$

Reaction with hydrogen peroxide first forms Compound I, formally oxidised by two electron equivalents over the ferric enzyme (where P· is a tryptophan radical in the case of CcP and the heme in APX). Compound I is then reduced to Compound II (oxidised by one electron equivalent) and then reduced to the ferric form with the release of water.

Scheme 1 Overall reaction of heme peroxidases.

In our study, we focused on yeast cytochrome *c* peroxidase and soybean ascorbate peroxidase. Cytochrome *c* peroxidase (C*c*P) is a mitochondrial electron transport chain enzyme with a presumed role in removing the damaging peroxides formed. Originally isolated from yeast in the 1940s (Altschul, Abrams, & Hogness, 1940), C*c*P has been studied extensively and there is a well-established protocol to grow large protein crystals (Yonetani, Chance, & Kajiwara, 1966). It was the first heme peroxidase to have its structure solved by X-ray crystallography in 1980

(Poulos et al., 1980). It was one of the first heme enzymes which oxidized intermediates were observed (George, 1953). Reaction with hydrogen peroxide initially forms the Compound I intermediate, which is metastable in the absence of an electron donor, in this reaction the distal histidine was shown to have a crucial role (Erman et al., 1993). Compound I of C*c*P decays with a half-life of a few hours at room temperature (Fulop et al., 1994)

Soy ascorbate peroxidase (APX) is one of the newest enzymes to have joined the scene of heme peroxidase research. Peroxidases are abundant in plants and scavenge hydrogen peroxide by oxidation of small organic molecules such as ascorbic acid. In particular, a structure of APX-ascorbic acid complex structure provided the first identification of a non-protein physiological substrate binding site for a heme peroxidase (Sharp, Mewies, Moody, & Raven, 2003). Upon reaction with hydrogen peroxide, the Compound I intermediate quickly decays to the less unstable Compound II state, the difference in stability the reaction intermediates of C*c*P and APX being ascribed to formation of the radical placed on an adjacent tryptophan in the case of C*c*P and on the heme for APX (see Moody and Raven (2018) for a fuller discussion). APX compound II has a half-life of around 500 s at room temperature.

The ferryl intermediates are of particular interest as the chemical nature of the ferryl heme group (Fe^{IV}-OH or $Fe^{IV}{=}O$) controls the reactivity of the enzymes (Scheme 2). The transient nature of these intermediates makes a structural study difficult. However, UV–vis spectroscopy has long been used to characterize and study these enzymes in solution and this can be adapted

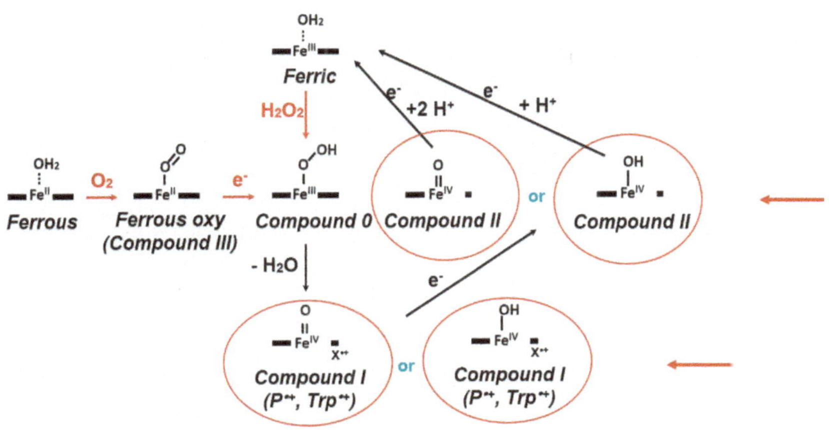

Scheme 2 The different stages in the reaction of C*c*P and APX, the possible Compound I and Compound II centres are circled.

to follow the reactions in the crystal. This idea was first used to demonstrate that crystals of chymotrypsin did indeed retain enzymatic activity (Rossi & Bernhard, 1970).

Establishing the nature of these transient intermediates had proven to be troublesome. Initially the majority of studies focused on measurement of the Fe—O distance or bond strength using various spectroscopic methods, a short distance being consistent with an Fe=O double bond, and a longer one indicating an Fe—OH single bond. These indirect methods produced ambiguous and conflicting results. In addition, most of the spectroscopic techniques suffer from the complication of laser induced photoreduction which introduced further complications in interpretations of the results.

The ability to react crystals of C*c*P with hydrogen peroxide to form the Compound I intermediate offered the possibility of measuring the distance directly, if diffraction data could be collected sufficiently quickly. Fulop et al. (1994), using a flow-cell with single-crystal spectroscopy to follow the reaction at 6 °C, collected data rapidly using the Laue method and 300–400 ms exposures, thus mitigating the effects of X-ray photoreduction. Depending on the protocol used they refined a value of 1.7–1.9 Å for the Fe-O distance (Fulop et al., 1994). The use of single-crystal spectroscopy and X-ray induced photoreduction with multiple passes synchrotron data collection was used to create a "movie" of the horse radish peroxidase reaction, interpreting an Fe-O distance of 1.7 Å for Compound I and 1.8 Å for compound II (Berglund et al., 2002). From a crystal of cryo-trapped C*c*P compound I, monitored before and after data collection by UV–vis and EPR spectroscopy an Fe-O distance of 1.87 Å was reported (Bonagura et al., 2003). However, subsequent work limiting X-ray exposure using multiple crystals, combined with single crystal spectrophotometry, showed that the Fe-O distance increased from 1.63 to 2.02 Å with photoreduction during data collection, and by using the same technique of minimal exposures and multiple crystals, cryo-trapped and unreduced Compound II of APX was seen to have an Fe-O distance of 1.84 Å. These results indicate Compound I is Fe=O, while Compound II is Fe-OH (Gumiero, Metcalfe, Pearson, Raven, & Moody, 2011; Gumiero, Murphy, Metcalfe, Moody, & Raven, 2010). However, there remained a suspicion that the Compound II structure was actually ferric hydroxide (Fe^{III}-OH) rather than the Fe^{IV} intermediate. Despite the use of SHELX full matrix refinement to give positional ESUs, there remains the caveat that the iron-oxygen distances observed might be distorted by the close proximity of electron-dense iron to the oxygen (Einsle et al., 2002; Fülöp et al., 2000; Seiffert et al., 2007).

4. Neutron structure of compound I of CcP

Using neutron crystallography to observe hydrogen (as deuterium) is the obvious way to resolve the true nature of the intermediates. Neutron diffraction data was first collected at LADI-III at ILL (Blakeley et al., 2010) on ferric (i.e., resting state) *Cc*P crystals that had been soaked in D_2O to allow exchange. This used the quasi Laue method, whereby 19 "stills" at different orientations and ~23 h exposure (each) were recorded using a wavelength range of 3.2–4.2 Å and cylindrical image-plate detector to give 2.4 Å data. The joint refinement including X-ray data clearly showed the distinctive banana-shaped density of D_2O bound to heme iron. It also confirmed that the distal histidine was in the neutral state. Once the parameters necessary to monitor and cryotrap Compound I in *Cc*P crystals had been established, neutron data was collected at the BIODIFF monochromatic neutron beam line at FRM-II (Ostermann & Schrader, 2015). This beam line having established the protocols needed to keep crystals at 100 K for the several hundred hours data collection takes to record a set of data from this intermediate. A total of 268 images using the rotation method (0.3° each) gave 2.5 Å data. After joint refinement with X-ray data, the neutron density of an oxygen could be seen clearly, establishing the identity of Compound I of *Cc*P as Fe=O (Fig. 1). Necessarily complimentary validation was required to establish the stability and identity of the intermediate in the crystal. Accordingly, the single-crystal UV–vis and EPR spectra were recorded and then re-measured after 20 days storage in liquid nitrogen. Not only did this work establish the identity of the ferryl oxygen intermediate, it also showed that the distal histidine is in the positively charged

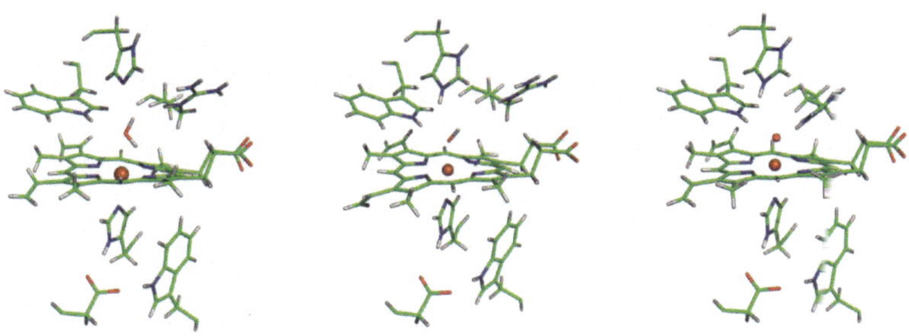

Fig. 1 Neutron crystal structure of (left) *Cc*P Ferric at room temperature, showing bound water in the distal pocket, (right) *Cc*P compound I at 100 K showing oxygen bound to form Fe=O, and (centre) APX compound II at 100 K showing the Fe-OH structure.

state, this and the revealed hydrogen-bond structure meant that the mechanism of Compound I formation had to be revised (Casadei et al., 2014; Groves & Boaz, 2014).

5. Neutron structure of compound II of APX

Unlike *Cc*P, previously grown crystals of APX tended to be small for neutron work (with a volume $\sim$0.001 mm^3 (Macdonald, Badyal, Ghamsari, Moody, & Raven, 2006), and so the hanging drop crystallization condition had to be systematically refined, this allowed crystals as large as $1.0 \times 0.6 \times 0.4$ mm to be grown. In order to trap Compound II in APX the ferric enzyme crystals (grown in D_2O) were reacted with *m*-chloroperbenzoic acid (CPBA), this first forms Compound I which rapidly decays to Compound II which is stable for long enough ($\sim$500 s) to trap in liquid nitrogen. This method avoids the possibility of also forming Compound III (the peroxo enzyme) by reaction with excess H_2O_2. The UV–vis and EPR spectra confirmed the intermediate formation and that it is stable in liquid nitrogen for at least 20 days. Furthermore, X-ray fluorescence scans of both ferric and Compound II crystals verified the higher oxidation state of the iron with the shift to higher energy. Initial tests on BIODIFF at FRM-II (Ostermann & Schrader, 2015) were followed by neutron data collection at 100 K using LADI-III at ILL (Blakeley et al., 2010), which has been modified to enable a cryogenic gas stream to flow over the crystal. Neutron diffraction data was collected using the quasi-Laue method and 15 exposures of 22.3 h at different orientations. To minimize the effects of photoreduction, only the first 10° of data from 10 different crystals was collected on the home source and merged to provide the X-ray data needed for joint refinement. The neutron map clearly showed cylindrical density corresponding to OD, with an Fe-O distance of 1.87 Å in close agreement with the earlier multi-crystal work (Gumiero et al., 2011). The work confirmed the identity of Compound II in APX is indeed Fe^{IV}-OH (Kwon et al., 2016) and that distal histidine is in the positive form, as it is in Compound I in *Cc*P (Fig. 1).

6. Summary

Neutron crystallography, combined with cryo-trapping and the use of a variety of spectroscopic techniques in the crystal and in solution has resolved long-standing uncertainties about the nature of the ferric

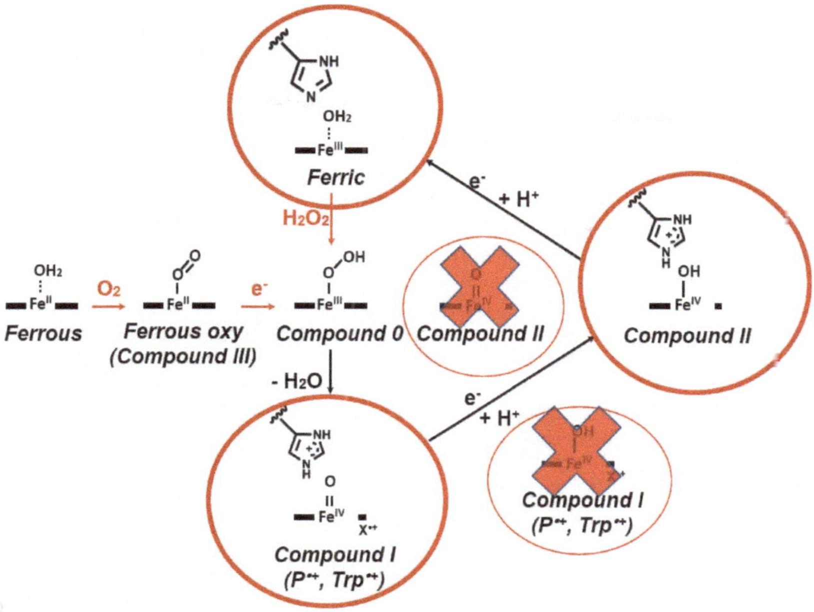

Scheme 3 The nature of the intermediates of the heme peroxidases resolved by neutron crystallography.

intermediates in the two heme peroxidases studied (Scheme 3), and this has revealed further details of the structures that meant a re-consideration of the details of the reaction mechanism was required. This affirms that the extrapolation of hydrogen positions from X-ray work alone may be less reliable than commonly imagined.

Acknowledgments

The research reported here has been supported by grants from the UK BBSRC (BB/K015656/1, BB/N015940/1, BB/M018598/1, BB/C00602X/1, BB/C001184/1, and 91/B19083), Leverhulme Trust, (F/00212/Q and RF-2016-135), The Wellcome Trust (WT094104MA) and Institut Laue-Langevin and by access to beamlines at Diamond Light Source, ESRF, Institut Laue-Langevin, and Heinz Maier-Leibnitz Zentrum. We are also grateful for the contributions of Andrea Gumiero, Cecilia Casadei and Jaswir Basran.

References

Altschul, A. M., Abrams, R., & Hogness, T. R. (1940). Cytochrome *c* peroxidase. *Journal of Biological Chemistry*, *136*(3), 777–794. Retrieved from http://www.jbc.org/content/136/3/777.short.

Berglund, G. I., Carlsson, G. H., Smith, A. T., Szöke, H., Henriksen, A., & Hajdu, J. (2002). The catalytic pathway of horseradish peroxidase at high resolution. *Nature*, *417*(6887), 463–468. https://doi.org/10.1038/417463a.

Blakeley, M. P., Teixeira, S. C., Petit-Haertlein, I., Hazemann, I., Mitschler, A., Haertlein, M., et al. (2010). Neutron macromolecular crystallography with LADI-III. *Acta Crystallographica Section D: Biological Crystallography*, *66*(Pt. 11), 1198–1205. https://doi.org/10.1107/s0907444910019797.

Bonagura, C. A., Bhaskar, B., Shimizu, H., Li, H., Sundaramoorthy, M., McRee, D. E., et al. (2003). High-resolution crystal structures and spectroscopy of native and compound I cytochrome *c* peroxidase. *Biochemistry*, *42*(19), 5600–5608. https://doi.org/10.1021/bi034058c.

Casadei, C. M., Gumiero, A., Metcalfe, C. L., Murphy, E. J., Basran, J., Concilio, M. G., et al. (2014). Heme enzymes. Neutron cryo-crystallography captures the protonation state of ferryl heme in a peroxidase. *Science*, *345*(6193), 193–197. https://doi.org/10.1126/science.1254398.

Einsle, O., Tezcan, F. A., Andrade, S. L., Schmid, B., Yoshida, M., Howard, J. B., et al. (2002). Nitrogenase MoFe-protein at 1.16 Å resolution: A central ligand in the FeMo-cofactor. *Science*, *297*(5587), 1696–1700. https://doi.org/10.1126/science.1073877.

Erman, J. E., Vitello, L. B., Miller, M. A., Shaw, A., Brown, K. A., & Kraut, J. (1993). Histidine 52 is a critical residue for rapid formation of cytochrome c peroxidase compound I. *Biochemistry*, *32*(37), 9798–9806. https://doi.org/10.1021/bi00088a035.

Fulop, V., Phizackerley, R. P., Soltis, S. M., Clifton, I. J., Wakatsuki, S., Erman, J., et al. (1994). Laue diffraction study on the structure of cytochrome c peroxidase compound I. *Structure*, *2*(3), 201–208. https://doi.org/10.1016/s0969-2126(00)00021-6.

Fülöp, V., Watmough, N. J., & Ferguson, S. J. (2000). Structure and enzymology of two bacterial diheme enzymes: Cytochrome cd1 nitrite reductase and cytochrome c peroxidase. *Adv. Inorg. Chem.*, *51*, 163–204.

George, P. (1953). The chemical nature of the second hydrogen peroxide compound formed by cytochrome c peroxidase and horseradish peroxidase. 2. Formation and decomposition. *Biochemical Journal*, *55*(2), 220–230. https://doi.org/10.1042/bj0550220.

Groves, J. T. (2014). Using push to get pull. *Nature Chemistry*, *6*(2), 89–91. https://doi.org/10.1038/nchem.1855.

Groves, J. T., & Boaz, N. C. (2014). Fishing for peroxidase protons. *Science*, *345*(6193), 142–143. https://doi.org/10.1126/science.1256754.

Gumiero, A., Metcalfe, C. L., Pearson, A. R., Raven, E. L., & Moody, P. C. (2011). Nature of the ferryl heme in compounds I and II. *Journal of Biological Chemistry*, *286*(2), 1260–1268. https://doi.org/10.1074/jbc.M110.183483.

Gumiero, A., Murphy, E. J., Metcalfe, C. L., Moody, P. C. E., & Raven, E. L. (2010). An analysis of substrate binding interactions in the heme peroxidase enzymes: A structural perspective. *Archives of Biochemistry and Biophysics*, *500*(1), 13–20. https://doi.org/10.1016/j.abb.2010.02.015.

Karlin, K. D. (2010). Model offers intermediate insight. *Nature*, *463*(7278), 168–169. https://doi.org/10.1038/463168a.

Kwon, H., Basran, J., Casadei, C. M., Fielding, A. J., Schrader, T. E., Ostermann, A., et al. (2016). Direct visualization of a Fe(IV)-OH intermediate in a heme enzyme. *Nature Communications*, 7, 13445. https://doi.org/10.1038/ncomms13445.

Kwon, H., Smith, O., Raven, E. L., & Moody, P. C. (2017). Combining X-ray and neutron crystallography with spectroscopy. *Acta Crystallographica Section D: Biological Crystallography*, *73*(Pt. 2), 141–147. https://doi.org/10.1107/S2059798316016314.

Macdonald, I. K., Badyal, S. K., Ghamsari, L., Moody, P. C., & Raven, E. L. (2006). Interaction of ascorbate peroxidase with substrates: A mechanistic and structural analysis. *Biochemistry*, *45*(25), 7808–7817. https://doi.org/10.1021/bi0606849.

Moody, P. C. E., & Raven, E. L. (2018). The nature and reactivity of ferryl heme in compounds I and II. *Accounts of Chemical Research*, *51*(2), 427–435. https://doi.org/10.1021/acs.accounts.7b00463.

Ostermann, A., & Schrader, T. (2015). BIODIFF: Diffractometer for large unit cells. *Journal of Large-Scale Research Facilities*, *1*, 1–4. https://doi.org/10.17815/jlsrf-1-19.

Patterson, W. R., & Poulos, T. L. (1995). Crystal structure of recombinant pea cytosolic ascorbate peroxidase. *Biochemistry*, *34*(13), 4331–4341. https://doi.org/10.1021/bi00013a023.

Pelletier, H., & Kraut, J. (1992). Crystal structure of a complex between electron transfer partners, cytochrome c peroxidase and cytochrome *c*. *Science*, *258*(5089), 1748–1755. https://doi.org/10.1126/science.1334573.

Poulos, T. L., Freer, S. T., Alden, R. A., Edwards, S. L., Skogland, U., Takio, K., et al. (1980). The crystal structure of cytochrome *c* peroxidase. *Journal of Biological Chemistry*, *255*(2), 575–580.

Rossi, G. L., & Bernhard, S. A. (1970). Are the structure and function of an enzyme the same in aqueous solution and in the wet crystal? *Journal of Molecular Biology*, *49*(1), 85–91. https://doi.org/10.1016/0022-2836(70)90377-3.

Seiffert, G. B., Ullmann, G. M., Messerschmidt, A., Schink, B., Kroneck, P. M., & Einsle, O. (2007). Structure of the non-redox-active tungsten/[4Fe:4S] enzyme acetylene hydratase. *Proceedings of the National Academy of Sciences of the United States of America*, *104*(9), 3073–3077. https://doi.org/10.1073/pnas.0610407104.

Sharp, K. H., Mewies, M., Moody, P. C., & Raven, E. L. (2003). Crystal structure of the ascorbate peroxidase-ascorbate complex. *Nature Structural & Molecular Biology*, *10*(4), 303–307. https://doi.org/10.1038/nsb913.

Sligar, S. G. (2010). Glimpsing the critical intermediate in cytochrome P450 oxidations. *Science*, *330*(6006), 924–925. https://doi.org/10.1126/science.1197881.

Yonetani, T., Chance, B., & Kajiwara, S. (1966). Crystalline cytochromes *c* peroxidase and complex ES. *Journal of Biological Chemistry*, *241*(12), 2981–2982. Retrieved from http://www.jbc.org/content/241/12/2981.abstract.

FSC
www.fsc.org
MIX
Papier | Fördert gute Waldnutzung
FSC® C08341

Kolibri 360 GmbH
Ferdinand-Jühlke-Straße 7
99095 Erfurt, Deutschland
produktsicherheit@kolibri360.de